DISCARD

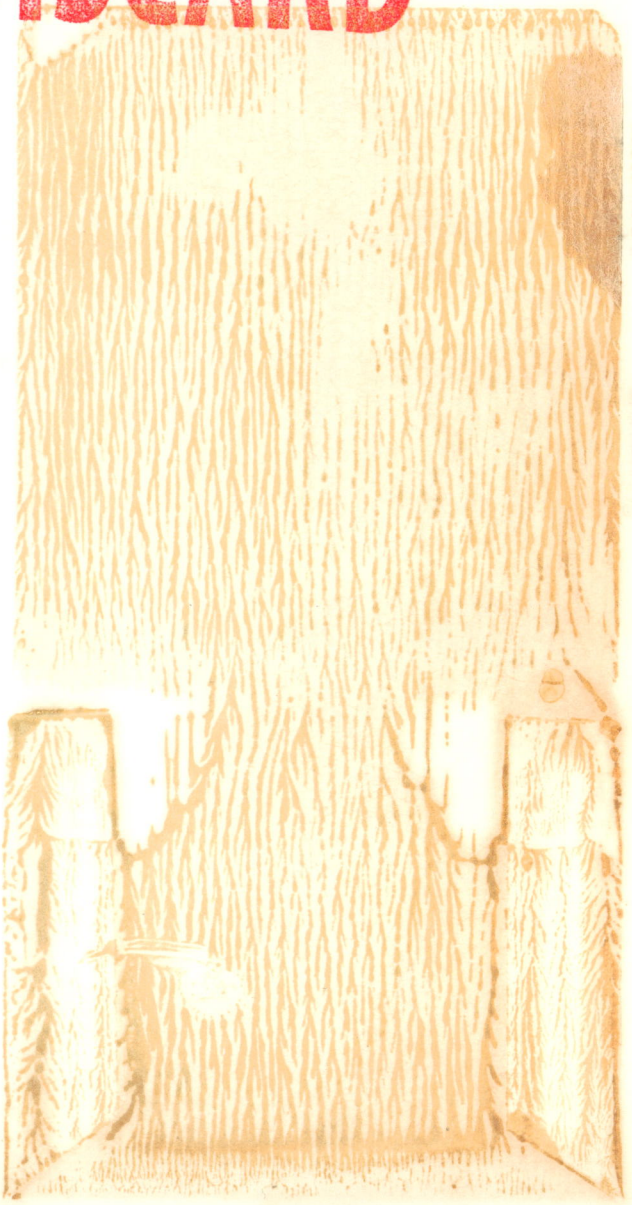

Basic Electronic Circuits Simplified

BY NELSON HIBBS

TAB BOOKS
Blue Ridge Summit, Pa. 17214

FIRST EDITION

FIRST PRINTING—DECEMBER 1972

Copyright ©1972 by TAB BOOKS

Printed in the United States
of America

Reproduction or publication of the content in any manner, without express permission of the publisher, is prohibited. No liability is assumed with respect to the use of the information herein.

Hardbound Edition: International Standard Book No. 0-8306-2622-X

Paperbound Edition: International Standard Book No. 0-8306-1622-5

Library of Congress Card Number: 72-87457

Dedicated to my daughters, Pepper and Kandy.

CONTENTS

1 BASIC LAWS OF ELECTRONICS 7
Ohm's Law—Kirchhoff's Law—Combining Ohm's and Kirchhoff's Laws—Von Helmholtz, Thevenin, Norton, and Millman Theorems—Thevenin Equivalent Circuits—Norton Equivalent Circuits

2 L, R, C AND TIME CONSTANTS 41
Reactance Formulas—The Inductor—The Capacitor—Multiple Inductors—Multiple Capacitors—Graphical Analysis—Step Function Waveforms—Time Constants—Series Integrator Circuits—Use of the LL00 Scale—Use of the LN-3 Scale—Use of the LL-3 Scale—For Epsilon Powers Less Than Minus One—Series Differentiator Circuits—Rise Time Calculations—Compensated Divider

3 DIODES 85
Construction of Junction Diodes—Zener Diode—Tunnel Diode—Back Diodes—Shockley Diode—Field Effect Diode—"Snap" Diode

4 ANALYZING TIME CONSTANTS OF DIODE CIRCUITS 109
Simple High Pass Filter (A Review)—Adding a Diode—Reversing Polarities—Getting a Bit More Complex—"Catching" Diodes Used in Both Directions—Tunnel-Diode Differentiator Feeding—Back-Diode Integrator—Field-Effect Diode Circuits—Sawtooth Generators

5 AMPLIFIER DEVICES 135
Triode Tube Analysis—Diodes and Pentodes—Equating Tubes and Transistors

6 AMPLIFIER CIRCUITS 197
Calculating Input and Output Resistance, and Voltage Gain—Triode Plate-Loaded Stage Voltage Gain For-

mulas—Thevenin Equivalent of Vacuum Tube as Voltage Generator—Norton Equivalent of Vacuum Tube as Current Generator—The "Metering Resistance"—Analysis of Transistor and FET Amplifiers Using "Transresistance"—Paraphase Amplifier—Differential Amplifier—Push-Pull Amplifier

7 COMPLETE ANALYSIS OF AN AMPLIFIER 231
Initial Methodology—Preliminaries and Self Test—Author's Analysis in Detail

8 FREQUENCY RESPONSE, AND OPERATIONAL AMPLIFIERS 252
Frequency Response—Gain Times Bandwidth—Determining Stray Capacities and Their Effects—Relating Bandwidth, Rise Time, and RC—Operational Amplifiers—Transistor Op Amp—Frequency-Compensated Op Amp—Emitter Follower-Common Emitter Op Amp—Push-Pull of Amps—In-Phase Feedback—Quiz

9 MORE ON FEEDBACK (OSCILLATORS) 280
Coaxial Transmission Line—Op Amp Becomes Oscillator — Sine-Wave Oscillator Example — Franklin Oscillator—Tuned Plate-Tuned Grid Oscillator—Armstrong Oscillator—Hartley Oscillator—Colpitts Oscillator—Grid-Plate Pierce Oscillator—Triode-Type Crystal-Controlled Oscillator—Neutralization—Growing Your Own Crystals

10 POWER, POWER SUPPLIES, AND SAFETY 300
Safety—Half-Wave Rectifier—Full Wave Rectifier—Series Regulator as Operational Amplifier — "Designing" a Power Supply — Solid-State Power Supply

APPENDIX 337
The Powers of Ten—Programmed Instruction of The Powers of Ten—Reactions—Directions—Answers—Conclusion

INDEX 349

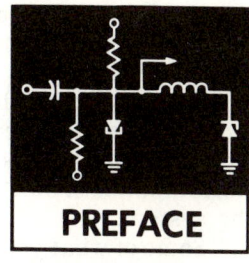

PREFACE

Our trade—electronics—is a fairly young one, really starting at around the turn of the 20th century with the development of the vacuum tube. Up until this time, it was predominantly a hobby for the rich and a curiosity for the scientifically inclined. The half century which followed allowed plenty of time for a standard, cut-and-dried approach to develop fairly universal training curriculums. Then came transistors and solid-state circuitry. These devices were treated as almost a completely separate subject. A new and different system for transistor circuit analysis was mathematically developed and scientifically proven to a far more accurate degree than the parts and circuits could be designed. The twenty years that followed 1950 saw an ever-increasing number of electronic equipments manufactured. Each new item seemed to require a new system of analysis all its own. Formulas for predicting field effect transistor operation bore no resemblance to those used to predict operation of regular transistorized circuitry and the formulas for transistor circuit analysis bore no resemblance to the now well-rooted vacuum-tube approach. Then each slightly different scientific field began to develop its own system for transistor analysis and half a dozen other approaches showed up, which left the poor technician caught in the middle with a vacuum-tube-algebra background, sadly in need of a full college degree to even understand most of the writings relative to these new devices.

However, glimmerings of light have begun to show through this cloud of previously required knowledge, and a slight modification along with a simplified expansion of the old vacuum-tube approaches is proving to be universally applicable to all circuitry. This is what this book is all about.

Please don't let the slight tongue-in-cheek presentation of this material cause you to treat it too lightly. My attempt here is to try to use common language and lightness of approach to promote more of an interest in learning the subject. All too often this type of material has been developed through extended use of shorthand of mathematics, so I will also attempt to use nothing beyond common, everyday algebra in developing the **tools of our trade**—which is how I like to think of all the useful conclusions and systems of thinking about modern circuitry.

The material in this book develops a system which largely bypasses the ultra-exact approach to circuit analysis. The roots of what the book presents are all soundly anchored to the fundamentals of electronics, and this, of course, is the proof of the system.

Some of what you have read before may be repeated here, but in a new and different way which enhances understanding.

I wish to express my appreciation to Tektronix, Inc., for their kind permission to reprint herein portions of several articles and several diagrams. I would also like to thank Mr. William Neill and Mr. Ron Olson for their encouragement and help in bringing my manuscript into reality.

Nelson W. Hibbs

Basic Laws of Electronics

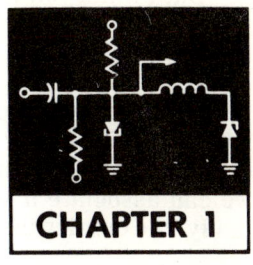

CHAPTER 1

Knowledge of electronics as a basis of industry is a comparatively new capability of man. I like to think of it as really beginning in 1827 when instructor of mathematics and physics, Georg Simon Ohm, published his pamphlet, "Mathematical Theory of the Galvanic Circuitry." You see, this was the publication that contained the basis of what we now call OHM'S LAW. Mr. Ohm (1787-1854) did not write E equals I times R. The terms, E, I, and R hadn't even been invented at that time. He did speak of electrical pressure, an electrical current that flowed, and the opposition to this exhibited by the circuit in which it flowed.

The scientists of his time who agreed with him were few and far between. For that matter, Mr. Ohm lost his job over the publication of his pamphlet. It actually took 14 years for the scientists of that time to grudgingly agree that Mr. Ohm had stated the laws of the electric circuit for the first time. (The Royal Society of London awarded him the Copley Medal in 1841.)

Progress was slow in those days. It took 40 more years to get these things of which Mr. Ohm wrote defined into basic units and terms of measure. It was the Electrical Congress in Paris, in 1881, which named electrical pressure (volts) after the Italian scientist, Volta, electrical current (amperes) after the Frenchman, Ampere, and the opposition to current flow after Mr. Ohm—the German—although I suppose we should call him a Bavarian, because that's where he was born. This Congress also defined each of these units of electrical measure in technical terms for the first time in history. We define them a little bit differently today and much more accurately, but Mr. Ohm's Law still holds true.

OHM'S LAW

Stated in terms used when I went to school, Ohm's Law was:

$$E = I \times R$$

But I think you will find it to your advantage to remember it as:

$$V = I \times R \qquad (Eq. 1\text{-}1)$$

What's that?

How do you keep these quantities straight? How do you REMEMBER them?

Well, I understand it is easier to remember something if we can associate it with something that's kind of crazy. Actually, the crazier the better. So let me suggest this;

Think of Old Man V-O-L-T-S

Pushing a Wheelbarrow of A-M-P-S

Up the Rocky Road of R-E-S-I-S-T-A-N-C-E

If you can remember this crazy picture, you'll never forget Ohm's Law. It'll keep straight in your mind (1) what does the pushing, (2) what gets pushed, and (3) what opposes the motion. We can take these three quantities, V, I, and R, and put them all in the same basket and this will keep our arithmetic straight as long as V is on top—like this:

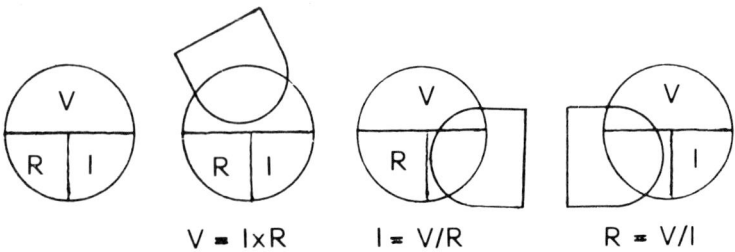

To find Voltage, cover V with a finger and note that I and R are on the same level and must be multiplied.

To find Current, cover I with a finger and note that V is over R; therefore divide Voltage by Resistance to obtain the answer.

To find Resistance, cover R with a finger and note that V is over I; therefore divide Voltage by Current.

KIRCHHOFF'S LAWS

The thinking of a Russian by the name of Kirchoff is the next thing to add to the picture for us. This gentleman came up with two ideas that are fundamental to the application of Ohm's Law to a real circuit.

Mr. Kirchhoff stated that the current flow is always the same in all parts of a series circuit (I_T I_1 I_2 ...). But, when we come to transmission-line theory, this idea can get us into trouble. I feel that it is better to think of Mr. Kirchoff's law for current as saying that the same quantity of electrons leave a given point in a circuit as enter it. And, this line of thought won't lead us astray.

The second law of the series circuit has to do with how the total voltage applied divides itself up among the numerous resistances that make up the circuit in question. The applied voltage divides among the various parts of the series circuit in direct proportion to the resistance of each part. Another way of saying the same thing is to note that the algebraic sum of all voltages in a closed loop is always equal to zero, or you could say that the sum of all the voltages in a series circuit is equal to that applied by the source. ($Vs = V1 + V2 + Vn...$).

Please note that I haven't said anything about what gets pushed around through our electrical circuit. I am going to assume that you are a believer in the Electron Theory for electrical current. In other words, it's the electron (or a quantity of them) in motion through a conductor that is the current measured in terms of the ampere. The electron will be thought of as being negative in nature. Thus a voltage will be thought of as being negative (—) when we have an excess quantity of electrons and positive (+) when we have a depletion (not enough) of electrons with respect to the neutral point we will call ground.

Now, I think our kit of tools is complete enough to start figuring some things out for ourselves about simple electrical circuits.

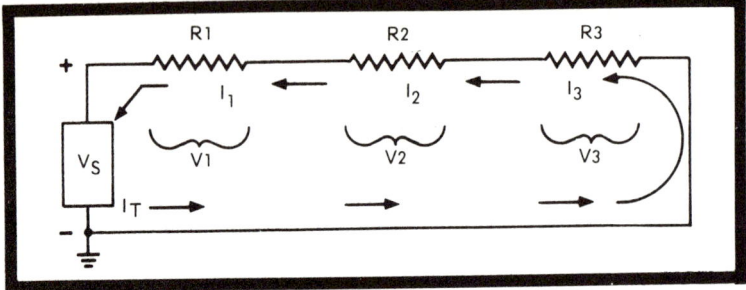

Fig. 1-1. Series DC circuit.

COMBINING OHM'S AND KIRCHHOFF'S LAWS

Please remember that all the rules you learned in algebra about equations will hold true for the equations we will write about our electrical circuits. Equals can be substituted for equals and if we do something to one side of the equation that changes the numerical value represented there, we must do the same thing to the other side. I'll do my best to remind you of these things as we go along.

Series DC Circuit

First, let's take a simple direct current (DC) series circuit made up of three resistors in series (one after the other) with a battery as a source of voltage. See Fig. 1-1.

Now, the idea is to apply Ohm's and Kirchhoff's conclusions to this circuit and see how many true equations we can write about it and what useable conclusions we can obtain.

From Kirchhoff; $\quad V_S = V_1 + V_2 + V_3 \quad$ (Eq. 1-2)

and $\quad I_T = I_1 = I_2 = I_3 \quad$ (Eq. 1-3)

Since Ohm's law must hold true for each part of the circuit as well as the whole thing, we can combine his throughts with Mr. Kirchhoff's and get;

$$R_T = \frac{V_S}{I_T} = \frac{V_1 + V_2 + V_3}{I_T}, \text{ or } \frac{V_1}{I_T} + \frac{V_2}{I_T} + \frac{V_3}{I_T},$$

$$\text{or } R_1 + R_2 + R_3 \qquad \text{(Eq. 1-4)}$$

In another form, Ohm's Law says V=I times R so we can write:

$$V_1 = I_1 \times R_1 = I_T \times R_1$$

$$V_2 = I_2 \times R_2 = I_T \times R_2$$

$$V_3 = I_3 \times R_3 = I_T \times R_3$$

and $V_S = I_T \times R_T$ or $I_T \times (R_1 + R_2 + R_3)$.

In the third form, Ohm's Law gives us I=V divided by R, thus we can say:

$$I_T = \frac{V_S}{R_T} = \frac{V_S}{R_1 + R_2 + R_3} \qquad \text{(Eq. 1-5)}$$

These are all useful conclusions, but let's pick one that will be of special use to us later on. Say, for instance, that we know the value of V_S and the values of R_1, R_2, and R_3, and we wish to predict the voltage which will appear across R_3. Ohm's Law tells us that V=I x R and we predicted V_3 would equal $I_T \times R_3$. But we don't know I_T. However, we can substitute equals for equals and we know

$$I_T = \frac{V_S}{R_1 + R_2 + R_3}$$

We'll use it.

And V_3 must equal

$$\frac{V_S}{R_1 + R_2 + R_3}$$

times R_3, or

$$\frac{R_3}{R_1 + R_2 + R_3} \times V_S. \qquad \text{(Eq. 1-6)}$$

In the same fashion, we can also say

$$V_2 = \frac{R_2}{R_T} \times V_S \text{ and } V_1 = \frac{R_1}{R_T} \times V_S.$$

Eq. 1-6 has become known as the voltage divider formula, and you'll use it again and again.

Quite often, we speak of the voltage at a given point in a circuit with respect to ground. So, let's look at this circuit (Fig. 1-1) in this fashion and note that the voltage on the left end of R_3 (with respect to ground or measured from ground) will just be the voltage drop across R_3. However, the voltage on the left end of R_2 with respect to ground will be quite dif-

11

ferent than the voltage across R_2. It will be the sum of the voltages across R_2 and R_3. The voltage on the left end of R_1 will be the supply voltage when measured with respect to ground, not just the voltage across R_1.

This voltage bit with respect to ground, though, is subject to change. You see, if we float our supply (disconnect the battery from ground) and put the ground in at the left end of R_3, we have changed the picture. The voltage on the left end of R_3 now is zero volts with respect to ground. The voltage across R_3 has not changed though, and the voltage on the right end of R_3 becomes negative with respect to ground and equal to the drop across R_3. The voltage on the left end of R_2 with respect to ground becomes equal to the drop across R_2. And, the voltage on the left of R_1 with respect to ground is equal to the sum of the drops across R_1 and R_2, not the supply voltage as it was before.

All of this just points up one thing and that is this: When we quote or predict a voltage at a given point in a circuit we must state it with reference to some other point in the circuit. It can be the voltage across a resistor (the voltage at one end with respect to the other end of the same resistor). The voltage at some point in a circuit can be stated with respect to ground or it can also be stated with respect to the supply voltage applied. We must keep our terms clear and precise. If we don't, we can get awfully mixed up.

Parallel DC Circuit

Taking our three known resistors and voltage supply in hand, let's hook them up in what we call a parallel circuit. The resistors are all side by side with the upper ends connected together and the lower ends all connected together. Then our battery is connected with its negative pole to the bottom of our circuit and positive pole to the top of our three resistors (Fig. 1-2).

And, again the idea is to apply Ohm's and Kirchhoff's formulas to this situation to find out what we can write in the way of equations about this circuit and, again, what useful conclusions we can obtain. Ohm's Law must hold true again for each part of the circuit as well as for the whole circuit. Note this time it is the current that gets divided up and not the voltage. We can write:

$$V_S = V_1 = V_2 = V_3$$

and

$$I_T = I_1 + I_2 + I_3$$ (Ref. Kirchhoff's Law for Current)

Since Ohm's Law says:

$$I = \frac{V}{R}, \text{ then } I_1 = \frac{V_S}{R_1}, I_2 = \frac{V_S}{R_2}, \text{ and } I_3 = \frac{V_S}{R_3}.$$

Since I_T equals the sum of the three currents, we can substitute equals for equals and come up with this:

$$I_T = \frac{V_S}{R_1} + \frac{V_S}{R_2} + \frac{V_S}{R_3} \text{ or } V_S \left(\frac{1}{R_1} + \frac{1}{R_2} + \frac{1}{R_3}\right).$$

One form of Ohm's Law said that R equaled V divided by I, so

$$R_T = \frac{V_S}{I_T}$$

And, although we don't really know the value of I_T, we do have a statement for it in terms of V_S and the three resistors. So, again, let's substitute equals for equals to get this statement:

$$R_T = \frac{V_S}{V_S \left(\frac{1}{R_1} + \frac{1}{R_2} + \frac{1}{R_3}\right)} \text{ or: } \frac{1}{\left(\frac{1}{R_1} + \frac{1}{R_2} + \frac{1}{R_3}\right)} \quad \text{(Eq. 1-7)}$$

since V_S cancels out and we are left with R_T strictly in terms of the individual resistors themselves. This is known as the reciprocal of the sum of the reciprocals formula for parallel resistance. Any number of resistors in parallel can be fitted into this formula since we just take the reciprocal of each resistor value, add them all up, and take the reciprocal of that

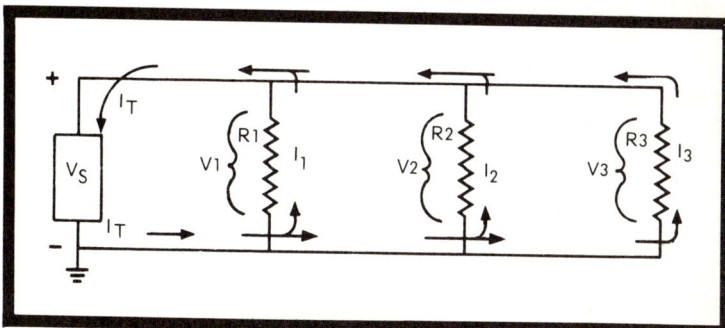

Fig. 1-2. Parallel DC circuit.

sum to obtain the total resistance of the parallel circuit. There's only one trouble with this. It usually takes a slide rule or a set of tables of reciprocals to achieve it. So let's take the simplest case we can (limit our circuit to just two resistors) and work it out a bit further to see if we can't simplify it a bit.

$$R_T = \frac{1}{\frac{1}{R_1} + \frac{1}{R_2}} = \frac{1}{\frac{R_2}{R_1 R_2} + \frac{R_1}{R_1 R_2}} = \frac{1}{\frac{R_2 + R_1}{R_1 R_2}}$$

And, if you remember your rule about what to do when you had to divide with a fraction, you'll recall that we invert it and multiply with it and get:

$$R_T = \frac{R_1 R_2}{R_1 + R_2} \qquad \text{(Eq. 1-8)}$$

And we don't need tables or a slide rule to work this one; we can do it longhand if need be. If there happens to be a third resistor as we had in our circuit to start with, we simply take the answer for the first two resistors and combine it with the third resistor in a second problem to get the total resistance of the three. Four resistors would be figured out in three problems and five resistors would take four problems to figure out. Personally, when it gets this complicated, I prefer a slide rule and the reciprocal formula.

VON HELMHOLTZ, THEVENIN, NORTON, AND MILLMAN THEOREMS

Now that we have developed the tools to handle the next advancement in thinking about circuitry, let's go back in history again for a bit.

Von Helmholtz Theorems

About the same time as Ohm was sweating out acceptance of his theories, there was another German by the name of von Helmholtz who was doing some original thinking of his own about electrical power sources. Von Helmholtz turned out to be the first man to list the special characteristics of the two basic ideal power sources; the Constant-Voltage source and the Constant-Current source. He pointed out that the Ideal Constant-Voltage source had to have zero resistance when electrically viewed back into its output leads, and that the Ideal Constant-Current source had to have infinite resistance when viewed from its output leads.

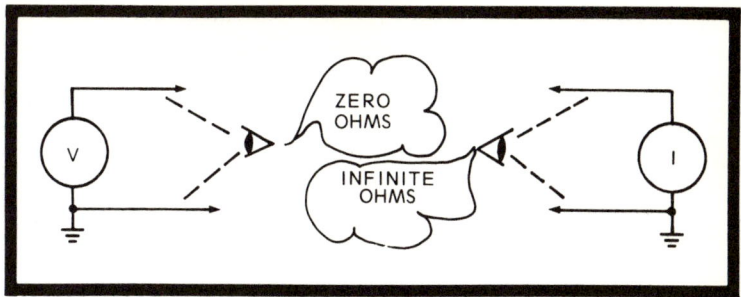

Fig. 1-3. Ideal constant voltage and constant current sources.

Then, about a generation later, Thevenin and Norton came along and brought these ideas into clearer focus for us. Thevenin conceived that, when viewed from two different points in the circuit, any linear network of resistance and voltage sources could be thought of as a single voltage source and a series resistance. Meanwhile, Norton did the same thing for the constant-current source when he deduced that any similar network, viewed in a similar manner, could be represented by a constant-current generator representing the sum of the short-circuit current between the two points and with the same resistance Thevenin saw, now seen in parallel with the current source.

These two ideas, the Helmholtz-Thevenin Theorem and the Helmholtz-Norton Theorem, have been sadly overlooked in most cases of curriculum development. Today they are much more commonly known as just Thevenin's Theorem and Norton's Theorem and Mr. von Helmholtz has mostly been left out of the picture. The original language in which each theorem was written, correct though it may have been, turned out to be a bit more than just confusing to the average student. So, these ideas did a lot of dust-gathering when they should have been in constant use. They simplify the complex so beautifully, though, that I'd like to suggest that we take them one at a time and go through them in detail. Thevenin, first, then Norton, and finally, Millman's development, which is sort of a compound of the two.

THEVENIN EQUIVALENT CIRCUITS

(This section is reprinted from Tektronix SERVICE SCOPE, October 1966.)

Apparently the first thing we need is a linear network of impedances and generators. To keep it simple, we will use

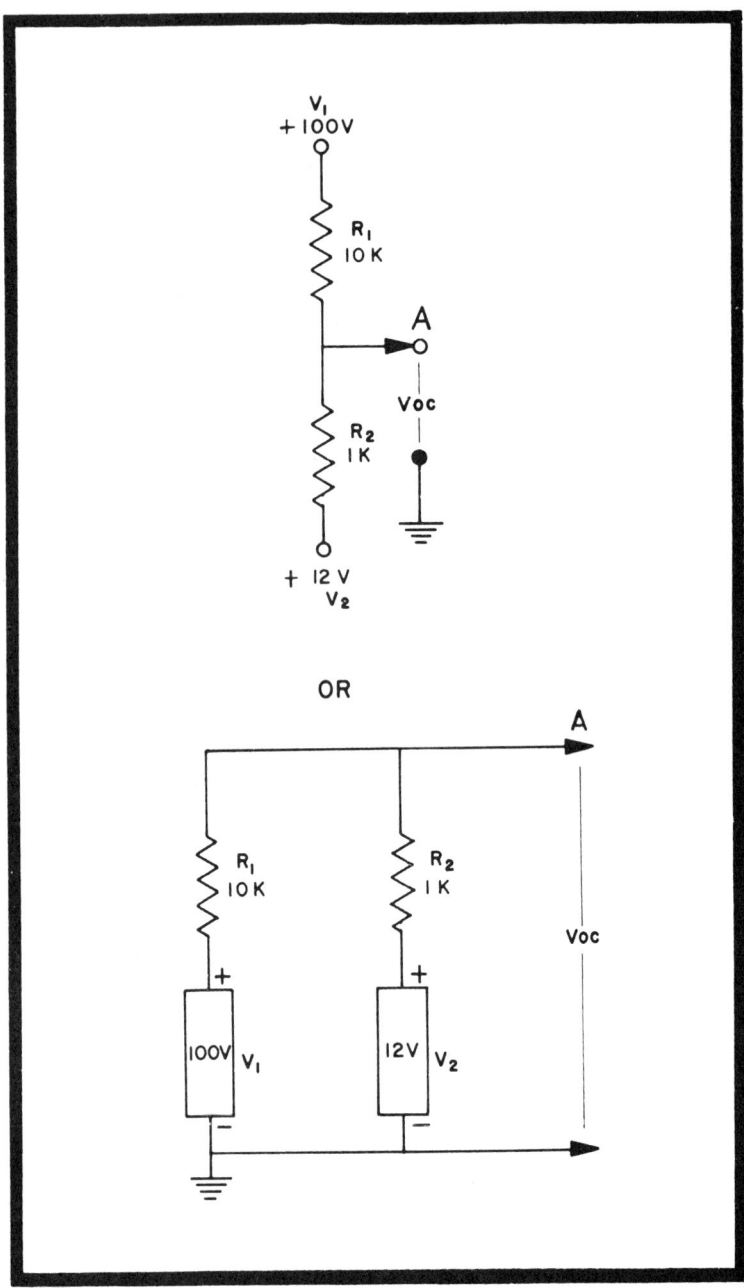

Fig. 1-4. Simple circuit of ideal voltage sources (no internal impedance) and resistors. (Courtesy Tektronix, Inc.)

resistors for the impedances and good solid voltage supplies for the generators. Our circuit might look like Fig. 1-4.

Thevenin's Theorem also says we must view this circuit from two points in the network. Let us select, for these two points, the ground lead and the common lead at point A. Next, Thevenin pointed out that we could make a substitution for this complex network. It would consist of a single voltage source (which he called V_{oc}) and a single series resistance (which he called Z_{th}), where Z equals Impedance.

Let us define V_{oc} and Z_{th}. Since the ground is one point of reference and the common lead is the other, V_{oc} becomes the voltage difference between these two points. Thus in the circuit in Fig. 1-4,

$$V_{oc} = V_2 + \frac{(V_1 - V_2) R_2}{R_1 + R_2}$$

Note: Voc equals open circuit voltage.
(reference - Eq. 1-6, voltage divider formula)

$$V_{oc} = 12 V + \frac{88 \times 1K}{10K + 1K} \quad \text{(Thevenin's open-circuit voltage)}$$

If we assume we are using ideal batteries for our "good solid voltage supplies" we will, of course, have zero impedance within the voltage sources. Looking back then into the circuit from our selected reference points, through the resistors to the zero impedance point, we will see an impedance made up of the parallel resistance of the two divider resistors and this impedance becomes Z_{th}. Thus in the circuit in Fig. 1-4,

$$Z_{th} = \frac{R_L \times R_2}{R_1 + R_2} = \frac{10K \times 1K}{10K + 1K} = \frac{10 \times 10^6}{11 \times 10^3} = .91K \text{ ohms}.$$

According to Thevenin, these two units, V_{oc} and Z_{th}, will be seen in series as in Fig. 1-5 when used as a substitute for the more complex circuit we started with in Fig. 1-4. Consequently, we call this combination of V_{oc} and Z_{th} a "Thevenin Equivalent Circuit."

Now let us put this idea into the practical framework of a real circuit.

Fig. 1-6 shows a transistor with a split collector load. Let us assume we have the collector curves for this transistor and we would like to draw in the load line to obtain an idea of the range of voltage and current this transistor might need to

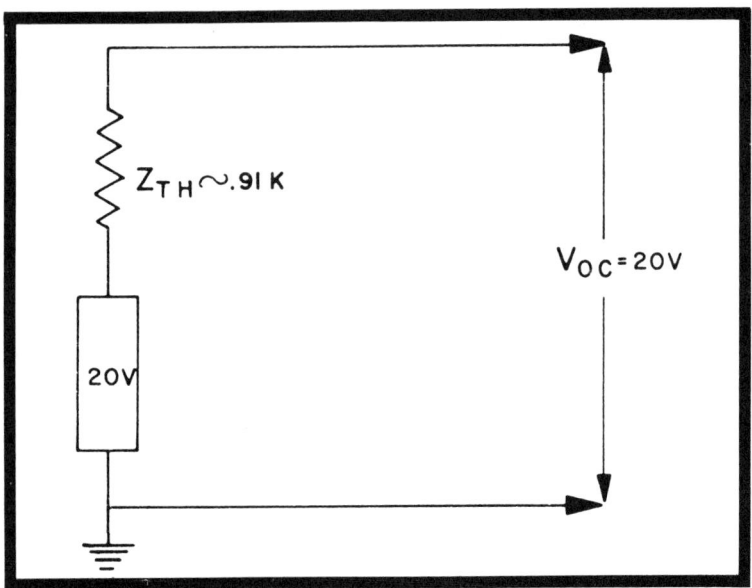

Fig. 1-5. Thevenin equivalent circuit of Fig. 1-4. (Courtesy Tektronix, Inc.)

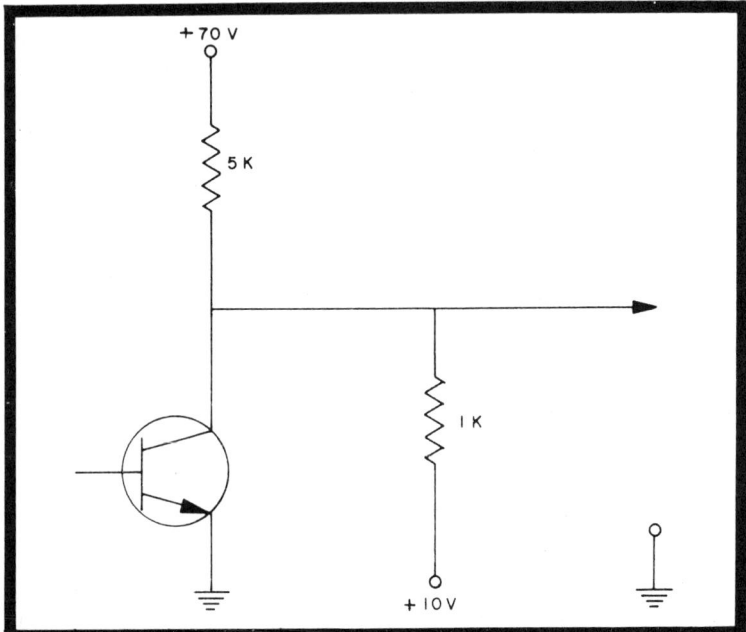

Fig. 1-6. Transistor with split collector load. (Courtesy Tektronix, Inc.)

work within. We now need to know what the effective V_{cc} is and how much resistance is in the actual effective load, R_L. Applying Thevenin's theorem, V_{cc} turns out to be the V_{oc}, and R_L becomes the Z_{th} of the theorem, thus the Thevenin Equivalent for the circuit in Fig. 1-6 would be the circuit shown in Fig. 1-7. We can now draw in the load line for the transistor, as shown in Fig. 1-8.

Naturally, the more complex linear networks will require a bit more figuring and will establish the reason for labeling Thevenin's voltage as V_{oc}, or open-circuit voltage, rather than calling it the "unloaded divider voltage" or something else. However, as you have just seen, the figuring only involves some very basic mathematics with which the electronic student or technician is (or should be) very familiar. There are other methods of analyzing complex linear circuits; but they all turned out to be more complex and time-consuming than this system, once I learned it.

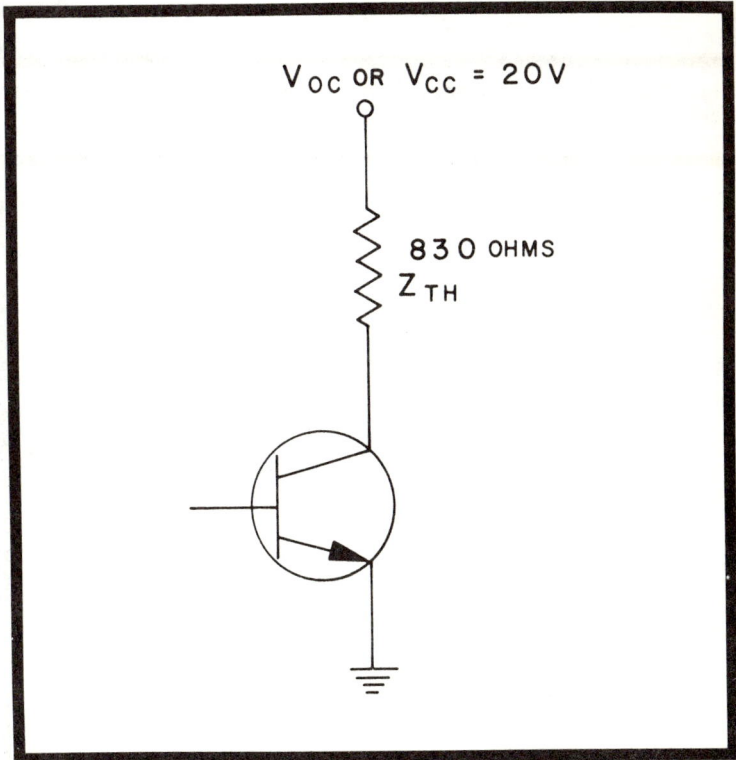

Fig. 1-7. Thevenin equivalent circuit of Fig. 1-6. (Courtesy Tektronix, Inc.)

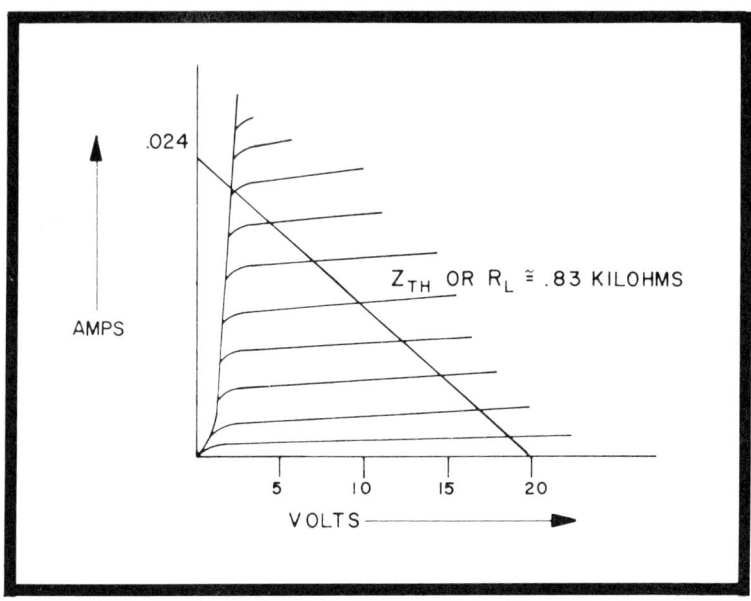

Fig. 1-8. Load line on collector curves for transistor in Fig. 1-6. (Courtesy Tektronix, Inc.)

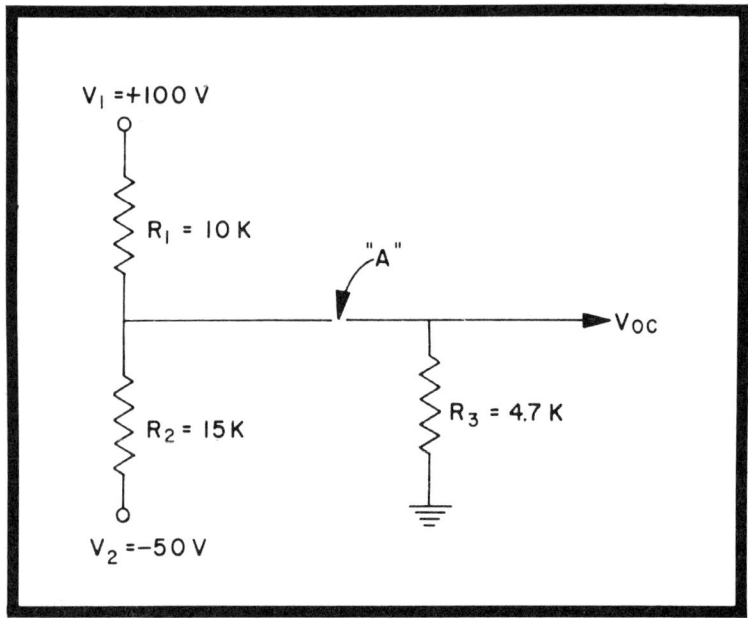

Fig. 1-9. A complex circuit for Thevenin analysis. (Courtesy Tektronix, Inc.)

As an example of a more complex circuit, look at the circuit in Fig. 1-9. The procedure, when using Thevenin's theorem and analyzing a complex circuit, is to progressively apply the theorem to portions of the circuit until all elements of the circuit have been considered. If in Fig. 1-9 then, we break the circuit at point "A", we can solve for V_{oc} and Z_{th} up to this point. In the interests of clarity, let us call the open-circuit voltage and impedance up to this point V_{oc_1} and Z_{th_1}, and the open-circuit voltage and the impedance of the entire circuit, V_{ocT} and Z_{thT}.
Thus:

$$V_{oc_1} = V_2 + \frac{(V_1 - V_2) R_2}{R_1 + R_2} \quad \text{(Ref. Eq. 1-6)}$$

$$= -50V + \frac{100V - (-50V) \times 15K}{15K + 10K}$$

$$= -50V + \frac{150V \times 15K}{25K}$$

$$= -50V + 90V = +40V.$$

And
$$Z_{th_1} = \frac{R_1 \times R_2}{R_1 + R_2}$$

$$= \frac{15K \times 10K}{15K + 10K}$$

$$= \frac{150 \times 10^6}{25 \times 10^3} = 6 \times 10^3 = 6K \text{ ohms}.$$

The Thevenin equivalent then, for that portion of the circuit in Fig. 1-9 up to point "A" is the one shown in Fig. 1-10.

(SHORT-CUT NOTE) Look at the two formulas that we just used. Note that each formula includes the expression R_2 divided by the sum of R_1 and R_2. We can figure this out once and use it twice without figuring it again as we actually did in the solution shown. In other words, we could have figured 15K over 25K as being equal to 3/5. Then figured 3/5 of the 150V to get the 90V we added to the reference voltage of —50V, and then used the same 3/5 to multiply the 10K (R_1) to get 6K for the solution of our problem. This procedure can speed things up for us.

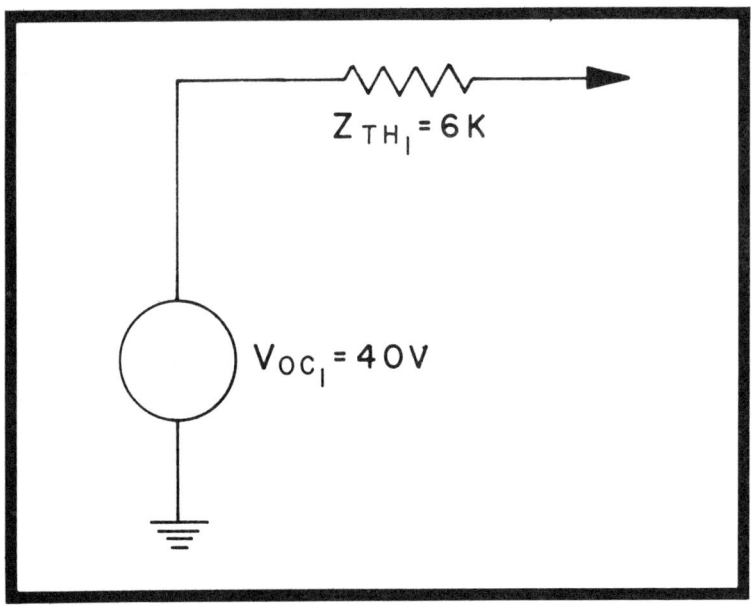

Fig. 1-10. Fig. 1-9 partially analyzed. (Courtesy Tektronix, Inc.)

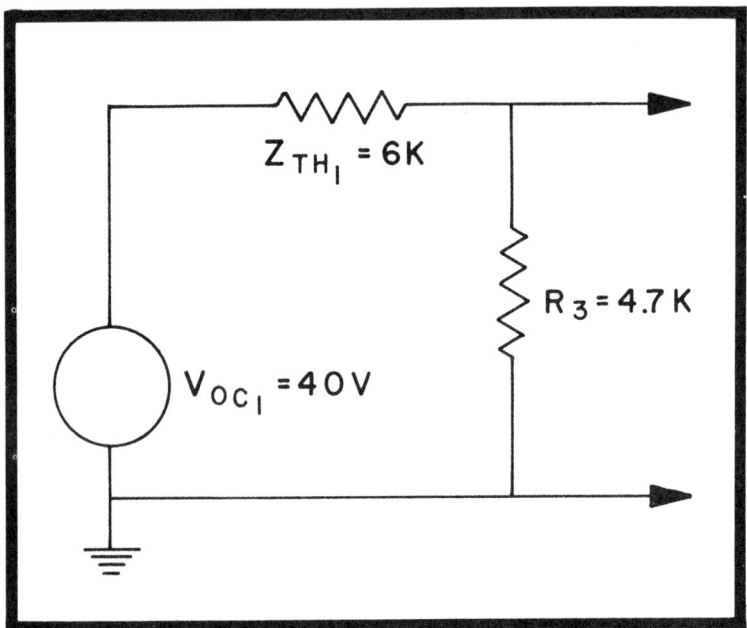

Fig. 1-11. Thevenin equivalent circuit placed in Fig. 1-9. (Courtesy Tektronix, Inc.)

We can now redraw the circuit in Fig. 1-9, replacing that portion of the circuit up to point "A" with its Thevenin equivalent. This gives us the circuit shown in Fig. 1-11. We can now apply Thevenin's Theorem to this circuit and obtain our original objective; i.e., a complete analysis of the circuit in Fig. 1-9.

Thus:

$$V_{oc_T} = \frac{V_{oc_1} \times 4.7K}{Z_{th_1} + 4.7K}$$

$$= \frac{40V \times 4.7K}{6K + 4.7K}$$

$$= 17.5V.$$

And

$$Z_{th_1} = \frac{Z_{th_1} \times R_3}{Z_{th_1} + R_3}$$

$$= \frac{6K \times 4.7K}{6K + 4.7K}$$

$$= 2.63K.$$

The open-circuit voltage (V_{oc}) and the impedance (Z_{th}) then for the entire circuit in Fig. 1-9 is 17.5V and 2.63K, respectively, and this circuit (shown in Fig. 1-12) can be substituted for the three-resistor, three-supply network we started with in any problem involving such a network.

From the foregoing, it should be apparent that in analyzing complicated circuits we open the circuit so that we consider only two supplies and their resistances at a time. Look at the circuit in Fig. 1-13. Here we would open the circuit at points "A", take V_1 and R_1, and V_2 and R_2, and simplify them into one voltage supply and its series resistance. To this we would add the next supply and its series resistance, apply the procedure of Thevenin and find this new equivalent, and so on, until we had simplified the entire circuit.

It is not difficult to use Thevenin's theorem once you understand it. It just takes practice, practice, practice!

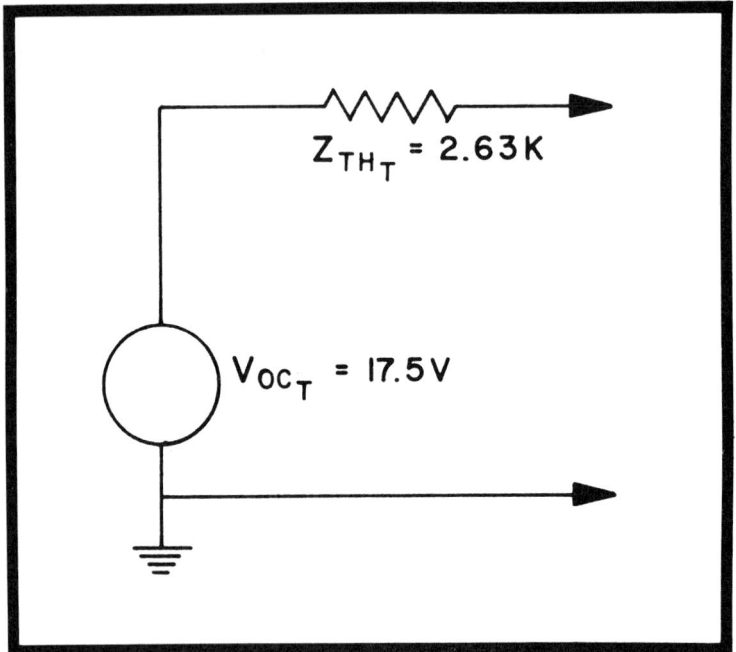

Fig. 1-12. Final Thevenin equivalent circuit for Fig. 1-9. (Courtesy Tektronix, Inc.)

NORTON EQUIVALENT CIRCUITS

The following discussion, used in training classes, is reprinted with permission of Tektronix, Inc.)

Since all sources of electrical power may be looked upon as either a **constant-voltage type plus a degree of imperfection** (Thevenin Equivalent) or a **constant-current type plus its degree of imperfection** (Norton Equivalent), and having discussed the Thevenin approach, let's now tackle the Norton approach.

Again we have an extension of Ohm's Law that is extremely useful when understood, but is very seldom stated in a form that the average student of electronics can understand. Norton's Theorem states, just like Thevenin, that any linear network of impedances and generators, if viewed from any two points in the network, can be replaced by an equivalent current source (I_{se}) and an equivalent impedance (Z_{th}) in shunt with this current source.

Once more, we need a linear network of impedances and generators. So to make this compatible with the previous discussion, let's use a circuit similar to the one we started with

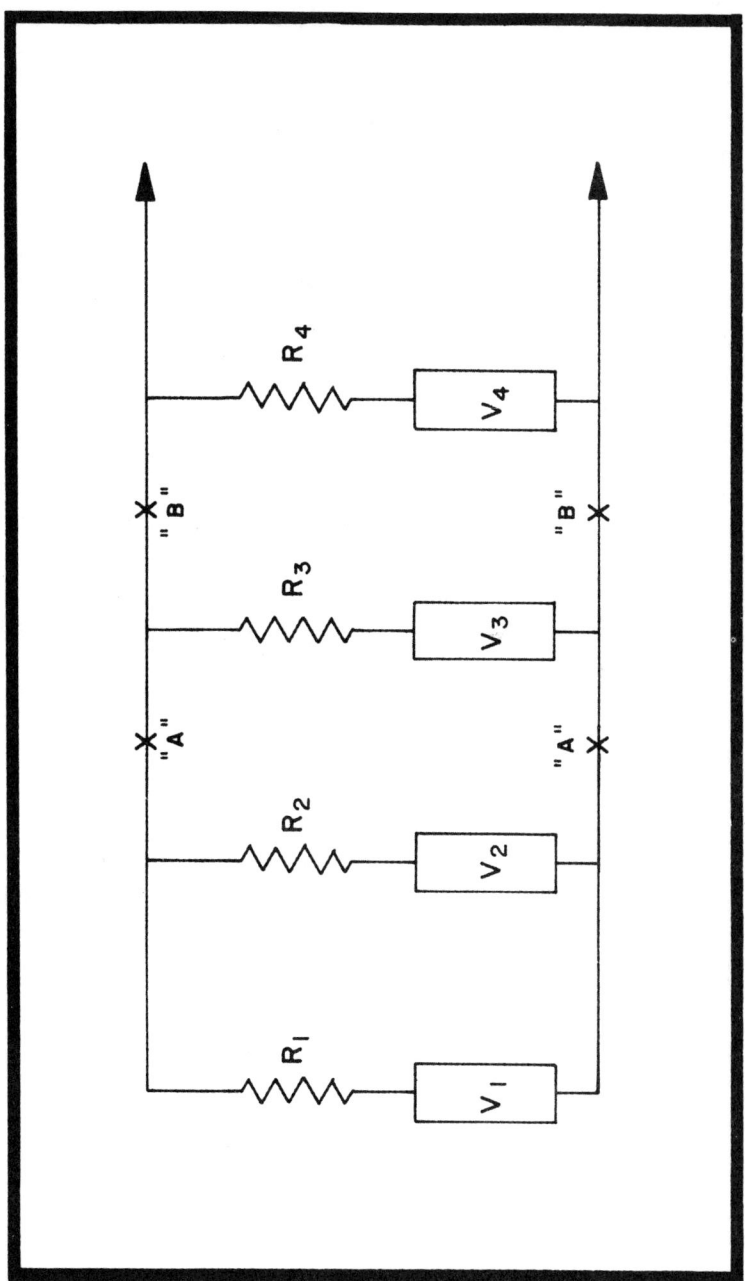

Fig. 1-13. Further example of simplifying for Thevenin equivalencies. (Courtesy Tektronix, Inc.)

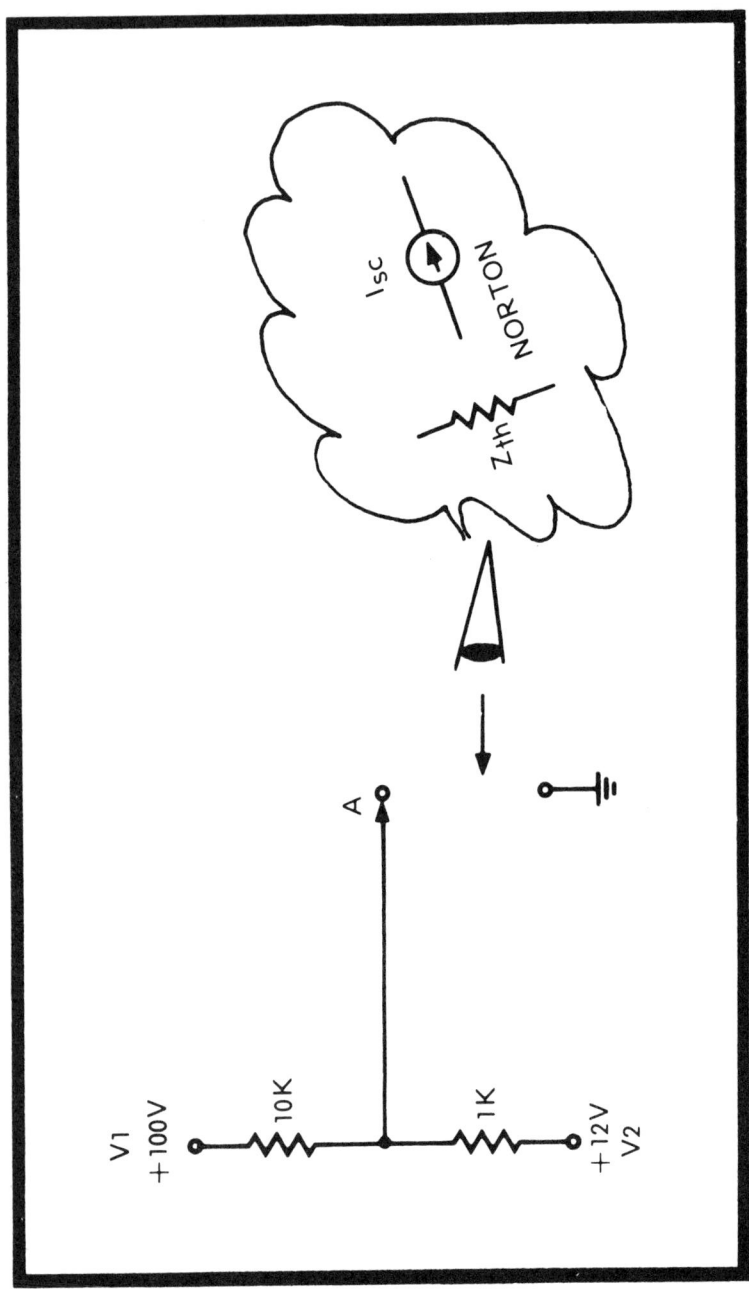

Fig. 1-14. A complex circuit for Norton analysis. (Courtesy Tektronix, Inc.)

before, Fig. 1-4. Our circuit will now look like the one in Fig. 1-14. Norton's Theorem says this circuit must be viewed from two points in the network. The two points we will use will be ground and the common lead (Point A in Fig. 1-14). Norton's conclusion was that we would see just one constant-current generator shunted by a single resistor (Z_{th}) existing between these two viewpoints.

Remember that the constant-current source is the opposite of the constant-voltage source which—when perfect—had zero back impedance, and the Z_{th} impedance was shown in series with it to show the imperfection. In this case, we have an infinite impedance represented in the generator, and if we are to look back into this circuit from the viewpoints of ground and the common lead, we will see a readable impedance. This impedance MUST be in parallel with the generator. This is shown in Fig. 1-15.

It will pay us to note here that a very efficient Thevenin Equivalent is a very poor constant-current source, and a very efficient Norton Equivalent is a very poor constant-voltage source. Remember, if current is constant, Ohm's Law points out that the resultant voltage will change directly as the load. In other words, if the load is changed, the voltage across it will be different. And failure of a power source to do this is simply its degree of imperfection when thought of as a Norton Equivalent.

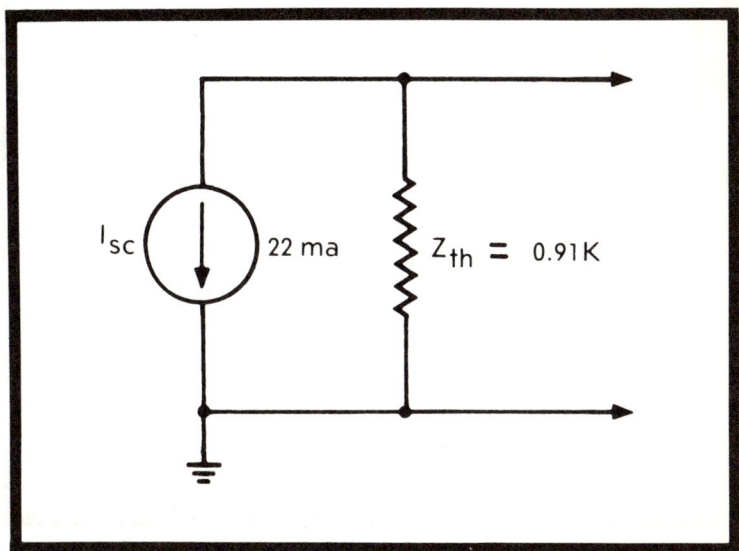

Fig. 1-15. Fig. 1-14 Norton equivalent. (Courtesy Tektronix, Inc.)

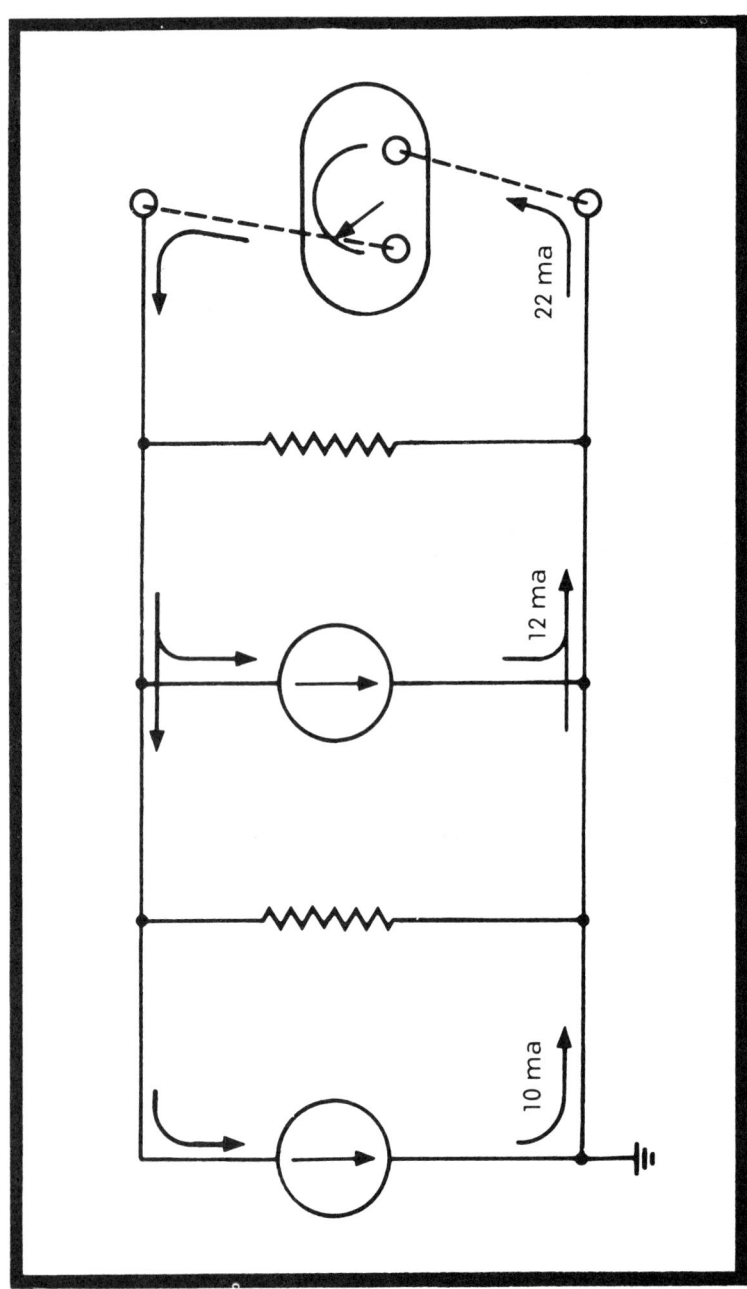

Fig. 1-16. Norton equivalent of Fig. 1-4 (the original Thevenin problem), showing paths of short-circuit currents. (Courtesy Tektronix, Inc.)

The subnotation for the current from our constant-current source and that for the impedance seen in parallel with it gives us our clue as to how to find their values.

Let's take current first. If our actual circuit is to be looked upon as that represented in Fig. 1-16, and we wish to know the real values of constant-current caused to flow by the generator, we would have to put a zero-impedance current-measuring meter onto this power source so that the meter would short out all the current in question through itself. Well, we can't actually do this, but we can figure the circuit in question just as though we could. Fig. 1-14 shows the circuit in question, so let's just use Ohm's Law to predict the current that would flow in a shorting bar placed across the output. Looking at the Norton Equivalent, we can see that all the current the two generators can supply will now add algebraically in the shorting bar.

Zero volts at point A gives us 100V over 10K for 10 ma and 12V over 1K for 12 ma. Looking at the direction of current flow for both currents (electron theory) in the shorting bar shows that when these two currents from the two constant-current generators are added algebraically, we will have a total I_{sc} (short-circuit current) of 10 plus 12 or 22 ma for our Norton Equivalent generator (Fig. 1-16).

Now, looking back into the composite circuit in Fig. 1-15, it is easy to see that R_1 and R_2 are again seen in parallel just as they were in the "Thevenin Equivalent Impedance" (Z_{th}). Their product over their sum is again 10K divided by 11K, or 0.91K ohms.

We now have values for both parts of the Norton Equivalent of the circuit we started with. The current generator causes electron current to flow into ground and pulls it up through its own back impedance when no load is applied to this source, which is a 22 ma current generator with 0.91K ohms impedance seen in parallel with it.

Pay attention to the fact here that if the entire 22 ma does flow through the shunting 910 ohms of impedance, we will have our open-circuit Thevenin voltage of ($22 \times 10^{-3} \times .91 \times 10^{+3}$ = 20 volts), that's right, 20 volts, just as we figured in the Thevenin discussion!

Right here, let's do as we did before. Put this concept to some constructive use. We'll take the same transistor circuit we did before, too; the one with the split load of 5K to +70 volts and 1K to +10 volts, now shown in Fig. 1-17.

The Norton Equivalent of this load will have a constant-current generator causing 70 volts over 5K, or 14 ma, along

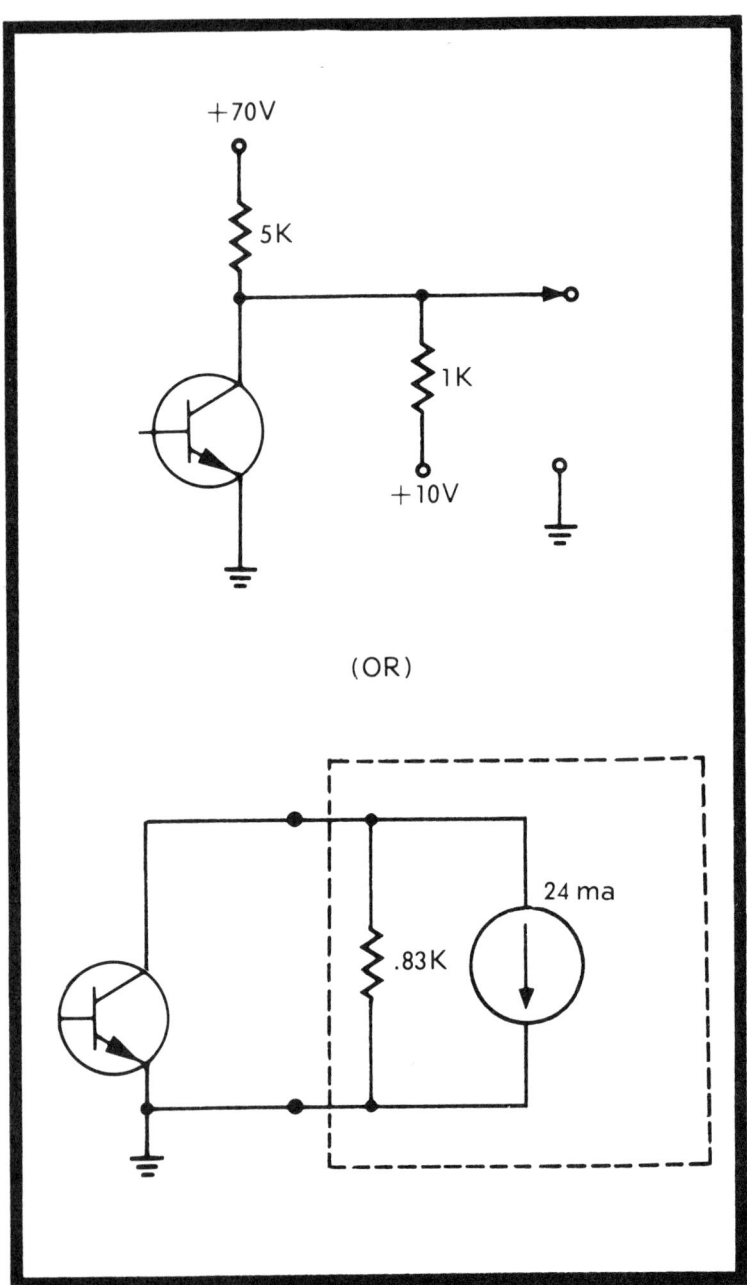

Fig. 1-17. Norton equivalent of Fig. 1-6. (Courtesy Tektronix, Inc.)

with 10 volts over 1K, or 10 ma. These two values are in the same direction, so they add and become the value for the constant-current source of 24 ma. The two resistances are seen in parallel again, for a combined value of 5K over 6K or 830 ohms. When we look at this Norton Equivalent in conjunction with the transistor (Fig. 1-17) we can easily see that the transistor can't possibly have any more current than 24 ma even if it were a complete short circuit, and as long as this is below the peak current rating for the device, we can't damage it by using it in the circuit. If this is a switching circuit, the 24 ma would be a good estimate of the amount of current change the supply would have to be capable of handling without shifting its own characteristics. And, if it won't handle it, we had better turn another transistor "on" as we turn this one "off" so the supply doesn't see the change of current at all.

Of course, with the transistor turned off, it is also easy to see that with this circuit we will never have a voltage greater than the Norton Generator in question can create by driving all its own current (24 ma) through its own back impedance (.83K ohms) across the transistor. This will be 24 x .83 or about 20 volts, and as long as this is less than BV_{ce} for the transistor, we are safe for this limitation.

Millman's Composite Theorem

The use of the Norton approach doesn't stop here, however. It's my personal feeling that it was probably Millman's use of the Norton approach or concept to the composite Thevenin type problem that lead Millman to develop his equation:

$$V_{out} = \frac{\frac{V_1}{R_1} + \frac{V_2}{R_2} + \frac{V_3}{R_3} + \frac{V_4}{R_4}}{\frac{1}{R_1} + \frac{1}{R_2} + \frac{1}{R_3} + \frac{1}{R_4}} \quad \text{(Eq. 1-9)}$$

when solving the problem represented in Fig. 1-18. This is the same circuit problem used to wind up the Thevenin discussion.

If we use the Norton approach of short-circuiting the output of this composite circuit to determine the current of the Norton Equivalent, you will note that we are solving the numerator of the Millman Equation. And, when we look back into this circuit to find the R_{th} and solve for the single value of

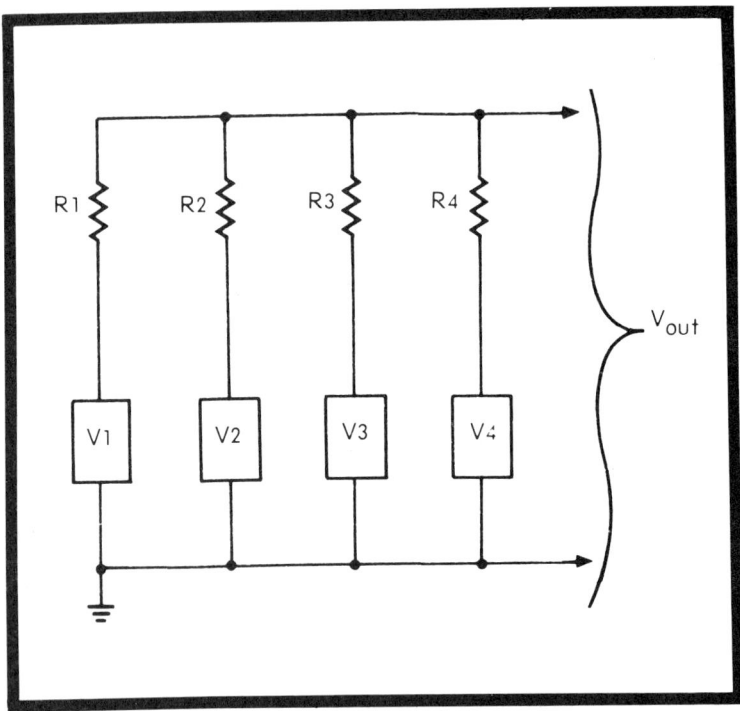

Fig. 1-18. Type of circuit where Millman's equation is used to solve for Vout.

all at one time, we have to take the reciprocal of the sum of the reciprocals, which is the solution of one over the denominator in Millman's Equation. Of course, it all comes back to Ohm's Law for final solution of V_{out} when we realize that Millman merely took his product of current and resistance to predict voltage:

$$(V_{out} = I_{sc(total)} \times R_{th(total)}). \quad \text{(Eq. 1-10)}$$

Now you have studied the basic tools upon which the remainder of this book is based. I suggest that if you are not very sure you understand the theorems, draw some simple circuits using different values and practice using the formulas until you consider yourself facile in their use; do this, and perhaps even review Chapter 1, before you proceed. And if you've been having a spot of trouble in manipulating Powers of Ten, go to the Appendix and use the programmed text you'll find there.

HERE'S A LITTLE GADGET THAT MIGHT HELP YOU WITH YOUR POWERS OF TEN AS YOU CONTINUE IN THIS BOOK. CAREFULLY CUT OUT THE TWO CIRCLES ON THIS PAGE AND THE NEXT PAGE, AND PASTE THEM ON CARDBOARD. ALSO CUT OUT TWO LITTLE SHADED AREAS OF THE SMALLER CIRCLE.

POKE A PIN THROUGH THE EXACT CENTER OF EACH OF THESE CIRCLES AND THEN MOUNT THE SMALLER ONE ON TOP OF THE LARGER ONE SO THAT IT TURNS FREELY, BEING SURE THAT THE PRINTED SIDE IS UP FOR EACH CIRCLE. THE BLACK DOT ON THE BIG CIRCLE SHOWS THE DECIMAL PLACEMENT FOR UNITS.

AS YOU TWIST THE LITTLE CIRCLE AROUND, THE "POWER" WINDOW WILL GIVE YOU THE POWER OF TEN TO RETURN THE DECIMAL TO THIS POINT, AND EVERY THIRD POWER OF TEN PREFIX WILL SHOW IN THE OTHER WINDOW.

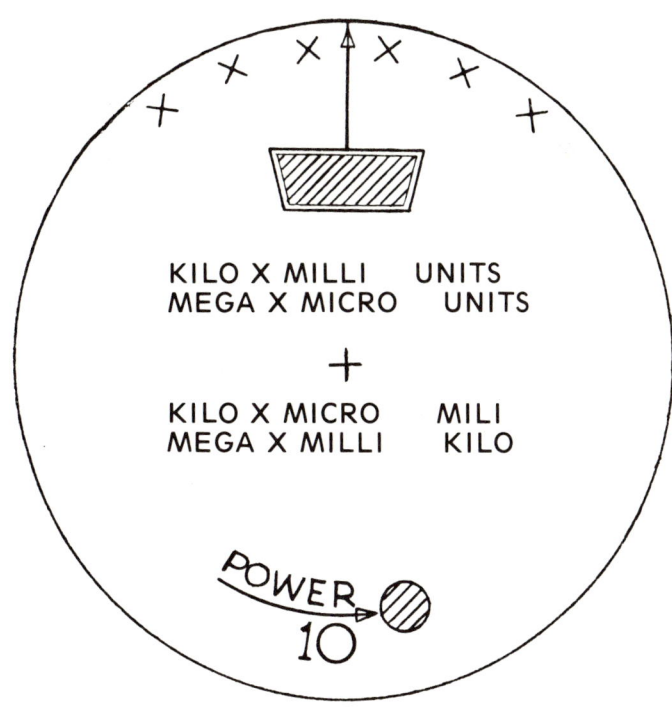

Back of this page left blank intentionally to permit cutting out circle.

33

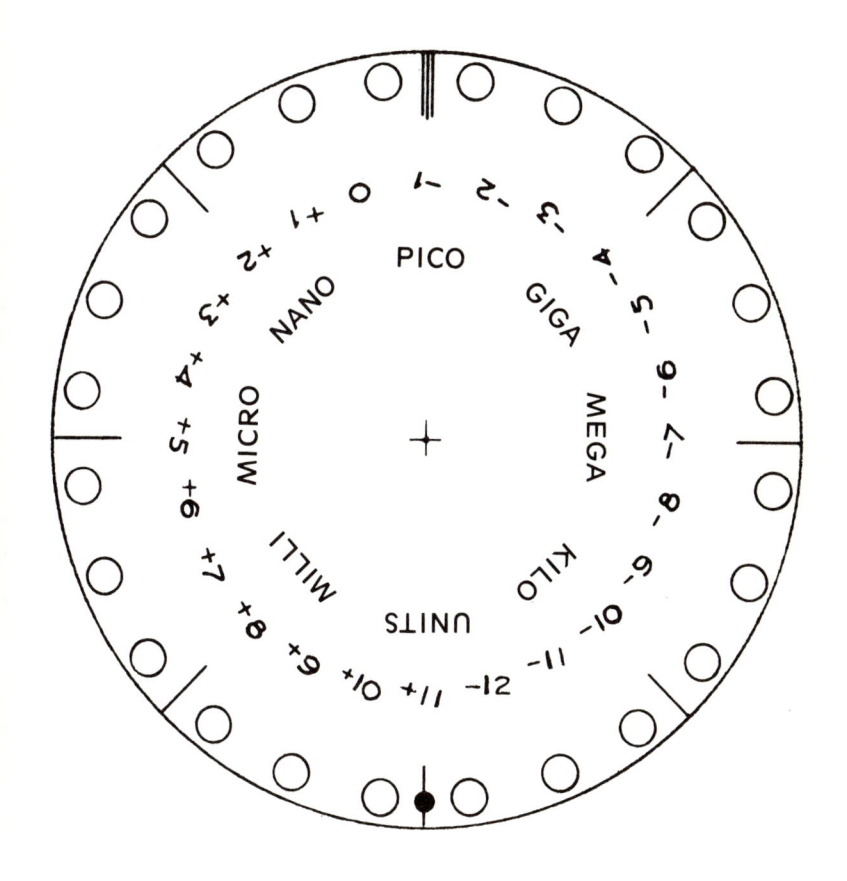

Back of this page left blank intentionally to permit cutting out circle.

HERE'S ANOTHER LITTLE GADGET THAT MIGHT HELP YOU KEEP OHM'S LAW STRAIGHT FOR YOURSELF AS WELL AS A FEW OF THE OTHER EQUATIONS USED IN THIS BOOK. ITS USE SHOULD BE SELF-EVIDENT AS YOU CONTINUE THROUGH THE BOOK.

CAREFULLY CUT THE TWO CIRCLES OUT, PASTE THEM ON CARDBOARD, AND PUT A PINHOLE IN THE EXACT CENTER OF EACH SO THAT YOU CAN MOUNT THEM ON THE SAME PIN, PRINTED SIDE UP.

CUT OUT THE SHADED HOLE IN THE SMALL CIRCLE SO THAT YOU CAN SEE THE EQUATION FOR THE NAME BRACKETED BY THE TWO ARROWS.

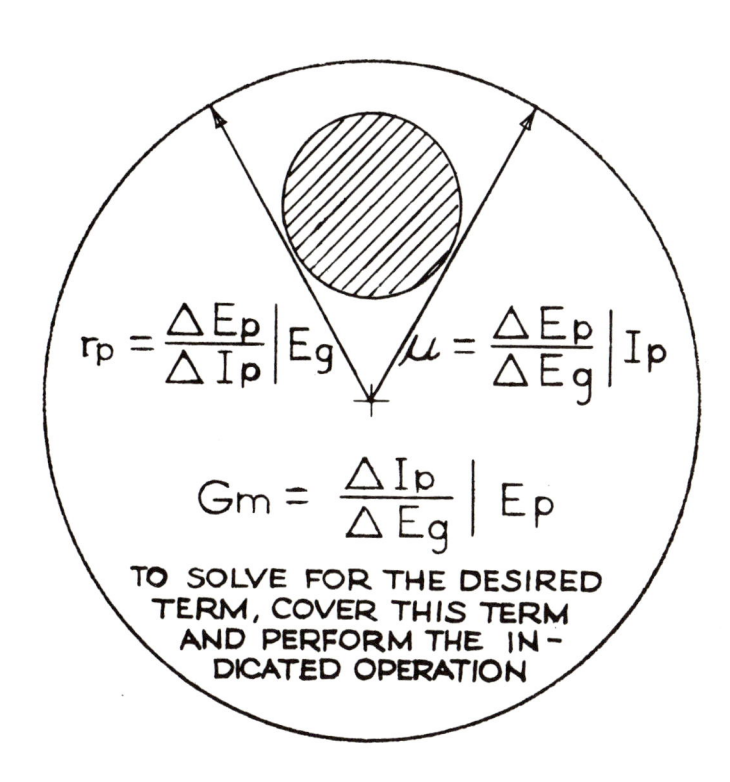

$$r_p = \frac{\Delta E_p}{\Delta I_p} \bigg| E_g \qquad \mu = \frac{\Delta E_p}{\Delta E_g} \bigg| I_p$$

$$G_m = \frac{\Delta I_p}{\Delta E_g} \bigg| E_p$$

TO SOLVE FOR THE DESIRED TERM, COVER THIS TERM AND PERFORM THE INDICATED OPERATION

Back of this page left blank intentionally to permit cutting out circle.

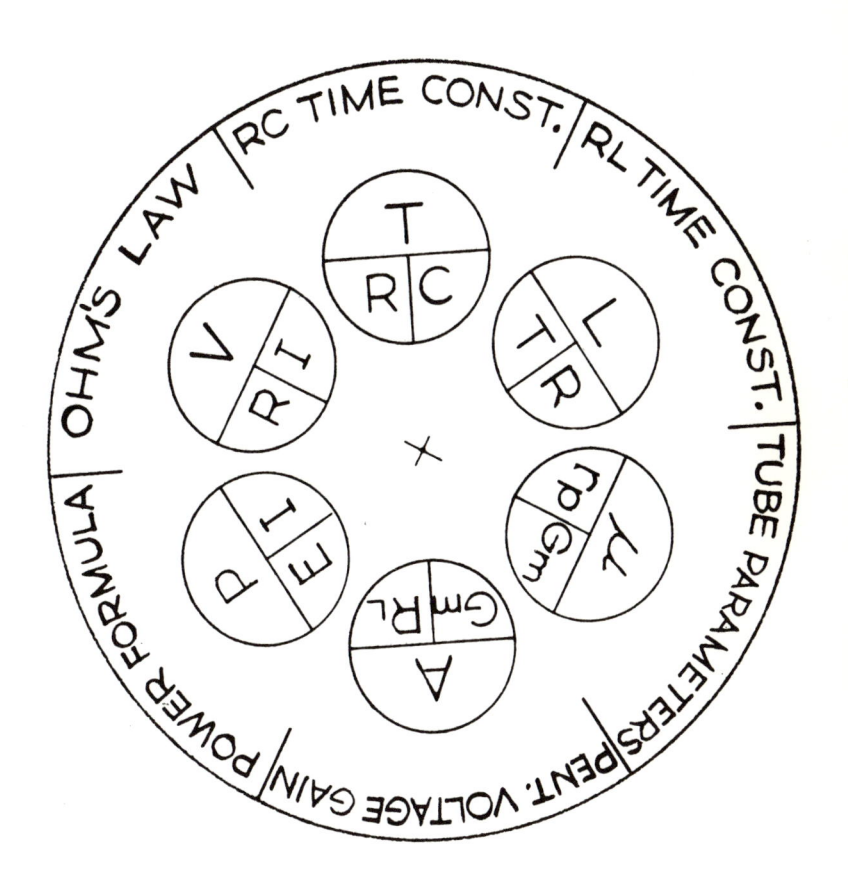

Back of this page left blank intentionally to permit cutting out circle.

L, R, C, and Time Constants

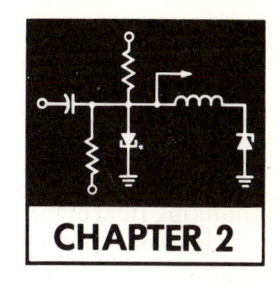

CHAPTER 2

For this chapter, let's add Mr. Henry's Inductor and Mr. Faraday's Capacitor to the general picture of electronics. And, since most texts treat these two items (L and C) relative to their reaction to sine wave signals, that is, AC signals of different frequencies, we shall, too. I will, however, make it as brief as possible before getting into a little bit different viewpoint. The viewpoint I'm speaking of is the Time-Constant concept. At first it sounds a lot more complicated, but if we have a copy of the curve, it's a whole lot easier. We'll get into this later though.

Right now let's concentrate on L (inductance measured in Henrys) and C (capacitance measured in Farads). These two items are normally thought of as being exact opposites, the capacitor being an open circuit to DC, and the inductor being a short circuit to DC. Neither one of these items act like this, though, under certain conditions. They both have other properties.

REACTANCE FORMULAS

You should know that a capacitor opposes change of voltage (such as a filter capacitor on a voltage supply) and that an inductor opposes change in current (such as a power supply choke coil). Also, you should remember that X_C equals 1 divided by $2\pi fC$ and that as either or both f and C get bigger, X_C proceeds to get smaller. Then with the inductor being the opposite, its formula is L equals $2\pi fL$ and as either or both f and L get bigger, X_L gets bigger, too. In other words, X_C is inversely related to frequency and the value of the device while X_L is directly related to frequency and the value of the device.

So what is a capacitor and what is an inductor? Well, let me describe their basic physical structures to you.

THE INDUCTOR

The Inductor (L) is a coil of conductive material (usually copper or aluminum wire). Inductors come in all sizes and are

41

measured in terms of the Henry. Now the Henry is a mighty big amount of inductance, so usually you will find a coil rated in millihenries (10^{-3} henrys or thousandths of a henry) or microhenries (10^{-6} henrys or millionths of a henry). It is enough to say here that a change of current within our conductor in the shape of an inductance (coil) generates a changing field of magnetic force that—as it cuts adjacent coils of the conductor—generates current opposing that which caused it in the first place. This just says that a coil will look like a high resistance to a current that is changing rapidly (high frequency) while it will look like a low resistance to a slowly changing current (low frequency). After all, this is just what the formula for X_L tells us. This opposition to current flow isn't really resistance either because it can't be directly added to any real resistance in the circuit. It's something a little bit different that we call Reactance. And for a coil, this is called Inductive Reactance, X_L.

THE CAPACITOR

The Capacitor (C) is made up of two conductive plates separated from each other by a layer of insulator material. It's actually an open circuit. These devices also come in all different kinds of styles and sizes. They are measured in terms of the Farad. And the Farad is an awful lot of capacity just like the Henry is a lot of Inductance. So, we will find lots of them in terms of microfarads (10^{-6} farads) and picofarads (10^{-12} farads).

Now, I used to think that it was possible to push extra electrons onto one plate of a capacitor without taking them off the other plate. And this line of thought got me into all kinds of trouble. It may be possible and it may not be possible, but I find it to my advantage to think of it as being impossible. I'm no physicist, but it stands to reason to me that the atomic structure of the material of the capacitor (as a whole single unit) has just so many protons within the nuclei that make up the atomic structure of the device and thus just so many electrons as well. Now some of these electrons are movable, provided the total electron population of the device isn't changed. Generators (batteries, power supplies, etc.) can be applied to capacitors to shift this electron population more to one side of the insulating material between the plates by way of the external circuit than the other. This gives us a voltage difference from one plate to the other, and this is the way I prefer to think of it. A voltage difference is developed. The capacitor is not **charged** or **discharged**. I don't like these words. They lead to arguments in semantics because they mean different things to different people. Don't use these two words in relation to capacitors if you can possibly stay away from them. They do nothing but confuse the issue.

MULTIPLE INDUCTORS

Now then, how do multiple inductors combine and how do multiple capacitors combine? Let's take them in turn.

The formula for a simple air-core inductor gives us the key to the very confusing issue of series or parallel inductors.

$$L = \frac{N^2 \times 3.192 \times A}{10^8 \times (\text{length})} \qquad \text{(Eq. 2-1)}$$

where L = the inductance of the coil, in henrys
3.192 = the permeability of air in perms per inch cube
A = the square inch cross-sectional area of the inside of the coil
N = the number of turns of the coil
(Length) = length of the coil in inches.

The formula states that L is directly proportional to the number of turns squared. This tells us that if we can double the number of turns of a coil without changing any of the rest of the measurements of the coil, we will get four times the inductance (i.e., 2^2 equals 4). However, this is impossible to achieve when hooking up two identical inductors in series. We would be bound to get something less than four times the value of one of them. BUT, just how much less, we're not exactly sure, because we have difficulty evaluating the degree of coupling between the two fields. Of course, if we can keep the two fields completely separated from each other, we can get the sum of the two, just as we would get half the inductance of a parallel pair of identical inductors, provided we could keep their fields separated. It is for this reason that combinations of inductors when commercially manufactured include some means of variable coupling so they can be tuned to the critical amount desired, rather than their being made accurately to begin with.

As for capacitors, they're a lot more easily handled and understood. However, before leaving inductors, I suppose I'd better at least list the basic formulas for the different combinations of inductors.

Parallel Inductance (no coupling): $L_T = \dfrac{L_1 \times L_2}{L_1 + L_2}$ (Eq. 2-2)

Parallel Inductance (fields aiding):

$$L_T = \dfrac{1}{\dfrac{1}{L_1 + M} + \dfrac{1}{L_2 + M}} \qquad \text{(Eq. 2-3)}$$

where M is the mutual inductance.

Parallel Inductance (fields opposing):

$$L_T = \dfrac{1}{\dfrac{1}{L_1 - M} + \dfrac{1}{L_2 - M}} \qquad \text{(Eq. 2-4)}$$

Series Inductances:

$$L_T = L_1 + L_2 + 2M \quad \text{(fields aiding)} \qquad \text{(Eq. 2-5)}$$

$$L_T = L_1 + L_2 - 2M \quad \text{(fields opposing)} \qquad \text{(Eq. 2-6)}$$

And, finally, Mutual Inductance; $M = \dfrac{L_A - L_B}{4}$ (Eq. 2-7)

Where L_A = total Inductance of 2 coils, fields aiding
and L_B = total Inductance of 2 coils, fields opposing.

Now for capacitors. These devices don't have any external fields and there is no mutual inductance or coupling to drastically louse up our calculations even though they may not be absolutely exact. You see, there are such things as temperature and humidity as well as maximum currents that different capacitors can handle, and these items can change our answers slightly. However, our answers will probably be much more accurate than the actual capacitors we might build into the physical circuit. These items are usually rated as being within 20 percent of their rated value. If you buy 10 percent capacitors, you'll have to pay more for them. AND, if you want 5 percent capacitors, you can get them, but a variable trimmer for the capacitor you wish to make exact is a lot less expensive.

The basic formula for the capacitance between two flat plates separated by a dielectric material gives us some clues as to how these items combine.

$$C = \dfrac{0.0885 \ k \ A}{d} \qquad \text{(Eq. 2-8)}$$

where C = capacitance in picofarads
k = the dielectric constant for the dielectric material used
A = the area of one plate in square centimeters
d = the distance between the plates in centimeters.

The two items in the formula that concern us most are A and d. Note that capacitance gets bigger as the areas of the plates get bigger. C and A are directly related to each other. If A is doubled, then C is doubled. Note now that C and d are indirectly related to each other. In other words, if the distance between plates gets bigger, then the value of C goes down by the same percentage. For example, doubling d decreases C by 50 percent. In still other words, if we hook up two capacitors in series, we are increasing the dielectric thickness (d in the formula) and the total capacity must go down. If, however, we hook up two capacitors in paralle, then we are increasing the area of the plates and the total capacity of this configuration must be greater. If you can remember this, it will help you keep the formulas for capacitors striaght.

MULTIPLE CAPACITORS

For capacitors in series;

$$C_{total} = \frac{1}{\frac{1}{C_1} + \frac{1}{C_2} + \frac{1}{C_3} + \text{etc.}} \qquad \text{(Eq. 2-9)}$$

Yes, series capacitors combine just like parallel resistors or parallel inductors. It's the same old reciprocal of the sum of the reciprocals. If it's just two capacitors in series, then you can use the product divided by the sum to get C_{total}. That is,

$$C_T = \frac{C_1 \times C_2}{C_1 + C_2} \qquad \text{(Eq. 2-10)}$$

For capacitors in parallel;

$$C_{total} = C_1 + C_2 + C_3 + \text{etc.} \qquad \text{(Eq. 2-11)}$$

And parallel capacitors just add like series resistors or series inductors with no mutual inductance. The same old basic formulas all over again.

As I mentioned in the first part of this chapter, capacitance and inductance have a quality that looks like resistance but isn't quite the same thing. This is true for alternating currents and voltages only. To DC, the capacitor is an open circuit and an inductance is a short circuit. You see, with an AC source, voltage is continually changing (up and down or back and forth—however you like to think of it) and

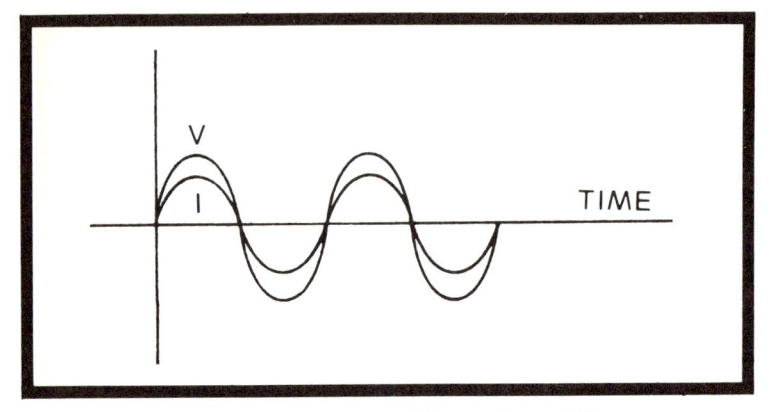

Fig. 2-1. Voltage and current with zero phase difference.

current is doing the same thing (Fig. 2-1). If these two things (i.e., current and voltage) don't reach peaks in the same direction at the same time, they are incapable of having the maximum peak power (product of the two) that they would have if they were exactly in step. The difference between these two levels of peak power is an apparent loss of power due to this quality in capacitors and inductors that looks like resistance but isn't. You see, if we could throw current and voltage back in step again, there would be no power loss. That is exactly what happens in a series resonant circuit. REMEMBER:

$$X_C = \frac{1}{2\pi fC} \qquad \text{(Eq. 2-12)}$$

and

$$X_L = 2\pi fL. \qquad \text{(Eq. 2-13)}$$

Each is a function of the 360 degree circle; therefore, it is the 6.28 (2π) factor in each reactance formula. You see, if you draw a circle (or at least go through the motions of doing so) and move the paper from right to left while moving your pencil up and down, you get a picture of what an AC signal might look

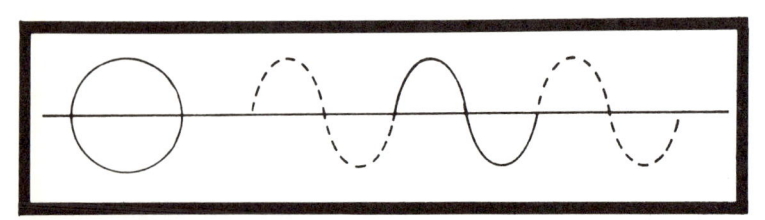

Fig. 2-2. Sine wave.

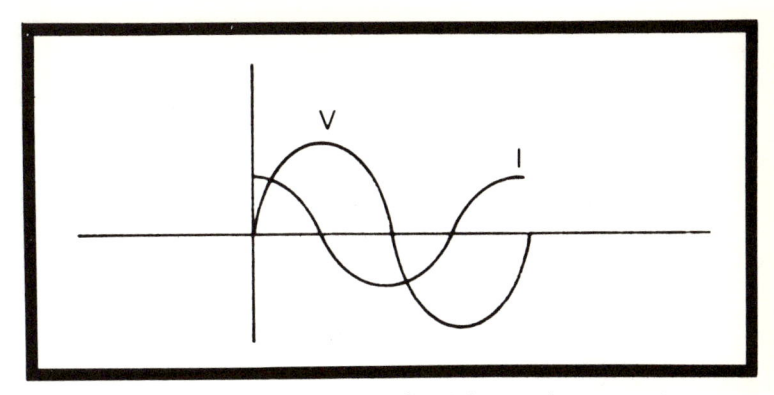

Fig. 2-3. Current leads voltage by 90 degrees (capacitive).

like. If the motions are absolutely uniform, you get what we call a sine wave. Or, as I see it, we get a waveform shaped like a bosom. Yes, there are numerous other more exact mathematical definitions for a sine wave, but this is the one I remember. (See Fig. 2-2.)

GRAPHICAL ANALYSIS

If an AC power source were hooked up across a pure capacitance with the bottom end of both items grounded, the power applied to the capacitor would be less than the source capability because current would lead voltage by 90 degrees (Fig. 2-3).

If the same power source were hooked up across a pure inductance with the bottom end of both items grounded, the same thing would happen but in reverse. The current would lag the voltage by 90 degrees (Fig. 2-4).

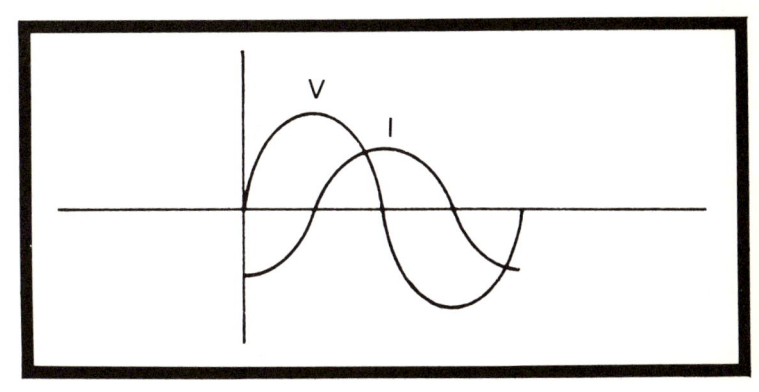

Fig. 2-4. Current lags voltage by 90 degrees (inductive).

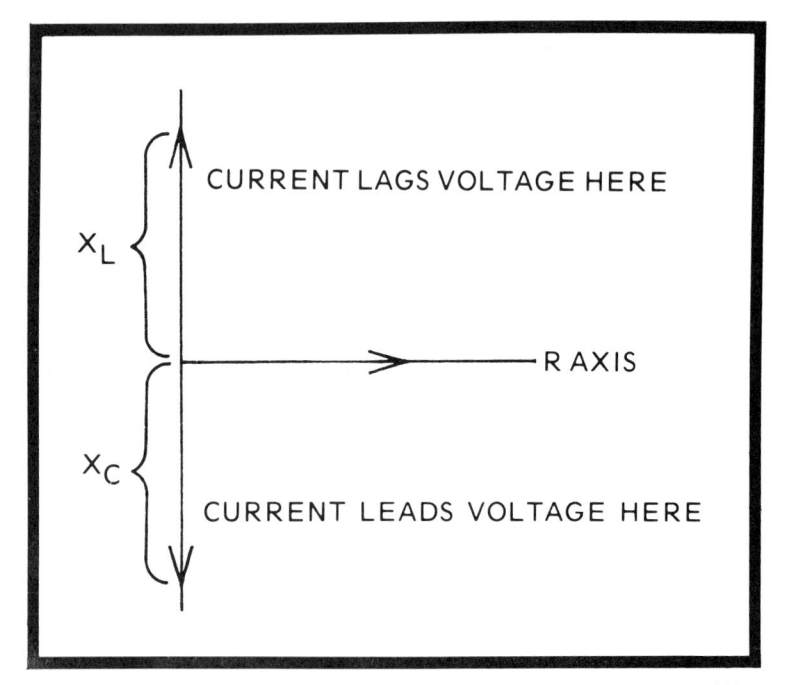

Fig. 2-5. Inductive reactance (X_L) and capacitive reactance (X_C) graphed with resistance axis.

For these reasons, for a circuit containing C, L, and R, when true resistance is graphed in a horizontal direction representing zero degrees of lead or lag for current with respect to voltage, we find X_L graphed in the vertical direction, while X_C is graphed in the downward direction, exactly opposite to the inductive reactance (Fig. 2-5). When we have X_L and X_C in a series circuit then, we will always take their difference to find their total effect. Therefore:

$$X_{total} = X_L - X_C$$

And, if X_L and X_C are equal, we will have a series-tuned circuit where the current is thrown out of phase with respect to the voltage by the first component and back in phase by the second component and no power will be used up by the two components (Fig. 2-1). This happens only at one frequency, but each combination of L and C has a frequency where this will happen. Thus, if we express the formula for $X_L - X_C$ in terms of the formula for each we will get;

$$2\pi f L = \frac{1}{2\pi f C}$$

and solve for f, we find that:

$$f \cdot f = \frac{1}{2\pi \cdot 2\pi \cdot C \cdot L}$$ and

by taking the square root of both sides of the equation we got after transposing, we obtain:

$$f = \frac{1}{2\pi\sqrt{CL}}$$ (Eq. 2-14)

and now we have a formula for finding the resonant frequency of any combination of L and C.

If, however, we are working with some frequency that is not the resonant frequency in a series circuit having L, C, and R, we have a different problem. We must go back to the two formulas for X_L and X_C and work them out (Eqs. 2-12, 2-13). These two values can then be graphed as in Fig. 2-5. Then using the scale, draw in the value for R at right angles as shown (Fig. 2-6). We find the difference between X_L and X_C

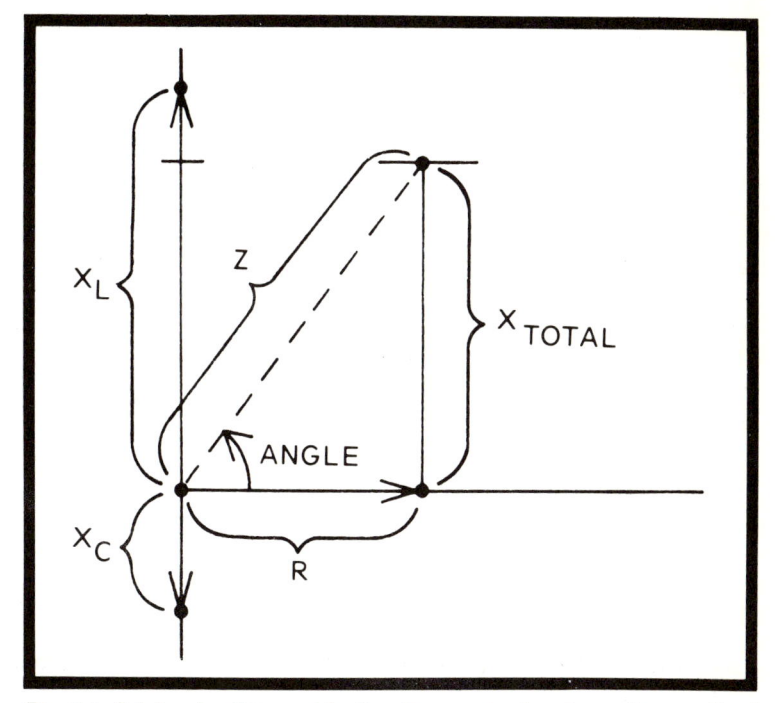

Fig. 2-6. Solving for Z_T graphically; the graph also shows the resultant current-to-voltage phase angle.

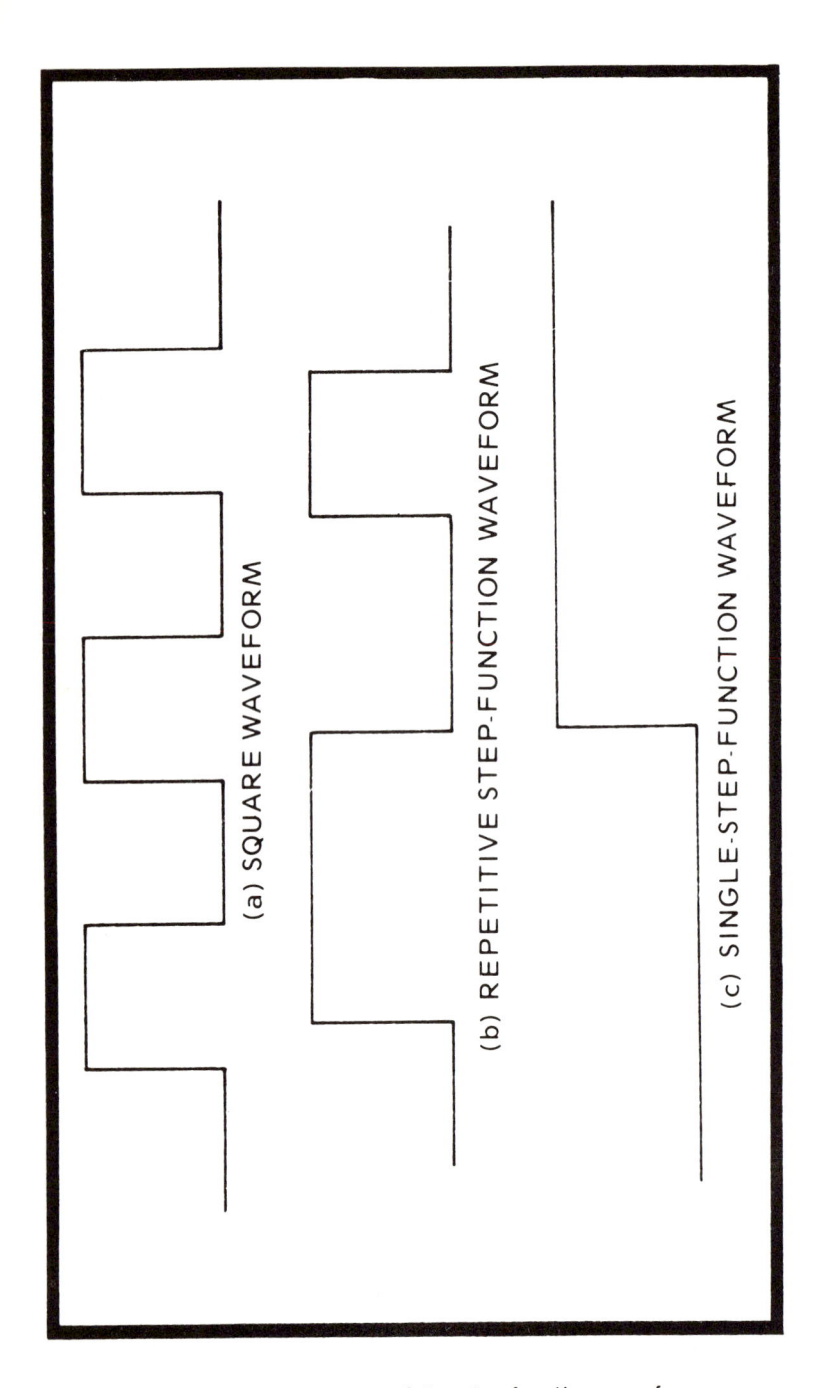

Fig. 2-7. Three variations of the step-function waveform.

and draw it in at the end of R. These two values now give us the right angle part of a triangle. The hypotenuse of this triangle becomes a graphic measurement of the total impedance of the circuit. This is called vector addition, where:

$$Z = \sqrt{(X_L - X_C)^2 + R^2} \qquad \text{(Eq. 2-15)}$$

(i.e., the square root of the sum of the squares.) It's nothing more nor less than the old Pythagorean Theorem you learned in plane geometry.

If the L, C, and R components of our circuit are in parallel, we figure the currents through the three devices as calculated by Ohm's Law in a new vector addition problem to get the total current that flows, and go back to Ohm's Law to find the total impedance by means of total current and the common voltage applied.

We now should be able to predict fairly well what will happen to a single-frequency sine wave when it encounters an R - C or an R - L combination of parts. But, how often do we have to do this nowadays? The answer to that question is, "Not very often." Present day circuitry is much more liable to have to handle composite frequencies; that is, waveforms made up of numerous frequencies, waveforms with shapes we do not want to change. And, if our circuitry does react oddly to some frequencies, it will change the shape of any waveform that contains those frequencies.

So, let's take a look-see at the most common wave shape used in this present day and age for this broad-band (many frequencies) analysis of a circuit. This is the square wave or step-function waveform, Fig. 2-7.

STEP-FUNCTION WAVEFORMS

The square wave is just what its name implies. It is a square-cornered waveform that pops up to a given level of voltage and then drops back down to the voltage it came from, over and over again. It will be up 50 percent of the time and down 50 percent of the time in equal increments. In this respect, it is different from the step waveform. The step waveform does not necessarily have a set frequency of repetition, but it does have a constant amplitude. In other words, it will go from one voltage to another one by means of an abrupt vertical step. The repetition rate with which this action is repeated is not necessarily constant, nor is the time that it stays at the upper voltage level.

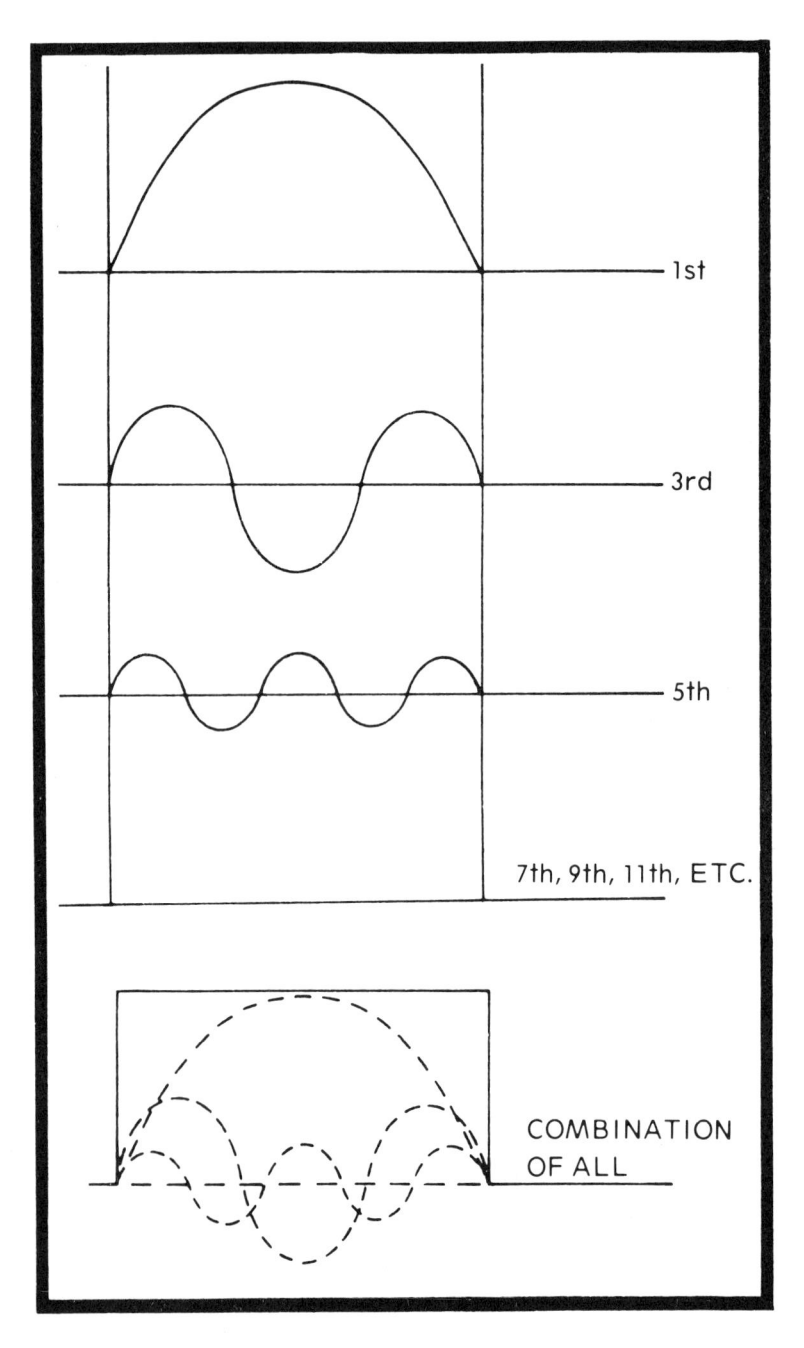

Fig. 2-8. Combination of odd harmonics to make a square wave.

Both waveforms, however, have a high-frequency content that is important to us. The sharper the leading-edge corner, the greater the high-frequency content of the waveform. This is important to us because we cannot test the high-frequency reaction of a circuit without using those frequencies.

Now, a square wave, of course, has the fundamental low frequency of its own basic repetition rate present. Then on top of this it contains odd harmonics (3rd, 5th, 7th, 9th, etc.) in phase with the original and at amplitudes reduced from the original by a factor represented by the harmonic that they happen to be with relation to the original (i.e., the 3rd harmonic will be 1/3 the amplitude of the original—the 5th harmonic will be 1/5 the voltage amplitude of the original, etc.) as in Fig. 2-8.

This sounds like an awfully complicated waveform, but in reality it can be roughly created by a simple make and break contact between some voltage and ground. I say roughly created because all circuits have some capacitance and-or inductance as well as resistance present. We haven't become that perfect yet. There is inductance in a piece of wire a quarter of an inch long and perfectly straight. And, as a friend of mine used to say, there's capacitance between pin heads at twenty miles. Also, it's a fact that there is no perfect conductor with absolutely no resistance (except that we are approaching it at cryogenic temperatures near absolute zero, but none of the things you and I are working on in the foreseeable future will operate at those temperatures, near -273 degrees Centigrade).

Usually, there is more capacitance than inductance, so we will consider capacitance and resistance first. The slight amount of inductance present will cancel a small amount of the capacitance present and we will be left with R and C as the main elements that will be limiting the voltage change that our perfect step generator will be thought of as creating.

The first thing to consider, I think, is just how does a capacitor change voltage with current flow? Well, just how does water level in a bucket change with a flow of water into it? If the flow of water is steady, then the rise of the water level is steady, and we have a set rate (ratio) of change per unit time. Of course, the bigger the bucket, the slower the rise; and the greater the quantity of water flow into the bucket, the faster the rise. So we can say then that change of level per unit of time is directly proportional to the quantity of water flow (current) and is indirectly proportional to the size of the bucket (size of the capacitor).

54

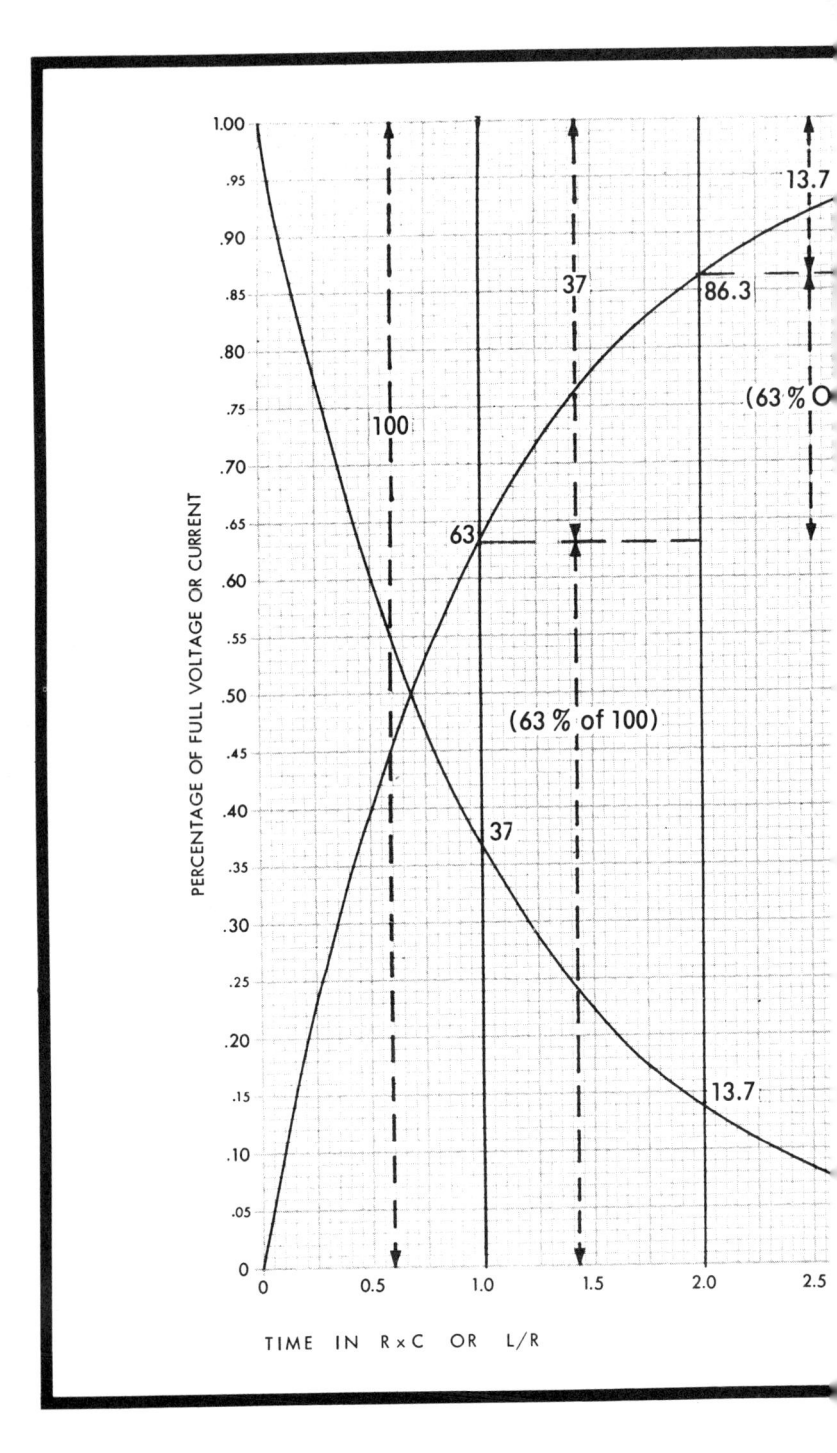

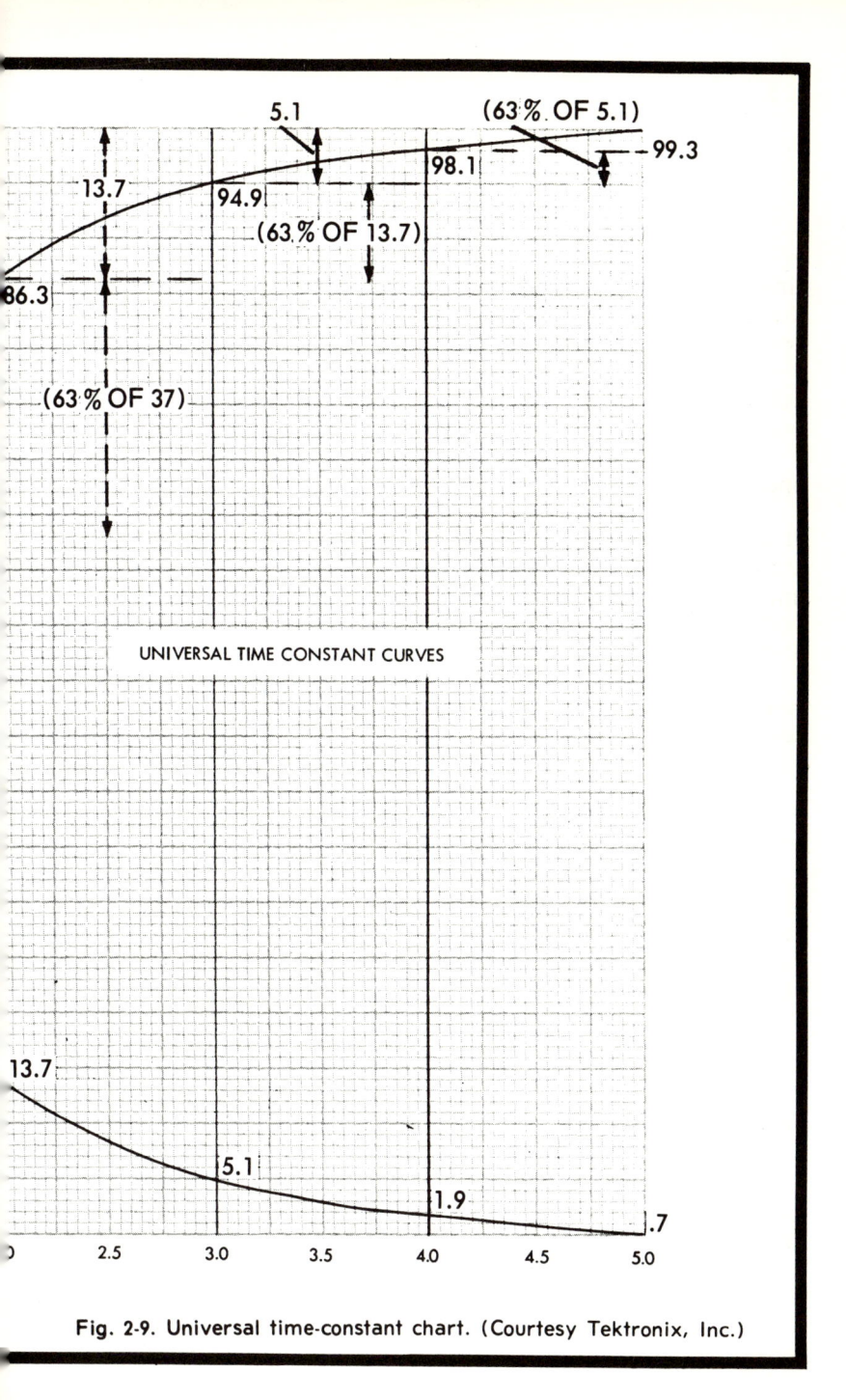

Fig. 2-9. Universal time-constant chart. (Courtesy Tektronix, Inc.)

This may seem like beating around the bush a bit, but I think it helps us to accept the equation:

$$\frac{dV}{dt} = \frac{I}{C}$$

This formula, change of voltage divided by the change of time to achieve it equals the steady current that flowed divided by the size of the capacitor, can be developed from the basics in this fashion also.

Q (the quantity of electrons necessary to change the voltage of a 1 farad capacitor 1 volt) has been defined in a couple of ways that are useful to us:

$$Q = C = V$$
and
$$Q = I \times dt$$

In still other words, Q is equal to the size of a given capacitor times a change of voltage achieved by a quantity of electrons moved from one plate to the other. And, Q is equal to a specific rate of electron flow times the time that it flowed. Both statements make sense if you think about it.

If the two Q values equal each other, then we should be able to say: C x dV equals I x dt. Now, if we transpose C and dt, we get:

$$\frac{dV}{dt} = \frac{I}{C} \qquad \text{(Eq. 2-16)}$$

As basic and important as this formula is, it isn't quite in the terms I want it in for our next step, so let's change it a bit more. Remember Ohm's Law? One of its forms stated that I = V/R. Let's use this in place of the I in the formula we just developed to get:

$$\frac{dV}{dt} = \frac{V}{CR} \qquad \text{(Eq. 2-17)}$$

TIME CONSTANTS

Now, if dV is equal to V, then the product of R and C must equal time. And, we say that it does. It is called the circuit Time Constant. This is the amount of time it would take for the capacitor in a series R-C circuit to take on the full applied voltage of a step function if the current maintained its original rate of flow. Of course, it doesn't really happen in a simple series R-C circuit by any manner or means. It takes extensive extra circuitry to achieve this, but that's another story for later on in this book.

Another way of defining the Time Constant (that which is the product of R and C in a circuit—or it could be L divided by R) if the circuit is capacitive in nature is to say that this is the time it takes the capacitor to change voltage by 63 percent of the applied step. For an inductive circuit, this is the time for 63 percent of the applied voltage step to be developed across the resistor in the circuit. And the voltage in question continues to change as time constants go by; it changes by 63 percent of the remaining portion of the step function at the start of each time constant.

Putting this in other words, at the end of the first time constant the voltage in question will be 63 percent of the full original step function. The voltage we will add to this to give the voltage at the end of the second time constant will be 63 percent of the remaining 37 percent of the step function, which will give us a total of 86.3 percent of the step at this time. To find the voltage at the end of the third time constant, we will take 63 percent of the 13.7 percent of the step still remaining to get a new total of 94.9 percent of the total step. At the end of the fourth time constant we'll have 98.1 percent of the total; and at the end of the fifth time constant, we will have 99.3 percent of the total step. (See Fig. 2-9.) This is the easiest way to figure the curve and, of course, there is a more complicated way; but we'll get to that after we figure out what is going on in this circuit we're talking about.

SERIES INTEGRATOR CIRCUITS

Consider the circuit in Fig. 2-10. The capacitor is on the bottom with the resistor in series with it and the step function voltage is being applied to the top of the resistor while we are looking at the voltage as it develops across the capacitor. At time zero (T_0), the instant the voltage changes, we cannot change the voltage across the capacitor. (Remember the capacitor opposes voltage change.) So the full amplitude of the step voltage appears across the resistor in the circuit. Ohm's Law says that a current has to flow in this case and it does. Current starts to shift from the upper plate of the capacitor to the lower plate by way of the external circuit (the generator which gave us the step voltage). Now the voltage from plate to plate of the capacitor begins to change. Of course, Kirchhoff's Law states that the voltage across the capacitor added to the voltage across the resistor has to equal the amplitude of the step function at all times. So, as the voltage across the

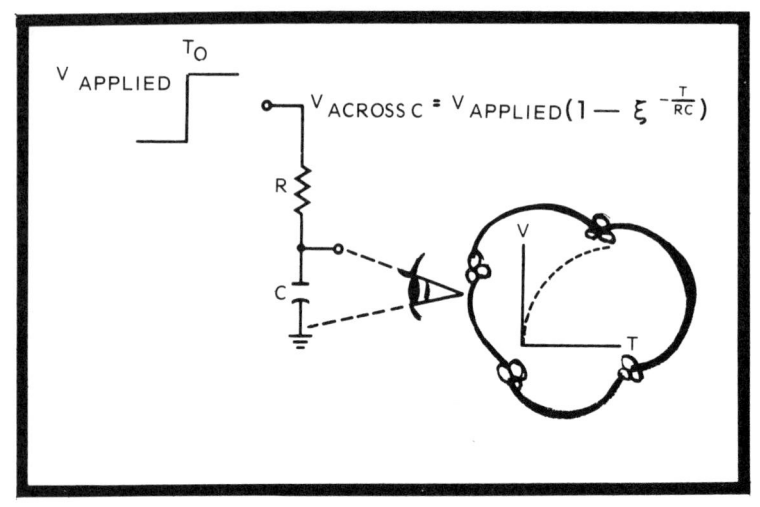

Fig. 2-10. The series R-C low pass filter or integrator circuit.

capacitor starts to go up, the voltage across the resistor starts to go down. Now note this: it is the voltage across the resistor that determines how much current flows in the circuit. So, the current shift starts to diminish also and the change of charge becomes less and less as each time constant goes by, until by the end of the fifth time constant it is general practice to say that all current has ceased to flow and there will be no further change.

What this tells us is that it will take five time constants for the capacitor in this circuit to take on the full voltage of the step function rather than one time constant as implied in the first definition of a time constant. For, you see, to make the capacitor take on the full step function voltage in one time constant, the current would have to be the same at all times and this circuit does not allow that to happen.

The inductive circuit that would give us the same wave shape at the output would look like that in Fig. 2-11. The inductor would be on top, with the resistor on the bottom and we would view the waveform across the resistor. (Remember, the inductor opposes change in current.) At T_0 in this circuit, the full step function voltage would appear across the inductor (L) and as the field of the inductor collapsed and allowed more and more current to flow, the voltage would build up across the resistor until (at the end of five time constants - L/R -) the full step function voltage would be across the resistor and maximum current would be flowing rather than no current at all, as in the capacitive circuit.

Since this waveform is almost like a ramp waveform (which would be the true shape of an integrated square wave) we call this circuit an integrator circuit. It isn't a true integrator, but it almost is.

Actually, this waveform follows the function of epsilon to the minus time after T_0 divided by R x C power, where epsilon is the base of the Naperian Log system equal to the base ten number 2.718. So, we must be taking the number

$$2.718^{-\frac{T}{RC}}$$

Now what does that mean? Well, if it's like a negative power of ten, I can change the negative to a positive by writing it as a reciprocal and note that it's related to

$$2.718^{\frac{T}{RC}}$$

Let's see, this should give us 63 percent or 0.63 at the end of one time constant or when T and R x C are equal, and 2.718 is raised to the first power or equal to itself. So, what is the reciprocal of 2.718? If our figures are correct it should come out to be equal to 0.37. And, we should be able to write the formula. At the end of one time constant,

$$V_{across\ C} = V_{applied} \times (1 - .37) = V_{applied} \times 63\%$$

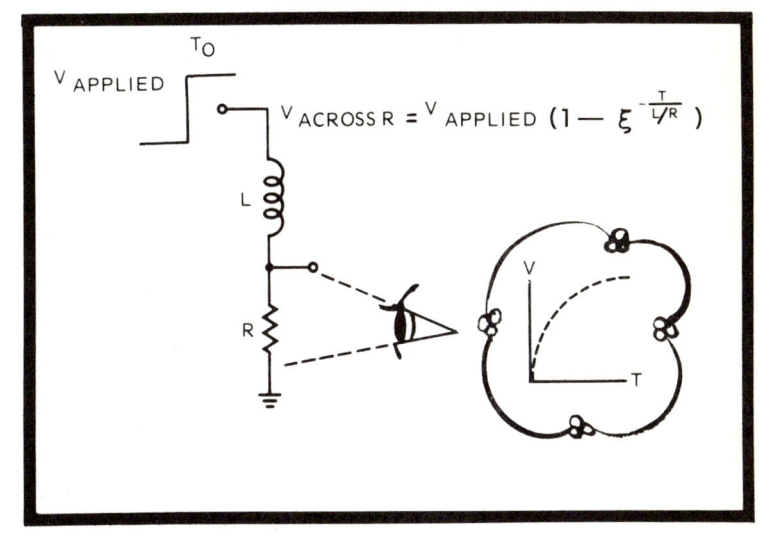

Fig. 2-11. The series R-L low pass filter or integrator circuit.

And the formula for the voltage across C at the end of any amount of time after T_0 will be:

$$V_{across\ C} = V_{applied} \times (1 - \varepsilon^{-\frac{T}{RC}})$$

Or, in the case of an inductive circuit:

$$V_{across\ R} = V_{applied} \times (1 - \varepsilon^{-\frac{T}{L/R}})$$

The key to the whole situation, you will note, is the time after T_0 divided by the time constant of the circuit. Now, I do not expect you to be able to raise the number 2.7183 (epsilon) to some decimal power. However, there are those who will expect you to be able to look it up. So I am including the table of Exponential Functions (Fig. 2-12) as part of this text. Assuming that you have found a value for time divided by time constant, you will look it up under the column labelled x. To find the value of epsilon raised to this minus power you simply go across to the column labelled e^{-x} and read the corresponding number. For instance, look up x equals 1.00. It's at the top of Fig. 2-12B. Reading across the page, you will find (three columns of numbers over) a column labelled e^{-x} and the number corresponding to the 1.00 value of x that you looked up is 0.367879. Rounding this number off to two significant numbers, you get 0.37. This is the number you subtract from 1.00 to get the 63 percent we started out with in this discussion.

The same conclusion can be reached by the use of your slide rule if it has a LL00 scale, the natural log scales Ln-0, Ln-1, Ln-2, and Ln-3 scales, or the LL0, LL1, LL2, and LL3 scales.

USE OF THE LL00 SCALE

You work this scale with reference to the A scale of your slide rule. If the power of epsilon were a minus 1, you would slide the cross-hair to the exact middle of the A scale in order to read the minus 1 power of epsilon now under the cross-hair of your slide rule on the LL00 scale. This would give you the number .37 which we subtracted from 1 to get 63 percent.

USE OF THE Ln-3 SCALE

The Natural Log scales work in conjunction with the D scale of your slide rule. Again, if the power of epsilon were a minus one, you would run the cross-hair of your rule to the exact left end of the D scale and read the value "0.37" or 1/e under the cross-hair on the Ln-3 scale.

EXPONENTIAL FUNCTIONS

x	e^x	$\text{Log}_{10}(e^x)$	e^{-x}	x	e^x	$\text{Log}_{10}(e^x)$	e^{-x}
0.00	1.0000	0.00000	1.000000	**0.50**	1.6487	0.21715	0.606531
0.01	1.0101	.00434	0.990050	0.51	1.6653	.22149	.600496
0.02	1.0202	.00869	.980199	0.52	1.6820	.22583	.594521
0.03	1.0305	.01303	.970446	0.53	1.6989	.23018	.588605
0.04	1.0408	.01737	.960789	0.54	1.7160	.23452	.582748
0.05	1.0513	0.02171	0.951229	**0.55**	1.7333	0.23886	0.576950
0.06	1.0618	.02606	.941765	0.56	1.7507	.24320	.571209
0.07	1.0725	.03040	.932394	0.57	1.7683	.24755	.565525
0.08	1.0833	.03474	.923116	0.58	1.7860	.25189	.559898
0.09	1.0942	.03909	.913931	0.59	1.8040	.25623	.554327
0.10	1.1052	0.04343	0.904837	**0.60**	1.8221	0.26058	0.548812
0.11	1.1163	.04777	.895834	0.61	1.8404	.26492	.543351
0.12	1.1275	.05212	.886920	0.62	1.8589	.26926	.537944
0.13	1.1388	.05646	.878095	0.63	1.8776	.27361	.532592
0.14	1.1503	.06080	.869358	0.64	1.8965	.27795	.527292
0.15	1.1618	0.06514	0.860708	**0.65**	1.9155	0.28229	0.522046
0.16	1.1735	.06949	.852144	0.66	1.9348	.28663	.516851
0.17	1.1853	.07383	.843665	0.67	1.9542	.29098	.511709
0.18	1.1972	.07817	.835270	0.68	1.9739	.29532	.506617
0.19	1.2092	.08252	.826959	0.69	1.9937	.29966	.501576
0.20	1.2214	0.08686	0.818731	**0.70**	2.0138	0.30401	0.496585
0.21	1.2337	.09120	.810584	0.71	2.0340	.30835	.491644
0.22	1.2461	.09554	.802519	0.72	2.0544	.31269	.486752
0.23	1.2586	.09989	.794534	0.73	2.0751	.31703	.481909
0.24	1.2712	.10423	.786628	0.74	2.0959	.32138	.477114
0.25	1.2840	0.10857	0.778801	**0.75**	2.1170	0.32572	0.472367
0.26	1.2969	.11292	.771052	0.76	2.1383	.33006	.467666
0.27	1.3100	.11726	.763379	0.77	2.1598	.33441	.463013
0.28	1.3231	.12160	.755784	0.78	2.1815	.33875	.458406
0.29	1.3364	.12595	.748264	0.79	2.2034	.34309	.453845
0.30	1.3499	0.13029	0.740818	**0.80**	2.2255	0.34744	0.449329
0.31	1.3634	.13463	.733447	0.81	2.2479	.35178	.444858
0.32	1.3771	.13897	.726149	0.82	2.2705	.35612	.440432
0.33	1.3910	.14332	.718924	0.83	2.2933	.36046	.436049
0.34	1.4049	.14766	.711770	0.84	2.3164	.36481	.431711
0.35	1.4191	0.15200	0.704688	**0.85**	2.3396	0.36915	0.427415
0.36	1.4333	.15635	.697676	0.86	2.3632	.37349	.423162
0.37	1.4477	.16069	.690734	0.87	2.3869	.37784	.418952
0.38	1.4623	.16503	.683861	0.88	2.4109	.38218	.414783
0.39	1.4770	.16937	.677057	0.89	2.4351	.38652	.410656
0.40	1.4918	0.17372	0.670320	**0.90**	2.4596	0.39087	0.406570
0.41	1.5068	.17806	.663650	0.91	2.4843	.39521	.402524
0.42	1.5220	.18240	.657047	0.92	2.5093	.39955	.398519
0.43	1.5373	.18675	.650509	0.93	2.5345	.40389	.394554
0.44	1.5527	.19109	.644036	0.94	2.5600	.40824	.390628
0.45	1.5683	0.19543	0.637628	**0.95**	2.5857	0.41258	0.386741
0.46	1.5841	.19978	.631284	0.96	2.6117	.41692	.382893
0.47	1.6000	.20412	.625002	0.97	2.6379	.42127	.379083
0.48	1.6161	.20846	.618783	0.98	2.6645	.42561	.375311
0.49	1.6323	.21280	.612626	0.99	2.6912	.42995	.371577
0.50	1.6487	0.21715	0.606531	**1.00**	2.7183	0.43429	0.367879

Fig. 2-12A. Table of exponential functions (sheet 1 of 6).

EXPONENTIAL FUNCTIONS (Continued)

x	e^x	$\text{Log}_{10}(e^x)$	e^{-x}	x	e^x	$\text{Log}_{10}(e^x)$	e^{-x}
1.00	2.7183	0.43429	0.367879	**1.50**	4.4817	0.65144	0.223130
1.01	2.7456	.43864	.364219	1.51	4.5267	.65578	.220910
1.02	2.7732	.44298	.360595	1.52	4.5722	.66013	.218712
1.03	2.8011	.44732	.357007	1.53	4.6182	.66447	.216536
1.04	2.8292	.45167	.353455	1.54	4.6646	.66881	.214381
1.05	2.8577	0.45601	0.349938	**1.55**	4.7115	0.67316	0.212248
1.06	2.8864	.46035	.346456	1.56	4.7588	.67750	.210136
1.07	2.9154	.46470	.343009	1.57	4.8066	.68184	.208045
1.08	2.9447	.46904	.339596	1.58	4.8550	.68619	.205975
1.09	2.9743	.47338	.336216	1.59	4.9037	.69053	.203926
1.10	3.0042	0.47772	0.332871	**1.60**	4.9530	0.69487	0.201897
1.11	3.0344	.48207	.329559	1.61	5.0028	.69921	.199888
1.12	3.0649	.48641	.326280	1.62	5.0531	.70356	.197899
1.13	3.0957	.49075	.323033	1.63	5.1039	.70790	.195930
1.14	3.1268	.49510	.319819	1.64	5.1552	.71224	.193980
1.15	3.1582	0.49944	0.316637	**1.65**	5.2070	0.71659	0.192050
1.16	3.1899	.50378	.313486	1.66	5.2593	.72093	.190139
1.17	3.2220	.50812	.310367	1.67	5.3122	.72527	.188247
1.18	3.2544	.51247	.307279	1.68	5.3656	.72961	.186374
1.19	3.2871	.51681	.304221	1.69	5.4195	.73396	.184520
1.20	3.3201	0.52115	0.301194	**1.70**	5.4739	0.73830	0.182684
1.21	3.3535	.52550	.298197	1.71	5.5290	.74264	.180866
1.22	3.3872	.52984	.295230	1.72	5.5845	.74699	.179066
1.23	3.4212	.53418	.292293	1.73	5.6407	.75133	.177284
1.24	3.4556	.53853	.289384	1.74	5.6973	.75567	.175520
1.25	3.4903	0.54287	0.286505	**1.75**	5.7546	0.76002	0.173774
1.26	3.5254	.54721	.283654	1.76	5.8124	.76436	.172045
1.27	3.5609	.55155	.280832	1.77	5.8709	.76870	.170333
1.28	3.5966	.55590	.278037	1.78	5.9299	.77304	.168638
1.29	3.6328	.56024	.275271	1.79	5.9895	.77739	.166960
1.30	3.6693	0.56458	0.272532	**1.80**	6.0496	0.78173	0.165299
1.31	3.7062	.56893	.269820	1.81	6.1104	.78607	.163654
1.32	3.7434	.57327	.267135	1.82	6.1719	.79042	.162026
1.33	3.7810	.57761	.264477	1.83	6.2339	.79476	.160414
1.34	3.8190	.58195	.261846	1.84	6.2965	.79910	.158817
1.35	3.8574	0.58630	0.259240	**1.85**	6.3598	0.80344	0.157237
1.36	3.8962	.59064	.256661	1.86	6.4237	.80779	.155673
1.37	3.9354	.59498	.254107	1.87	6.4883	.81213	.154124
1.38	3.9749	.59933	.251579	1.88	6.5535	.81647	.152590
1.39	4.0149	.60367	.249075	1.89	6.6194	.82082	.151072
1.40	4.0552	0.60801	0.246597	**1.90**	6.6859	0.82516	0.149569
1.41	4.0960	.61236	.244143	1.91	6.7531	.82950	.148080
1.42	4.1371	.61670	.241714	1.92	6.8210	.83385	.146607
1.43	4.1787	.62104	.239309	1.93	6.8895	.83819	.145148
1.44	4.2207	.62538	.236928	1.94	6.9588	.84253	.143704
1.45	4.2631	0.62973	0.234570	**1.95**	7.0287	0.84687	0.142274
1.46	4.3060	.63407	.232236	1.96	7.0993	.85122	.140858
1.47	4.3492	.63841	.229925	1.97	7.1707	.85556	.139457
1.48	4.3929	.64276	.227638	1.98	7.2427	.85990	.138069
1.49	4.4371	.64710	.225373	1.99	7.3155	.86425	.136695
1.50	4.4817	0.65144	0.223130	**2.00**	7.3891	0.86859	0.135335

Fig. 2-12B. Table of exponential functions (sheet 2 of 6).

EXPONENTIAL FUNCTIONS (Continued)

x	e^x	$\text{Log}_{10}(e^x)$	e^{-x}	x	e^x	$\text{Log}_{10}(e^x)$	e^{-x}
2.00	7.3891	0.86859	0.135335	**2.50**	12.182	1.08574	0.082085
2.01	7.4633	.87293	.133989	2.51	12.305	1.09008	.081268
2.02	7.5383	.87727	.132655	2.52	12.429	1.09442	.080460
2.03	7.6141	.88162	.131336	2.53	12.554	1.09877	.079659
2.04	7.6906	.88596	.130029	2.54	12.680	1.10311	.078866
2.05	7.7679	0.89030	0.128735	**2.55**	12.807	1.10745	0.078082
2.06	7.8460	.89465	.127454	2.56	12.936	1.11179	.077305
2.07	7.9248	.89899	.126186	2.57	13.066	1.11614	.076536
2.08	8.0045	.90333	.124930	2.58	13.197	1.12048	.075774
2.09	8.0849	.90768	.123687	2.59	13.330	1.12482	.075020
2.10	8.1662	0.91202	0.122456	**2.60**	13.464	1.12917	0.074274
2.11	8.2482	.91636	.121238	2.61	13.599	1.13351	.073535
2.12	8.3311	.92070	.120032	2.62	13.736	1.13785	.072803
2.13	8.4149	.92505	.118837	2.63	13.874	1.14219	.072078
2.14	8.4994	.92939	.117655	2.64	14.013	1.14654	.071361
2.15	8.5849	0.93373	0.116484	**2.65**	14.154	1.15088	0.070651
2.16	8.6711	.93808	.115325	2.66	14.296	1.15522	.069948
2.17	8.7583	.94242	.114178	2.67	14.440	1.15957	.069252
2.18	8.8463	.94676	.113042	2.68	14.585	1.16391	.068563
2.19	8.9352	.95110	.111917	2.69	14.732	1.16825	.067881
2.20	9.0250	0.95545	0.110803	**2.70**	14.880	1.17260	0.067206
2.21	9.1157	.95979	.109701	2.71	15.029	1.17694	.066537
2.22	9.2073	.96413	.108609	2.72	15.180	1.18128	.065875
2.23	9.2999	.96848	.107528	2.73	15.333	1.18562	.065219
2.24	9.3933	.97282	.106459	2.74	15.487	1.18997	.064570
2.25	9.4877	0.97716	0.105399	**2.75**	15.643	1.19431	0.063928
2.26	9.5831	.98151	.104350	2.76	15.800	1.19865	.063292
2.27	9.6794	.98585	.103312	2.77	15.959	1.20300	.062662
2.28	9.7767	.99019	.102284	2.78	16.119	1.20734	.062039
2.29	9.8749	.99453	.101266	2.79	16.281	1.21168	.061421
2.30	9.9742	0.99888	0.100259	**2.80**	16.445	1.21602	0.060810
2.31	10.074	1.00322	.099261	2.81	16.610	1.22037	.060205
2.32	10.176	1.00756	.098274	2.82	16.777	1.22471	.059606
2.33	10.278	1.01191	.097296	2.83	16.945	1.22905	.059013
2.34	10.381	1.01625	.096328	2.84	17.116	1.23340	.058426
2.35	10.486	1.02059	0.095369	**2.85**	17.288	1.23774	0.057844
2.36	10.591	1.02493	.094420	2.86	17.462	1.24208	.057269
2.37	10.697	1.02928	.093481	2.87	17.637	1.24643	.056699
2.38	10.805	1.03362	.092551	2.88	17.814	1.25077	.056135
2.39	10.913	1.03796	.091630	2.89	17.993	1.25511	.055576
2.40	11.023	1.04231	0.090718	**2.90**	18.174	1.25945	0.055023
2.41	11.134	1.04665	.089815	2.91	18.357	1.26380	.054476
2.42	11.246	1.05099	.088922	2.92	18.541	1.26814	.053934
2.43	11.359	1.05534	.088037	2.93	18.728	1.27248	.053397
2.44	11.473	1.05968	.087161	2.94	18.916	1.27683	.052866
2.45	11.588	1.06402	0.086294	**2.95**	19.106	1.28117	0.052340
2.46	11.705	1.06836	.085435	2.96	19.298	1.28551	.051819
2.47	11.822	1.07271	.084585	2.97	19.492	1.28985	.051303
2.48	11.941	1.07705	.083743	2.98	19.688	1.29420	.050793
2.49	12.061	1.08139	.082910	2.99	19.886	1.29854	.050287
2.50	12.182	1.08574	0.082085	**3.00**	20.086	1.30288	0.049787

Fig. 2-12C. Table of exponential functions (sheet 3 of 6).

EXPONENTIAL FUNCTIONS (Continued)

x	e^x	$\text{Log}_{10}(e^x)$	e^{-x}	x	e^x	$\text{Log}_{10}(e^x)$	e^{-x}
3.00	20.086	1.30288	0.049787	**3.50**	33.115	1.52003	0.030197
3.01	20.287	1.30723	.049292	3.51	33.448	1.52437	.029897
3.02	20.491	1.31157	.048801	3.52	33.784	1.52872	.029599
3.03	20.697	1.31591	.048316	3.53	34.124	1.53306	.029305
3.04	20.905	1.32026	.047835	3.54	34.467	1.53740	.029013
3.05	21.115	1.32460	0.047359	**3.55**	34.813	1.54175	0.028725
3.06	21.328	1.32894	.046888	3.56	35.163	1.54609	.028439
3.07	21.542	1.33328	.046421	3.57	35.517	1.55043	.028156
3.08	21.758	1.33763	.045959	3.58	35.874	1.55477	.027876
3.09	21.977	1.34197	.045502	3.59	36.234	1.55912	.027598
3.10	22.198	1.34631	0.045049	**3.60**	36.598	1.56346	0.027324
3.11	22.421	1.35066	.044601	3.61	36.966	1.56780	.027052
3.12	22.646	1.35500	.044157	3.62	37.338	1.57215	.026783
3.13	22.874	1.35934	.043718	3.63	37.713	1.57649	.026516
3.14	23.104	1.36368	.043283	3.64	38.092	1.58083	.026252
3.15	23.336	1.36803	0.042852	**3.65**	38.475	1.58517	0.025991
3.16	23.571	1.37237	.042426	3.66	38.861	1.58952	.025733
3.17	23.807	1.37671	.042004	3.67	39.252	1.59386	.025476
3.18	24.047	1.38106	.041586	3.68	39.646	1.59820	.025223
3.19	24.288	1.38540	.041172	3.69	40.045	1.60255	.024972
3.20	24.533	1.38974	0.040762	**3.70**	40.447	1.60689	0.024724
3.21	24.779	1.39409	.040357	3.71	40.854	1.61123	.024478
3.22	25.028	1.39843	.039955	3.72	41.264	1.61558	.024234
3.23	25.280	1.40277	.039557	3.73	41.679	1.61992	.023993
3.24	25.534	1.40711	.039164	3.74	42.098	1.62426	.023754
3.25	25.790	1.41146	0.038774	**3.75**	42.521	1.62860	0.023518
3.26	26.050	1.41580	.038388	3.76	42.948	1.63295	.023284
3.27	26.311	1.42014	.038006	3.77	43.380	1.63729	.023052
3.28	26.576	1.42449	.037628	3.78	43.816	1.64163	.022823
3.29	26.843	1.42883	.037254	3.79	44.256	1.64598	.022596
3.30	27.113	1.43317	0.036883	**3.80**	44.701	1.65032	0.022371
3.31	27.385	1.43751	.036516	3.81	45.150	1.65466	.022148
3.32	27.660	1.44186	.036153	3.82	45.604	1.65900	.021928
3.33	27.938	1.44620	.035793	3.83	46.063	1.66335	.021710
3.34	28.219	1.45054	.035437	3.84	46.525	1.66769	.021494
3.35	28.503	1.45489	0.035084	**3.85**	46.993	1.67203	0.021280
3.36	28.789	1.45923	.034735	3.86	47.465	1.67638	.021068
3.37	29.079	1.46357	.034390	3.87	47.942	1.68072	.020858
3.38	29.371	1.46792	.034047	3.88	48.424	1.68506	.020651
3.39	29.666	1.47226	.033709	3.89	48.911	1.68941	.020445
3.40	29.964	1.47660	0.033373	**3.90**	49.402	1.69375	0.020242
3.41	30.265	1.48094	.033041	3.91	49.899	1.69809	.020041
3.42	30.569	1.48529	.032712	3.92	50.400	1.70243	.019841
3.43	30.877	1.48963	.032387	3.93	50.907	1.70678	.019644
3.44	31.187	1.49397	.032065	3.94	51.419	1.71112	.019448
3.45	31.500	1.49832	0.031746	**3.95**	51.935	1.71546	0.019255
3.46	31.817	1.50266	.031430	3.96	52.457	1.71981	.019063
3.47	32.137	1.50700	.031117	3.97	52.985	1.72415	.018873
3.48	32.460	1.51134	.030807	3.98	53.517	1.72849	.018686
3.49	32.786	1.51569	.030501	3.99	54.055	1.73283	.018500
3.50	33.115	1.52003	0.030197	**4.00**	54.598	1.73718	0.018316

Fig. 2-12D. Table of exponential functions (sheet 4 of 6).

EXPONENTIAL FUNCTIONS (Continued)

x	e^x	$\text{Log}_{10}(e^x)$	e^{-x}	x	e^x	$\text{Log}_{10}(e^x)$	e^{-x}
1.00	2.7183	0.43429	0.367879	**1.50**	4.4817	0.65144	0.223130
1.01	2.7456	.43864	.364219	1.51	4.5267	.65578	.220910
1.02	2.7732	.44298	.360595	1.52	4.5722	.66013	.218712
1.03	2.8011	.44732	.357007	1.53	4.6182	.66447	.216536
1.04	2.8292	.45167	.353455	1.54	4.6646	.66881	.214381
1.05	2.8577	0.45601	0.349938	**1.55**	4.7115	0.67316	0.212248
1.06	2.8864	.46035	.346456	1.56	4.7588	.67750	.210136
1.07	2.9154	.46470	.343009	1.57	4.8066	.68184	.208045
1.08	2.9447	.46904	.339596	1.58	4.8550	.68619	.205975
1.09	2.9743	.47338	.336216	1.59	4.9037	.69053	.203926
1.10	3.0042	0.47772	0.332871	**1.60**	4.9530	0.69487	0.201897
1.11	3.0344	.48207	.329559	1.61	5.0028	.69921	.199888
1.12	3.0649	.48641	.326280	1.62	5.0531	.70356	.197899
1.13	3.0957	.49075	.323033	1.63	5.1039	.70790	.195930
1.14	3.1268	.49510	.319819	1.64	5.1552	.71224	.193980
1.15	3.1582	0.49944	0.316637	**1.65**	5.2070	0.71659	0.192050
1.16	3.1899	.50378	.313486	1.66	5.2593	.72093	.190139
1.17	3.2220	.50812	.310367	1.67	5.3122	.72527	.188247
1.18	3.2544	.51247	.307279	1.68	5.3656	.72961	.186374
1.19	3.2871	.51681	.304221	1.69	5.4195	.73396	.184520
1.20	3.3201	0.52115	0.301194	**1.70**	5.4739	0.73830	0.182684
1.21	3.3535	.52550	.298197	1.71	5.5290	.74264	.180866
1.22	3.3872	.52984	.295230	1.72	5.5845	.74699	.179066
1.23	3.4212	.53418	.292293	1.73	5.6407	.75133	.177284
1.24	3.4556	.53853	.289384	1.74	5.6973	.75567	.175520
1.25	3.4903	0.54287	0.286505	**1.75**	5.7546	0.76002	0.173774
1.26	3.5254	.54721	.283654	1.76	5.8124	.76436	.172045
1.27	3.5609	.55155	.280832	1.77	5.8709	.76870	.170333
1.28	3.5966	.55590	.278037	1.78	5.9299	.77304	.168638
1.29	3.6328	.56024	.275271	1.79	5.9895	.77739	.166960
1.30	3.6693	0.56458	0.272532	**1.80**	6.0496	0.78173	0.165299
1.31	3.7062	.56893	.269820	1.81	6.1104	.78607	.163654
1.32	3.7434	.57327	.267135	1.82	6.1719	.79042	.162026
1.33	3.7810	.57761	.264477	1.83	6.2339	.79476	.160414
1.34	3.8190	.58195	.261846	1.84	6.2965	.79910	.158817
1.35	3.8574	0.58630	0.259240	**1.85**	6.3598	0.80344	0.157237
1.36	3.8962	.59064	.256661	1.86	6.4237	.80779	.155673
1.37	3.9354	.59498	.254107	1.87	6.4883	.81213	.154124
1.38	3.9749	.59933	.251579	1.88	6.5535	.81647	.152590
1.39	4.0149	.60367	.249075	1.89	6.6194	.82082	.151072
1.40	4.0552	0.60801	0.246597	**1.90**	6.6859	0.82516	0.149569
1.41	4.0960	.61236	.244143	1.91	6.7531	.82950	.148080
1.42	4.1371	.61670	.241714	1.92	6.8210	.83385	.146607
1.43	4.1787	.62104	.239309	1.93	6.8895	.83819	.145148
1.44	4.2207	.62538	.236928	1.94	6.9588	.84253	.143704
1.45	4.2631	0.62973	0.234570	**1.95**	7.0287	0.84687	0.142274
1.46	4.3060	.63407	.232236	1.96	7.0993	.85122	.140858
1.47	4.3492	.63841	.229925	1.97	7.1707	.85556	.139457
1.48	4.3929	.64276	.227638	1.98	7.2427	.85990	.138069
1.49	4.4371	.64710	.225373	1.99	7.3155	.86425	.136695
1.50	4.4817	0.65144	0.223130	**2.00**	7.3891	0.86859	0.135335

Fig. 2-12E. Table of exponential functions (sheet 5 of 6).

EXPONENTIAL FUNCTIONS (Continued)

x	e^x	$\text{Log}_{10}(e^x)$	e^{-x}	x	e^x	$\text{Log}_{10}(e^x)$	e^{-x}
3.00	20.086	1.30288	0.049787	3.50	33.115	1.52003	0.030197
3.01	20.287	1.30723	.049292	3.51	33.448	1.52437	.029897
3.02	20.491	1.31157	.048801	3.52	33.784	1.52872	.029599
3.03	20.697	1.31591	.048316	3.53	34.124	1.53306	.029305
3.04	20.905	1.32026	.047835	3.54	34.467	1.53740	.029013
3.05	21.115	1.32460	0.047359	3.55	34.813	1.54175	0.028725
3.06	21.328	1.32894	.046888	3.56	35.163	1.54609	.028439
3.07	21.542	1.33328	.046421	3.57	35.517	1.55043	.028156
3.08	21.758	1.33763	.045959	3.58	35.874	1.55477	.027876
3.09	21.977	1.34197	.045502	3.59	36.234	1.55912	.027598
3.10	22.198	1.34631	0.045049	3.60	36.598	1.56346	0.027324
3.11	22.421	1.35066	.044601	3.61	36.966	1.56780	.027052
3.12	22.646	1.35500	.044157	3.62	37.338	1.57215	.026783
3.13	22.874	1.35934	.043718	3.63	37.713	1.57649	.026516
3.14	23.104	1.36368	.043283	3.64	38.092	1.58083	.026252
3.15	23.336	1.36803	0.042852	3.65	38.475	1.58517	0.025991
3.16	23.571	1.37237	.042426	3.66	38.861	1.58952	.025733
3.17	23.807	1.37671	.042004	3.67	39.252	1.59386	.025476
3.18	24.047	1.38106	.041586	3.68	39.646	1.59820	.025223
3.19	24.288	1.38540	.041172	3.69	40.045	1.60255	.024972
3.20	24.533	1.38974	0.040762	3.70	40.447	1.60689	0.024724
3.21	24.779	1.39409	.040357	3.71	40.854	1.61123	.024478
3.22	25.028	1.39843	.039955	3.72	41.264	1.61558	.024234
3.23	25.280	1.40277	.039557	3.73	41.679	1.61992	.023993
3.24	25.534	1.40711	.039164	3.74	42.098	1.62426	.023754
3.25	25.790	1.41146	0.038774	3.75	42.521	1.62860	0.023518
3.26	26.050	1.41580	.038388	3.76	42.948	1.63295	.023284
3.27	26.311	1.42014	.038006	3.77	43.380	1.63729	.023052
3.28	26.576	1.42449	.037628	3.78	43.816	1.64163	.022823
3.29	26.843	1.42883	.037254	3.79	44.256	1.64598	.022596
3.30	27.113	1.43317	0.036883	3.80	44.701	1.65032	0.022371
3.31	27.385	1.43751	.036516	3.81	45.150	1.65466	.022148
3.32	27.660	1.44186	.036153	3.82	45.604	1.65900	.021928
3.33	27.938	1.44620	.035793	3.83	46.063	1.66335	.021710
3.34	28.219	1.45054	.035437	3.84	46.525	1.66769	.021494
3.35	28.503	1.45489	0.035084	3.85	46.993	1.67203	0.021280
3.36	28.789	1.45923	.034735	3.86	47.465	1.67638	.021068
3.37	29.079	1.46357	.034390	3.87	47.942	1.68072	.020858
3.38	29.371	1.46792	.034047	3.88	48.424	1.68506	.020651
3.39	29.666	1.47226	.033709	3.89	48.911	1.68941	.020445
3.40	29.964	1.47660	0.033373	3.90	49.402	1.69375	0.020242
3.41	30.265	1.48094	.033041	3.91	49.899	1.69809	.020041
3.42	30.569	1.48529	.032712	3.92	50.400	1.70243	.019841
3.43	30.877	1.48963	.032387	3.93	50.907	1.70678	.019644
3.44	31.187	1.49397	.032065	3.94	51.419	1.71112	.019448
3.45	31.500	1.49832	0.031746	3.95	51.935	1.71546	0.019255
3.46	31.817	1.50266	.031430	3.96	52.457	1.71981	.019063
3.47	32.137	1.50700	.031117	3.97	52.985	1.72415	.018873
3.48	32.460	1.51134	.030807	3.98	53.517	1.72849	.018686
3.49	32.786	1.51569	.030501	3.99	54.055	1.73283	.018500
3.50	33.115	1.52003	0.030197	4.00	54.598	1.73718	0.018316

Fig. 2-12F. Table of exponential functions (sheet 6 of 6).

USE OF THE LL3 SCALE

The log log scales are similar to the natural log scales in that they work in conjunction with the D scale of your slide rule. However, the similarity ends here in that the log log scales are for the positive power of epsilon. Once you get an answer on the LL3 scale (for instance) for a power located on the D scale, you have to get its reciprocal. In other words, you have to divide it into 1 before you can subtract it from 1 to get the same old 63 percent that we developed earlier.

FOR EPSILON POWERS LESS THAN MINUS ONE

If the time after T_0 divided by the time constant of the circuit is less than minus 1 but greater than .1, you will use the left side of the A scale for readout on the LL00 scale or you will use the same old D scale for the Natural Log and Log log scales but find your answers under the hair-line on the Ln-2 scale or the LL2 scale, respectively. Remember though, if you are finding your answer on the LL2 scale, you still have to find its reciprocal before you have the same answer you read directly off the other two types of scales.

This gives us two methods of determining the instantaneous voltage of a step function into what we call a "low pass filter" circuit: One, the mathematical tables of Exponential Functions and two, your slide rule with any one of the three types of scales related to epsilon.

There is another. The easiest of the bunch. That's why I've saved it till last. Just read the answer off a graph!

Fig. 2-13 is a graph of the change related to epsilon. All we have to do is figure the same old **time after T_0** divided by the **time constant** of the circuit. And, assuming that this gives us a number between zero and five, locate it on the lower horizontal axis of this graph. (Note: if it's greater than five, for all practical purposes, the capacitor has taken on 100 percent of the change of voltage represented by the step function.) Now move in an exactly vertical direction up the graph to where you intersect the RC curve. From here you move directly to the left to read the percentage of the total step function taken on by the capacitor at this time. No slide rule to set wrong or interpret wrongly, no math tables to fumble through and maybe read wrong; just a simple graph to interpret. And, here is a very excellent one in Fig. 2-13 that can be interpreted quite closely. (Reprinted with permission from Tektronix Inc.)

The next part of our time constant story has us change the circuitry a little bit. So far all we've looked at is the voltage

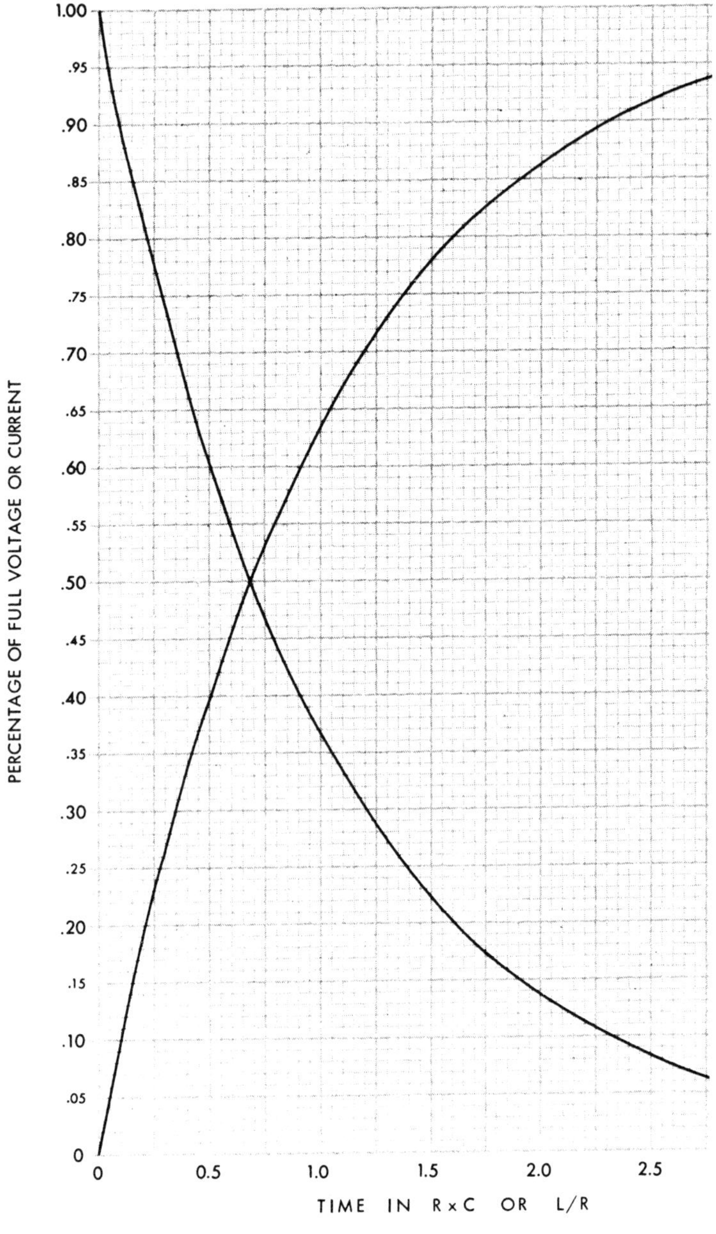

1.00
.95
.90
.85
.80
.75
.70
.65
.60
.55
.50
.45
.40
.35
.30
.25
.20
.15
.10
.05
0
PERCENTAGE OF FULL VOLTAGE OR CURRENT
0
0.5
1.0
1.5
2.0
2.5
TIME IN R x C OR L/R

UNIVERSAL TIME CONSTANT CURVES

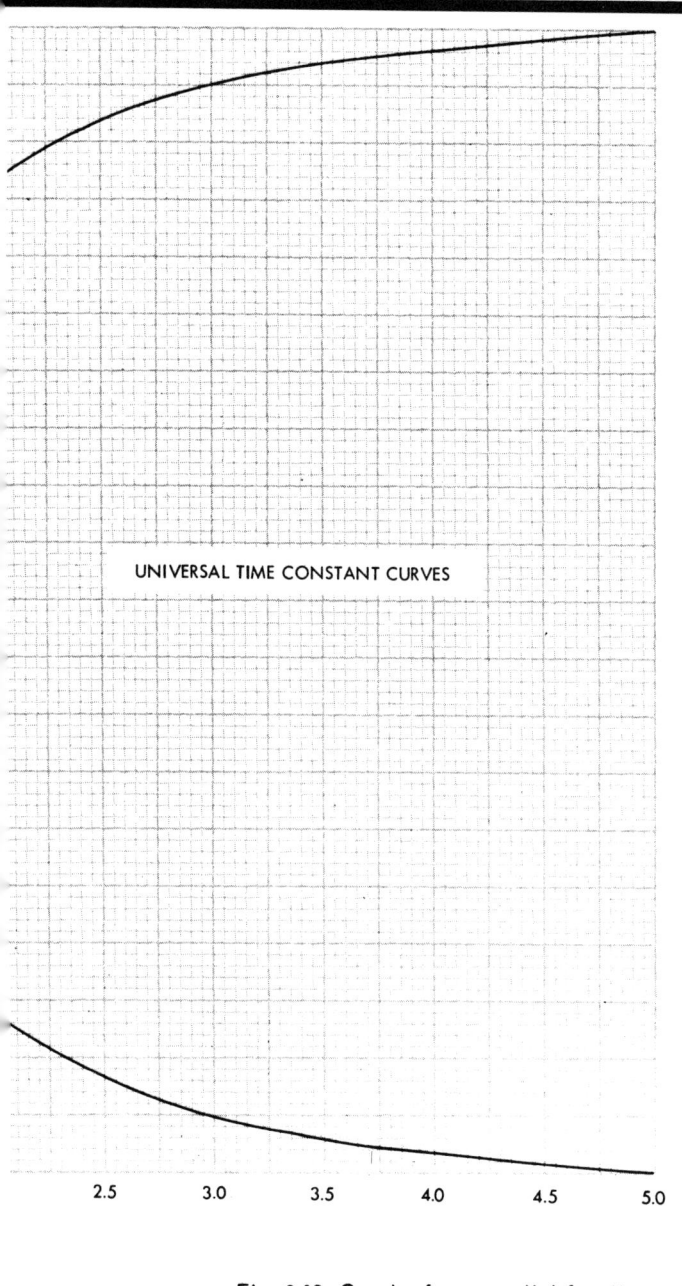

Fig. 2-13. Graph of exponential functions.

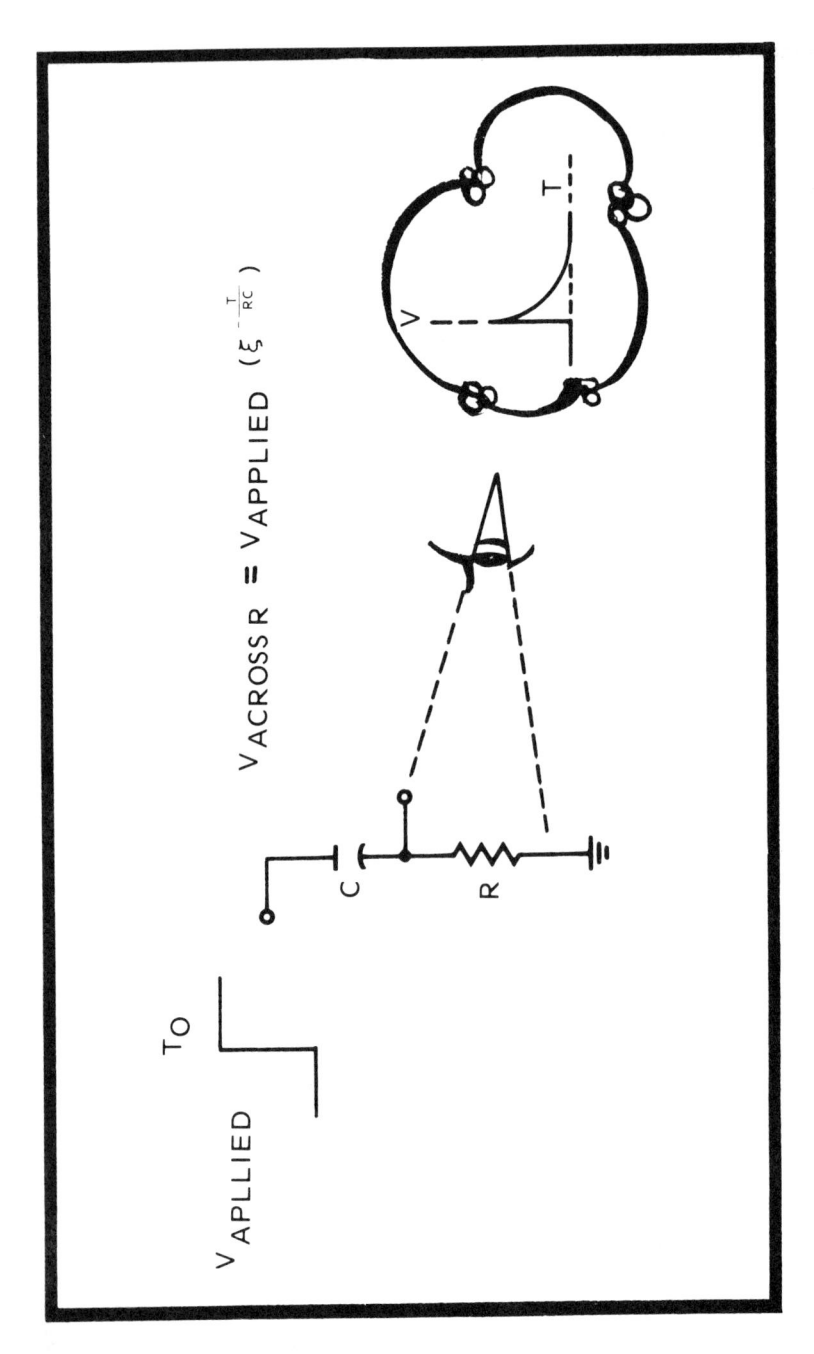

Fig. 2-14. The high pass R-C filter or differentiator circuit.

across the capacitor in an R-C series circuit with the capacitor grounded, and the voltage across the resistor in an L-R circuit with the resistor grounded.

Also, you will note, we have found a way to figure the instantaneous voltage at any time after T_0 between .1 time constant to 5 time constants. For anything beyond 5 time constants, we say 100 percent of the change has taken place. For anything less than .1 time constant, the epsilon related numbers have to be carried out to so many places that they become awkward to handle. So I recommend using a different formula. Go back to the formula:

$$\frac{dV}{dT} = \frac{I}{C}$$

To be absolutely truthful, the current (I) is not perfectly stable during this time, but the change is so small that we can safely ignore it. This will give us a change of voltage on the capacitor that will be easy to calculate. For the L R circuit, you will just have to be satisfied with interpreting the curve down in this region, even though it does end up being just a rough estimate.

SERIES DIFFERENTIATOR CIRCUITS

Now let's take the R-C circuit and change the position of the parts to that shown in Fig. 2-14. We now find ourselves looking at a circuit with the capacitor on top and the resistor on the bottom, and we will predict the shape of the voltage waveform across the resistor as the step function is applied to the capacitor.

Again, remember we cannot change the voltage across the capacitor instantaneously. If there was a zero voltage difference between the plates of the capacitor before the step function came along, there will be no voltage between the two plates immediately after the step function has occurred. This means that the full step voltage amplitude will be immediately across the resistor. Now as the capacitor accumulates electrons on the bottom plate and loses them from the top plate, it begins to develop a voltage difference, plate to plate, and the voltage across the resistor begins to diminish along the same old time-constant curve. However, it's turned up-side-down this time. We call this a differentiated waveform. And you'll find that the formula for predicting the instantaneous voltage across R any time after T_0 is a bit simpler than the one we used before. You no longer have to subtract the value of epsilon to the minus T over RC power from one. You just

71

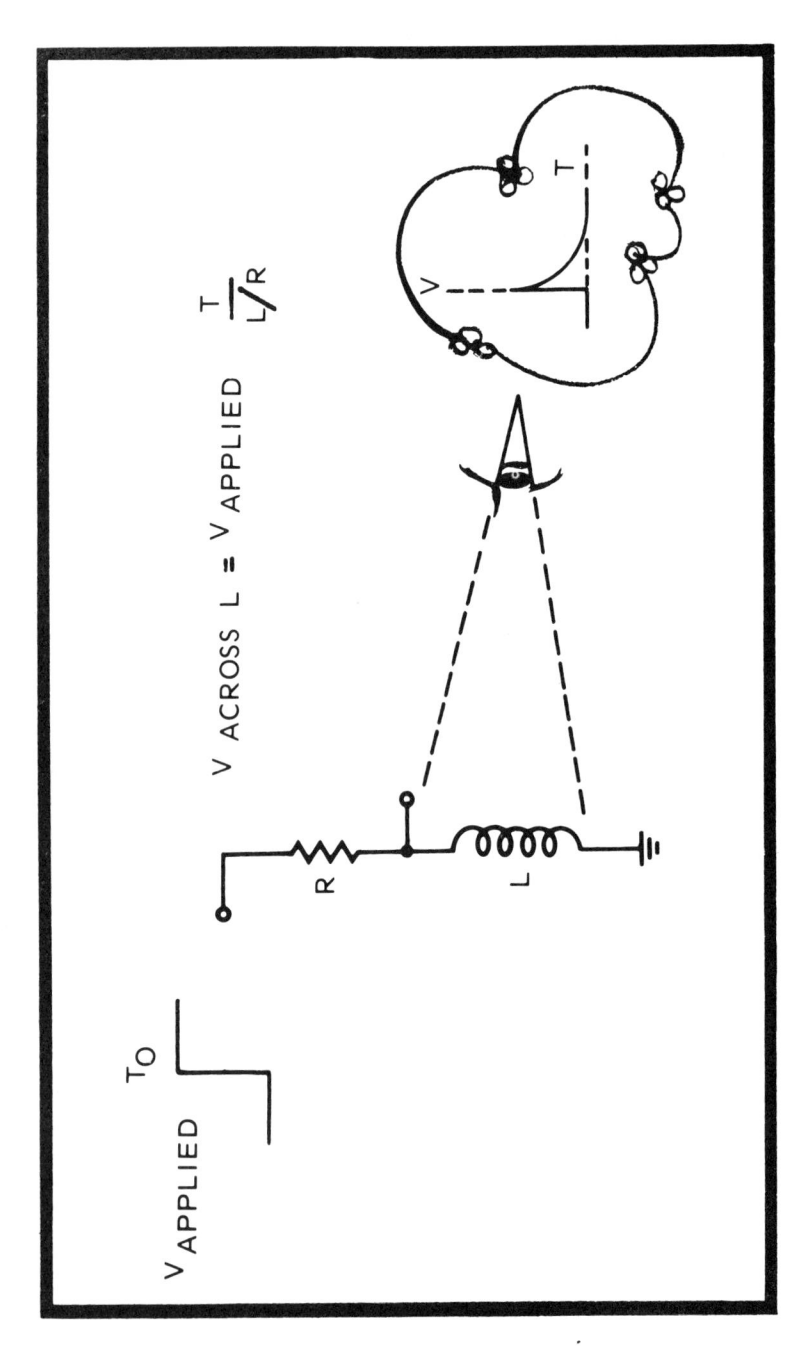

Fig. 2-15. The high pass R-L filter or differentiator circuit.

multiply V_{applied} by epsilon to the minus T over RC power, and the formula is:

$$V_{across\ R} = V_{applied} \times \varepsilon^{-\frac{T}{RC}}$$

Note that if the power of epsilon is minus five or more, the voltage across R is zero for all practical purposes. And, if the minus power of epsilon is less than 0.1, the voltage across R is equal to the voltage applied.

The combination of R and L that will give us the same waveshape as the combination of C and R just considered has R on top and the L on the bottom. We find ourselves looking at the waveshape of voltage developed across L as shown in Fig. 2-15.

Recall again that an inductor opposes a change of current and if there was no current flowing in the inductor before T_0, there will be no current flowing in the inductor immediately after T_0 either. This puts the full voltage of the step function across the inductor since none is lost by having a current flow through the resistor in series with it. As soon as the field of the inductor begins to allow current to seep through, Ohm's Law comes into effect and part of the applied voltage begins to develop across the resistor. This voltage subtracts from the original voltage across L and this voltage in turn begins to diminish along the same old time constant curve as before until (at the end of 5 time constants) there is no voltage left across the inductor and we have the same waveshape as the previous R-C differentiator circuit gave us.

Both of these circuits take the step function applied and differentiate it. They are both what we call High Pass Filters. In other words, the output will be the high-frequency content of the input waveform only.

The epsilon related formula for the R-L differentiator circuit is:

$$V_{across\ L} = V_{applied} \times \varepsilon^{-\frac{T}{L/R}}$$

Note again, that for any negative power of epsilon numerically larger than 5, the voltage across L is zero; and for any negative power less than 0.1, the voltage across L equals the voltage applied. This last is not absolutely true, but for all practical purposes it is.

There are some other limitations that we must consider relative to the truth of this situation we have been discussing. If you will remember, I said, "Let us assume that we have a perfect step generator." And there is no such thing! All step

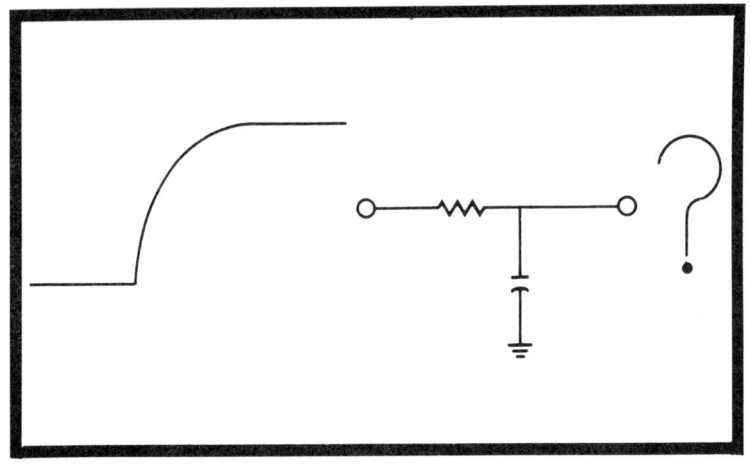

Fig. 2-16. The exponential waveform inputted to a low pass filter.

generators develop an exponential curve of some time duration. Some of them just do it in shorter times than others.

So, the next step is to look into the situation we have when an exponential (the rolled-off waveform that came out of the low pass filter of Fig. 2-10) comes through another low pass filter circuit. See Fig. 2-16. Quite obviously, the output will rise to the maximum input level sooner or later. But that's the question. How long is it going to take it to do just that, and how do we figure it?

RISE TIME CALCULATIONS

This requires a new definition, a definition of the measurement of a step waveform with an exponential rise that we call Rise Time (T_r). The **Rise Time** of such a waveform is the time it takes the waveform to go from 10 percent of the rise to 90 percent of the rise (the center 80 percent of the total change). We can relate this T_r to the time constant of the circuit by means of the RC curve, Fig. 2-9. Note on the curve that the 10 percent point occurs at 0.1 RC time and that the 90 percent point on the curve occurs at 2.3 RC time. This says that:

$$T_r = (2.3 - .1) \times RC, \text{ or } 2.2 \; RC$$

And, a little algebra says that:

$$RC = \frac{T_r}{2.2}$$

74

And when we have two such circuits, one driving the other, we have a time of rise total that is equal to the square root of the sum of the squares. That is:

$$T_{r\ out} = \sqrt{T_{r1}^2 + T_{r2}^2}$$

On this basis then, I think it's fair to say that the rise time of the step generator should be at least ten times faster than the low pass circuit we are driving with the generator if we are going to measure a time of rise that comes close to approximating that of the circuit under test. Let me say that again in other words; if you are going to use a step waveform to test a low pass circuit, make sure that the generator pulse rise time is less than one tenth of 2.2 times the time constant of the circuit you are testing.

Now we also have to consider what happens when we use an exponential waveform to drive a high pass RC circuit. Note that this graph (Fig. 2-17) drawn with a horizontal axis in terms of a logarithmic scale of the generator time constant rather than a linear presentation of the time constant that we've been used to looking at in the other graph. Fig. 2-17 shows us that the requirements on the time constant of the generator become tighter yet in relation to the time constant of the circuit being driven. You see, N is the ratio of the time constant of the circuit under test to the time constant of the generator. Actually, if V_{out} is going to reach the full amplitude of $V_{applied}$ in the first few fractions of a moment after T_0 (as we figured it would since the capacitor does not change voltage immediately), then the time constant of the generator must be less than one one hundredth (.01) of the time constant of the circuit being driven. For ratios less than 100 to 1, I recommend interpreting relative to different values of N.

This looks almost like we are driving the high-pass filter with a ramp waveform which is a case in point all to itself. What do you suppose would happen if we drove a high-pass circuit with a ramp waveform that had a rise time more than five times greater than the circuit time constant?

Let's see if we can't roughly figure this out for ourselves. Refer to Fig. 2-18. The ramp starts out at T_0 and begins to develop a voltage across the resistor. Current begins to flow out of and into the capacitor and it begins to develop a voltage difference, plate to plate. So the transferral of the ramp voltage to the resistor becomes less than the change of the ramp but still continues to grow and more and more current

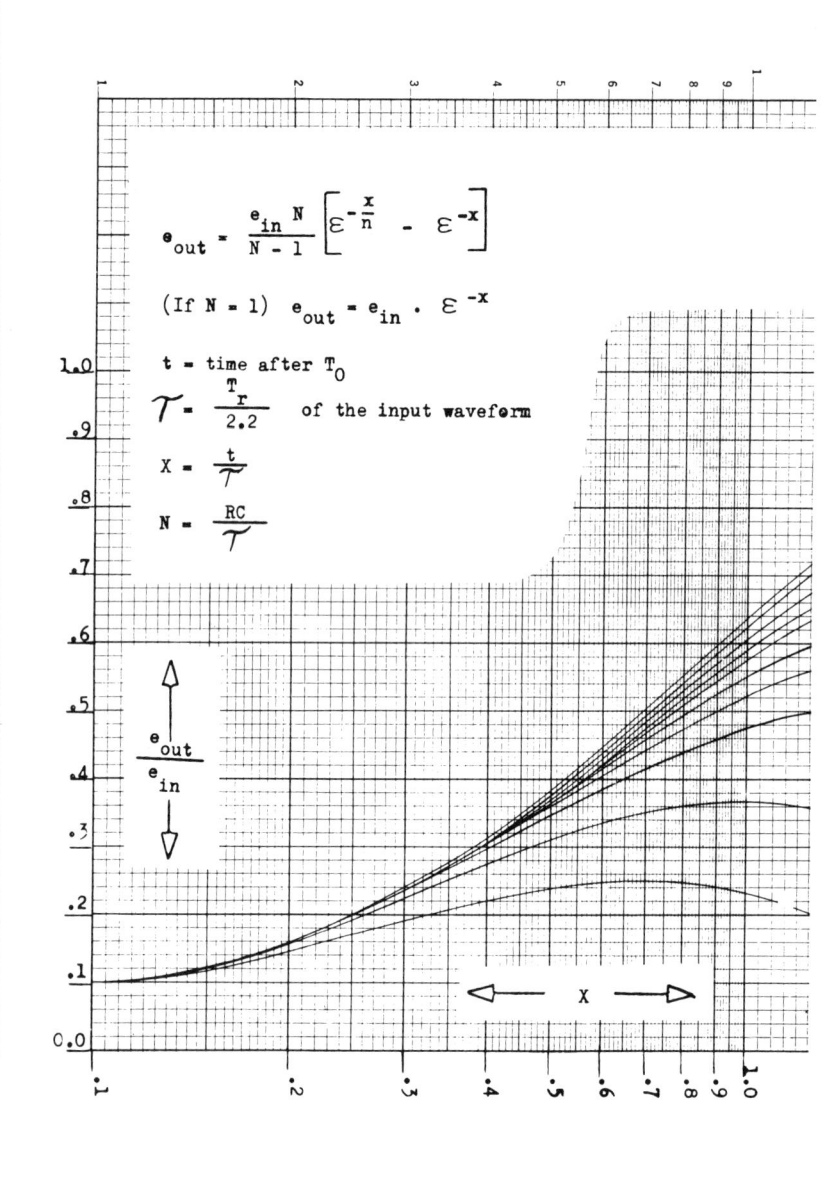

$$e_{out} = \frac{e_{in} N}{N-1}\left[\varepsilon^{-\frac{x}{n}} - \varepsilon^{-x}\right]$$

(If $N = 1$) $e_{out} = e_{in} \cdot \varepsilon^{-x}$

t = time after T_0

$\mathcal{T} = \dfrac{T_r}{2.2}$ of the input waveform

$X = \dfrac{t}{\mathcal{T}}$

$N = \dfrac{RC}{\mathcal{T}}$

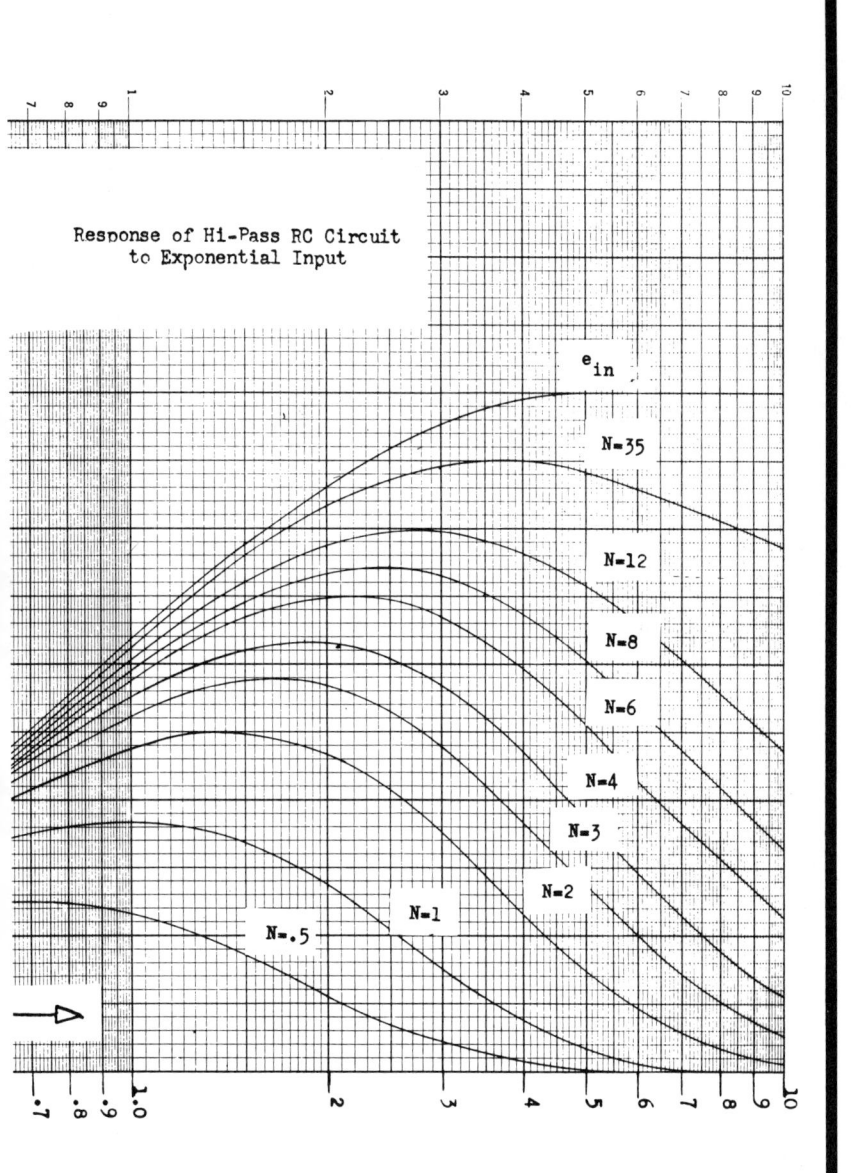

Fig. 2-17. Response of high pass R-C circuit to exponential input.

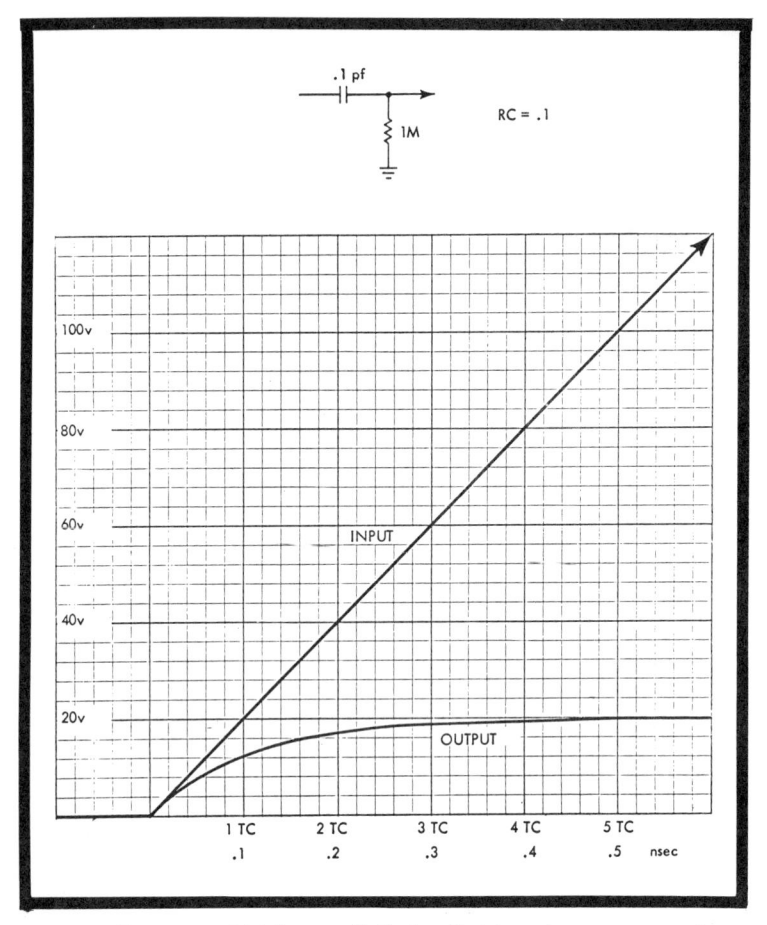

Fig. 2-18. Response of high pass R-C circuit driven by ramp waveform. (Courtesy Tektronix, Inc.)

flows through the resistor, into the capacitor, and a like amount of current flows out of the top of the capacitor. This does reach a limit, but where and when?

Well, look at it this way. The ramp represents a very definite change of voltage per unit of time. According to the picture, the ramp is changing 20 volts in one RC period of time.

Now then do you remember how a capacitor changes voltage with current? We had a formula that stated that the change of voltage across a capacitor per unit of time was equal to a specific magnitude of current divided by the size of the capacitor (Eq. 2-16). What do you suppose happens when the magnitude of current through the resistor reaches a big enough value with respect to the size of the capacitor to match the ramp? That's right! The capacitor is now changing voltage just as fast as the ramp, and the voltage across the resistor stops changing. Under this situation, the voltage across R can never become greater than the change of voltage of the ramp in one RC unit of time for the circuit. In this case, the ramp will never develop more than 20 V across the resistor. The voltage will reach the 20-volt level in five time constants of the circuit and the waveform will follow the exponential curve. In a practical sense, we do have to be careful not to exceed the breakdown voltage of the capacitor in this circuit. The capacitor does take on the lion's share of the ramp voltage applied.

COMPENSATED DIVIDER

Another very interesting circuit that can be figured strictly in terms of its R and C components is the compensated divider. See Fig. 2-19. A pure resistive divider usually drives a circuit that has a certain amount of input capacity, probably very little, but it is there. And, as frequency goes up, this capacity, representing a certain amount of X_C, begins to short out the lower resistor in the circuit, which changes the amount of total signal the driven circuit can see. So we have to do something about it.

What we do is this: we place a high frequency divider in parallel with the resistive divider such that the high

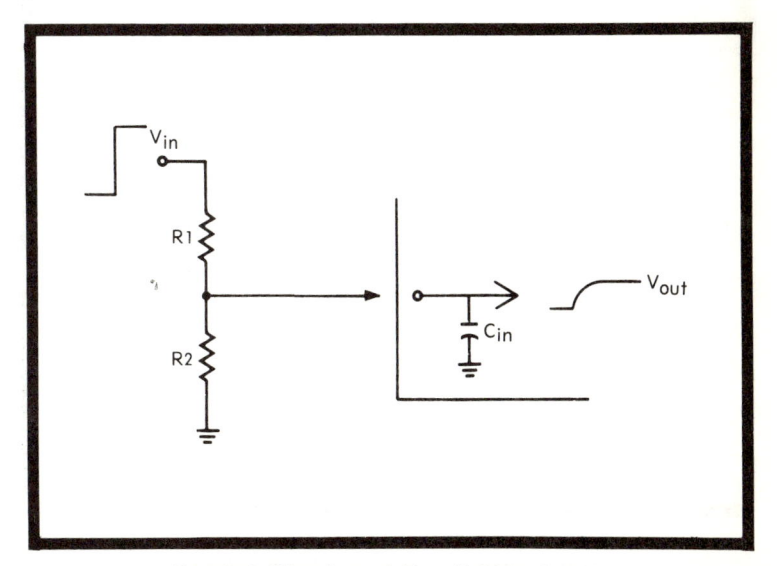

Fig. 2-19. Simple resistive divider circuit.

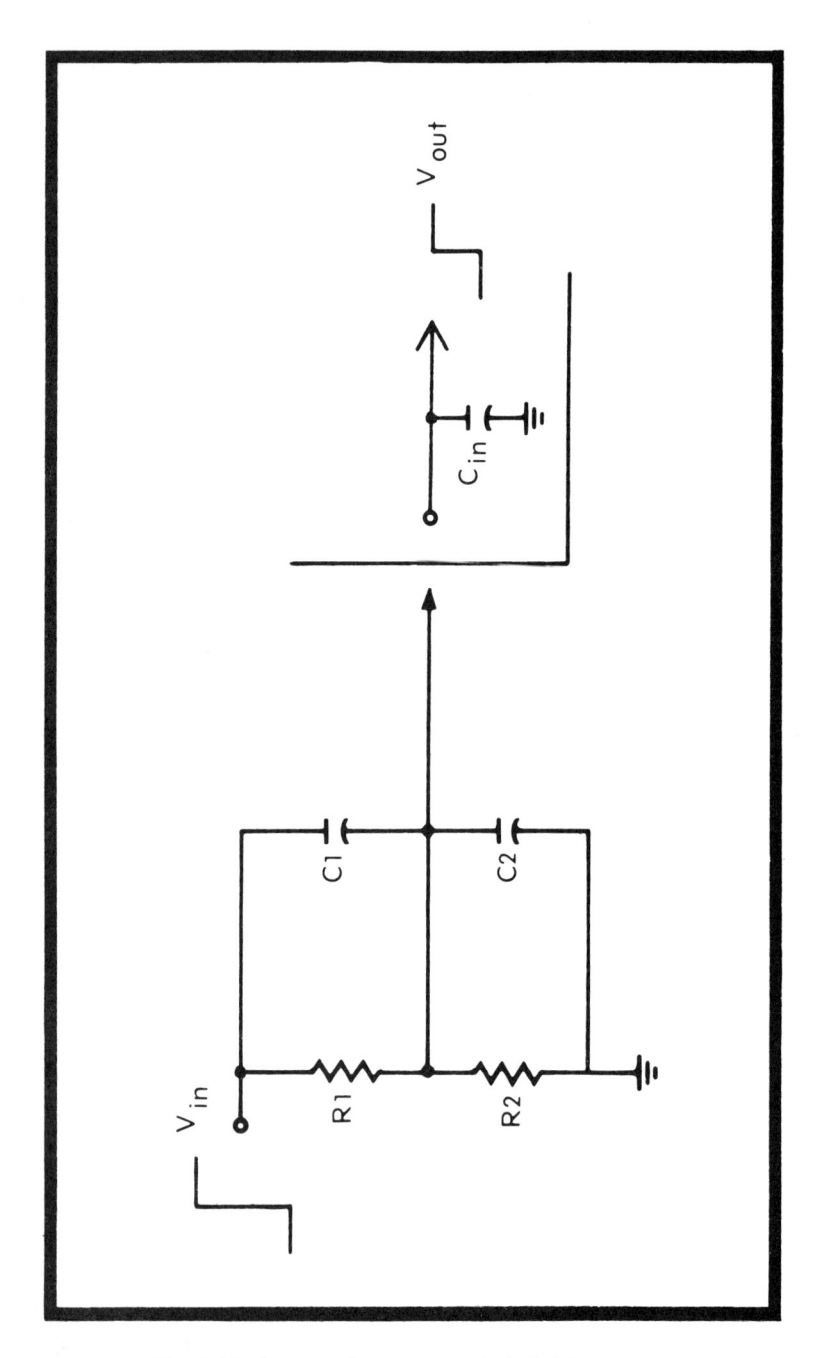

Fig. 2-20. A correctly compensated divider circuit.

frequencies are divided to the same degree as the resistors divide the low frequencies (Fig. 2-20).

This isn't quite as difficult as it sounds. The hardest job is figuring out how to go about it, so let's do that together.

We know that the magnitude of V_{out} according to the resistors is:

(low frequency) $$V_{out} = \frac{R_2}{R_1 + R_2} \times V_{in}$$

And I said that the X_C values of the two capacitors must divide the V_{in} to exactly the same degree. This allows me to say:

(hi frequency) $$V_{out} = \frac{\left(\frac{1}{2\pi f C_2}\right)}{\left(\frac{1}{2\pi f C_1}\right) + \left(\frac{1}{2\pi f C_2}\right)} \times V_{in}$$

And (hi frequency) V_{out} equals (low frequency) V_{out}. This then says that X_{C1} equals R_1, and X_{C2} equals R_2. therefore, the ratio of the resistors must be matched by the X_{C1} to X_{C2} ratio:

$$\frac{R_1}{R_2} = \frac{X_{c1}}{X_{c2}}$$

or, $$\frac{R_1}{R_2} = \frac{\left(\frac{1}{2\pi f C_1}\right)}{\left(\frac{1}{2\pi f C_2}\right)}$$

Now, do you remember what you do when you divide with a fraction? You invert (turn it up-side-down) and multiply with it:

so, $$\frac{R_1}{R_2} = \left(\frac{1}{2\pi f C_1}\right)\left(\frac{2\pi f C_2}{1}\right) = \left(\frac{2\pi f C_2}{2\pi f C_1}\right)$$

There! Now we have the form of the formula that can tell us something. Note that we can simplify it even more as long as we realize that no matter what frequency is coming into this divider, it hits both capacitors at the same time and we will

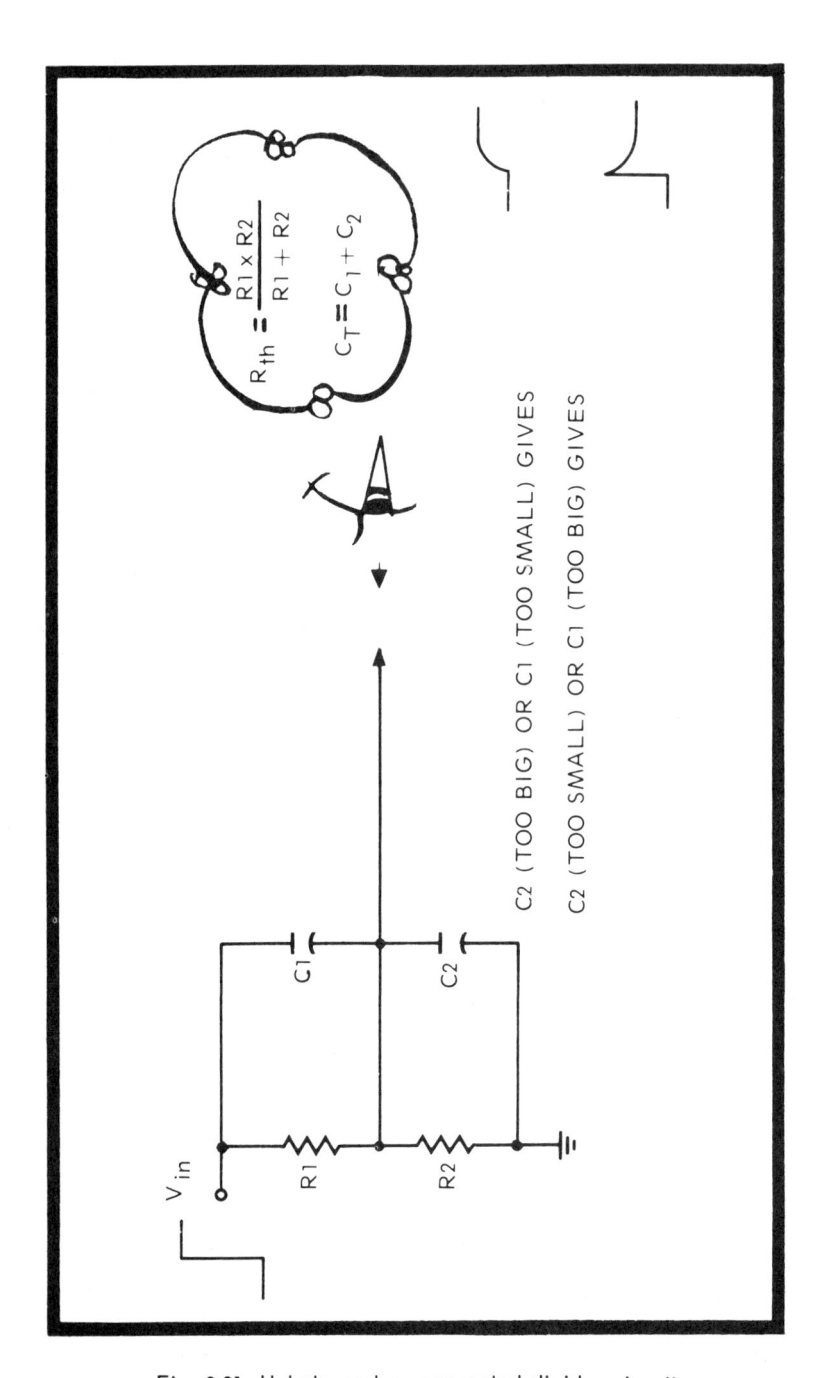

Fig. 2-21. Unbalanced compensated divider circuit.

always have identical ($2 \pi f$) values and, consequently, we can cancel them. This gives us the formula:

$$\frac{R_1}{R_2} = \frac{C_2}{C_1} \quad \text{(an inverse ratio of values)}$$

which says (after we transpose R_2 and C_1) that

$$R1 \times C1 = R2 \times C2$$

This then says that, if we have a resistive divider with R_1 twice the value of R_2 and V_{out} equal to one third V_{in}, C2 must be twice the value of C_1 if the hi frequency V_{out} is to equal one third hi frequency V_{in}.

Pay attention to this, though; we have ignored the value of C_{in} in Fig. 2-20. And, if adding C_{in} to C_2 changes its value by any appreciable amount, then the real value of C_2 should be considered to be the sum of the two. For this reason, in actual circuitry where this is critical, C_1 is usually a variable capacitor (or it may be C_2 that is variable) so it may be tuned to the exact value it should have.

What happens if this is not true? Well, that should be easy for you to figure by this time. See Fig. 2-21. If $R_1 \times C_1$ does not equal $R_2 \times C_2$ then the output waveshape will be distorted.

If C_2 is too large (or C_1 too small) its X_{C2} value will be too small in comparison with X_{C1} and as the step waveform comes out it will not be high enough to match the resistor-divided V_{out} that will be reached after 5 RC time constants have gone by and we will have a rolled-off exponential waveform.

If C_2 is too small (or C_1 too large) its X_{C2} value will be too great in comparison with X_{C1} and as the step waveform comes out it will be higher than the resistor-divided V_{out} that will be reached after 5 RC time constants have gone by, and we will have an over-shoot in the waveform.

As far as the output RC time constant is concerned, you should be able to see that looking back into V_{out}, you will see the Thevenin Equivalent of the two resistors (their product over their sum) and the simple sum of the two capacitors in parallel (both will have to change charge at a common point). This Thevenin Equivalent resistance and the total capacitance form the R and C for the circuit, determining the time constant of the recovery curve to the R_1—R_2 divider level.

Don't get the idea that the unbalanced compensated divider is wrong. It's only wrong when it should be a balanced compensated divider. An engineer may deliberately design a circuit so that it distorts the incoming waveform. Remember, the purpose of the circuit must always be taken into consideration.

There is one more note that I should add to this discussion and perhaps this is as good a place as any. It has become common practice among manufacturers to write the simple numerical value of a capacitor in on a schematic with little or no indication of the power of ten related to it. You are expected to understand that if the number is less than 1.00 (i.e., .01 or .001) that the value is in terms of microfarads, and if the value is greater than 1.0 (i.e. 100 or maybe 1000) that the value is in terms of picofarads. This doesn't leave any reason for not labeling a 10 microfarad filter capacitor without its power of ten label (uf). So usually in the power supply you will find large filter capacitors adequately labelled, but in other circuitry where values are usually less than 1 uf, there will usually be no power of ten terminology associated with the value of a capacitor. You just have to remember: less than 1 is microfarads and more than 1 is picofarads.

Diodes

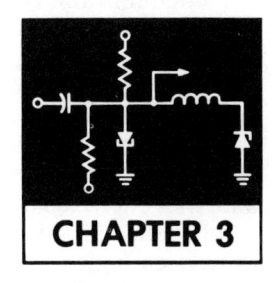

CHAPTER 3

A modern electronics dictionary defines a DIODE as, "A two-terminal device which will conduct electricity more easily in one direction than in the other." And, although this is true, diodes have been specialized far beyond any implication contained in these words.

The history of the diode starts with man's first practical investigation into what electricity really is. It was the man who named the VOLT itself who first discovered the diode function of two dissimilar metals when placed in contact with each other.

One of the earliest forms of the diode is still used today. It is the copper-oxide rectifier, made from copper and cuprous oxide; when firmly pressed together, these two materials will pass electrons from the copper to the cuprous oxide with much greater ease than in the other direction.

Another of the early diodes still in use today is the selenium rectifier. This diode consists of a thin layer of selenium deposited onto an aluminum plate, with a highly conductive layer of another metal coated over this. Electron current in this device goes from the highly conductive layer to the selenium and then to the aluminum plate. Several of these plates are usually stacked together to make a single diode. You see, the larger the plates, the greater the forward-current capability of the diode; and the more plates used, the greater the reverse breakdown voltage characteristic. This simple versatility made such diodes a great favorite for a long time.

A bit later it was discovered that a sharp point of metal in contact with a plane of metal had this same diode characteristic to some extent; and the most common use of this type of diode was the old Cat's Whisker on the Galine Crystal in the now very old-fashioned crystal sets. The galine, formed in nature, is lead sulfide and the point of wire making contact to it formed what is now known as a Shottky barrier type of junction. Under scientific construction today, a similar junction forms one of the fastest turn-on turn-off diodes we have, known as a "hot carrier" diode.

Vacuum-tube type diodes of course were first discovered by Edison when he found what is now called the "Edison Effect" from his light globe experiments. Though it really took J. A. Fleming to put them to use, circa 1905, when he got his first patents covering his development of the vacuum device. Then, a very short time later, a vacuum tube which made use of the point emission theory and the Edison Effect came into being and was known as the B H Tube. It actually worked quite well.

It didn't take long to give the vacuum tube type diodes a heated cathode like they have today.

And, shortly after came mercury rectifiers and other gas-filled envelopes that would glow when sufficient voltage was applied to them, and neon bulbs were born. They, too, are a form of diode.

It was some of these gas-filled diodes that became the first constant-voltage devices. The gas inside would ionize when enough voltage was applied (then the "striking" voltage would drop back down a bit) and as long as a certain amount of current continued to flow, the gas would remain ionized and the device would maintain a constant voltage across itself.

Up until just a few years ago, that was the diode story. But, then came refinements of atomic theory, semiconductors, and the birth of the transistor. And, believe me, diode development has progressed just about as fast as transistor development. Understanding of atomic theory, energy levels, and semiconductors, brought into being the P-N junction. This too is a diode, but with such variations that their application and resultant usefulness have been expanded tremendously.

CONSTRUCTION OF JUNCTION DIODES

The understanding of how all so-called solid-state devices work is wrapped up in the scientific explanations of how to make the best of insulators into partial conductors.

I think the best illustration of this can be achieved with pure water. You see, pure water has an atomic structure that is the same as the best of insulators. Scientists call it a pure Octet structure. And, this structure is what must be changed to make this insulator material into a partial conductor. In the case of pure water, we can **dope** it with a few grains of salt to achieve this alteration.

For changing the insulator materials that solid-state diodes are made from (that is, silicon and germanium) into semiconductors of a more specific nature, scientists are a bit more picky and choosy in what they add, but the resultant doping action is very similar. Gallium causes what we usually see illustrated as positive conduction, while arsenic causes what we see as negative conduction.

In other words, we now have two different types of conducting material. One of these has the positive charge or hole (the absence of an electron) that moves through the material, while in the other it is the negative charge or extra electron that moves and results in conduction. These two types of conduction have caused the subclassification of the basic insulator material into "P" (for positive) and "N" (for negative) material, depending upon what kind of "salt" was added.

The limited conductivity of "P" material can be illustrated using plus signs (+) as though they (the holes) were the current that was moving through the material from positive to negative.

In the case of "N" material, the extra electrons (—) are illustrated as though they were the thing that moves through the material from negative to positive.

Solid-state diodes are now made up of a piece of "P" material joined to a piece of "N" material within the same crystalline structure. The line of demarcation between these two types of material is called the junction. Thus, we can now define a solid-state diode as being a P-N junction.

If we apply voltages to this device (remembering that like charges repel each other) such that the holes in the "P" material are forced to meet the negative charges in the "N" material at or across the junction, the device will act as a conductor and electron current will actually pass from one end to the other. If, however, we apply voltages in the opposite direction (remembering that unlike charges attract each other), you can see that the current carriers are pulled away from the junction, leaving nothing but pure insulator in between and the device does not conduct. Electron current can no longer pass through it. And, consequently, it obeys the original definition of the diode.

The current vs voltage graph that our present day measuring devices (transistor curve tracers—an oscilloscope variation) would show us for such an action looks like that seen in Fig. 3-1 where the curves of three different types of diodes are shown. The horizontal sensitivity of the instrument that drew these graphs for all three diode types was .2 volts per division and the vertical sensitivity was set at 1 ma per division.

Note the differences shown for the different materials that the diodes were made from. A diode made of doped germanium conducts 2 ma of current at about 0.2 volts of forward

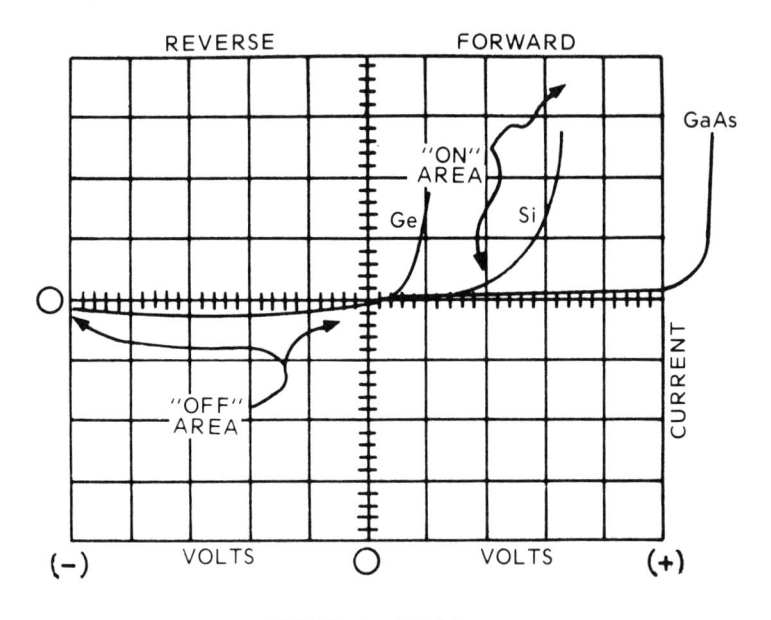

HORIZ: 0.2 V/DIV.
VERT: 1 MA/DIV.

GERMANIUM (Ge)   +.2 volt
SILICON (Si)   +.6 volt
GALLIUM ARSENIDE (GaAs)   +1.2 volts

NORMAL OPERATING REGION:
Over the whole curve.

CATHODE —[N | P]— ANODE

(−) ⟶ (+)

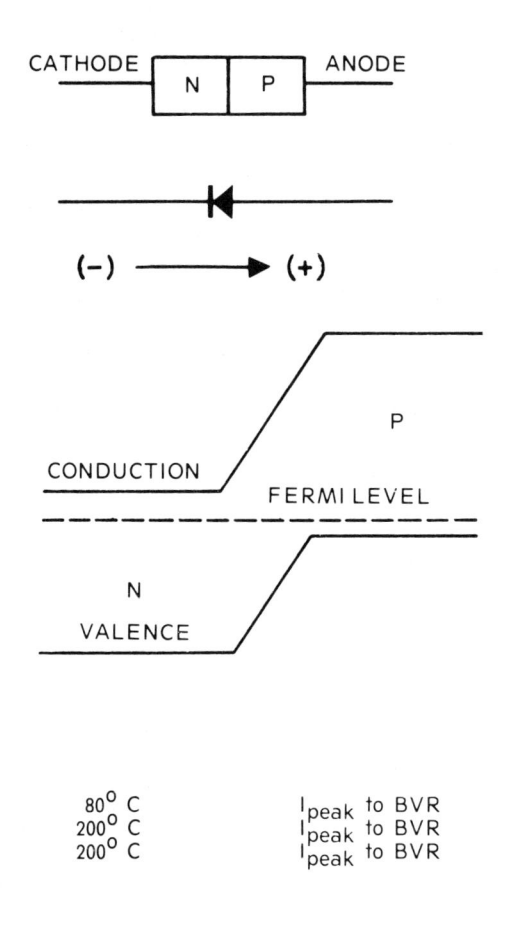

CONDUCTION

P

FERMI LEVEL

N
VALENCE

80° C I_{peak} to BVR
200° C I_{peak} to BVR
200° C I_{peak} to BVR

Fig. 3-1. Diode current vs voltage graph.

bias. A diode made of doped silicon requires a little over 0.6 volts to conduct the same amount of current. And, the gallium arsenide diode has to have about 1.2 volts of forward bias to cause it to conduct 2 ma. These voltages are characteristics of the basic material each diode is made from and identifies it.

There is nothing in the diode symbol used in schematics to differentiate between these different diodes. You just have to count on the parts list or the individual part number to do this for you. Each of these three diodes has the cathode made of "N" material and the anode made of "P" material as shown to the right of the graph in Fig. 3-1. The cathode is the junction line in the schematic symbol and the anode is the arrowhead pointing to the source of electrons. This last statement is just about universally true. The arrowhead in the solid-state schematic symbol points to the source of electrons under the FORWARD-BIASED condition of the junction. However, we do have to remember that some devices are not used for their forward-bias characteristics. There are several that are used under REVERSE-BIAS conditions and we'll find out about these as we come to them.

ZENER DIODE

One such device used for its reverse-bias characteristics is the zener diode. If you attach the cathode (or "N" material) side of the junction to ground and then apply a positive voltage to the anode (or "P" material) side of the junction, you would get a graph similar to that shown in Fig. 3-2. In the forward-bias condition you would get the normal curve of a silicon diode, while in the reverse-bias condition, you would get practically no current through the device until a point of breakdown had been reached. Please understand that we can apply a large enough voltage (or "push") to almost any material to make it conduct, if only for a fraction of a second. And when this happens, the heat generated usually burns up the material). However, the silicon junction will withstand quite a bit of heat (close to 180 degrees Centigrade), and the heat generated in watts is equal to the product of the voltage across the junction and the current through it. And, if the heat so generated can be radiated into the air around the diode fast enough to keep the junction temperature below 180 degrees Centigrade, it is safe to operate the device in this condition.

Well, Zener diodes are so designed. While a curve for a regular diode will slope off to what we might call the "southwest" after breakdown is reached, a Zener diode curve will head straight "south" at this point (Fig. 3-2). It was Mr. Zener who researched this phenomenon, and now this sharp corner is called a Zener Knee. He found that if we increase the amount of the doping stuff we add to the basic material to make it conduct, we can cause this sharp knee to appear at a lower and lower voltage.

Now, there is a whole family of Zener diodes and they can be purchased for the specific voltage at which they break down. Referring to Fig. 3-2, note what this graph tells us. It is saying that this device will guarantee a specific voltage over a fairly wide range of different current levels.

Mr. Zener discovered that this type of conduction wasn't exactly the same thing as thermal breakdown conduction of the normal junction. It is now called AVALANCHE conduction in order to differentiate it from THERMAL breakdown.

Engineers did get smart and come up with a different symbol for this device. The straight line is still the cathode and the arrowhead is still the "P" material of the junction, but the straight line of the junction has been changed into something resembling a Z (for Zener). This tells us that this gadget is predominantly used for carrying electron current in the direction of the arrow and thus makes use of its reverse-bias characteristics (the Zener knee at breakdown). So, let me suggest that we don't argue over which is the anode and which is the cathode. Let's just understand that this device is being used backwards to take advantage of its bias characteristics.

TUNNEL DIODE

Semiconductor material can be doped heavier yet, so that the Zener knee shows up well below the 10-volt level. And, if such a device were doped even heavier yet, we might end up with what Dr. Esaki named a "tunnel" diode. This diode conducts right off the bat under reverse bias, as shown in Fig. 3-3. It also conducts immediately upon application of forward bias, but to a limited degree. Then as we increase voltage a bit more, its conductivity decreases to where it can be thought of as joining its normal diode conduction curve near 0.2 volt. Then as more forward bias is applied, it proceeds to conduct again, reaching a level of forward peak current near 0.45 volt. These devices are rated by the amount of current they will conduct at the first peak reached under forward bias conditions. Some will conduct as high as 50 milliamps before going into the so called valley region near the 0.2 volt part of the curve.

A ratio of the peak current (near the 0.05V part of the curve) to the valley current (near the 0.2V part of the curve)

91

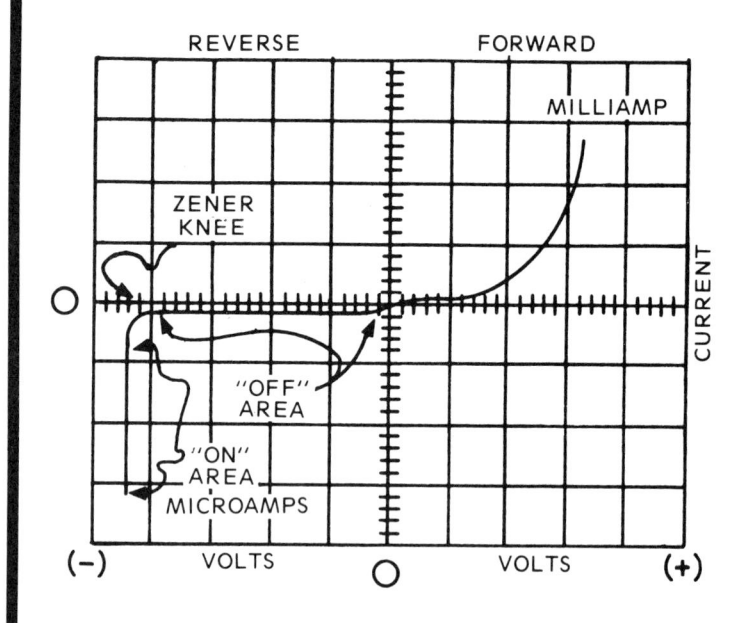

```
         REVERSE        |        FORWARD
                        |
                        |              MILLIAMP
                        |
    ZENER               |
    KNEE                |
                        |
  O ─────────────────── | ───────────────────
        ↑               |
        │  "OFF"        |
        │  AREA         |
        │               |
      "ON"              |
      AREA              |
    MICROAMPS           |
  (−)      VOLTS        O       VOLTS      (+)
```

Usually made of Silicon

Rated by their Zener Knee Voltage and peak power dissipation.

Can carry much greater forward currents than in reverse, but predominantly used for their Zener Knee and reverse-current-carrying capability.

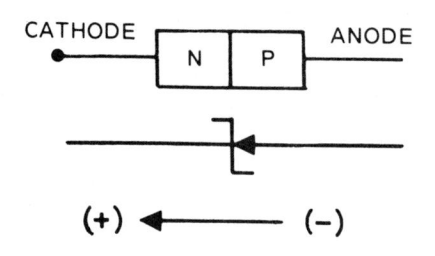

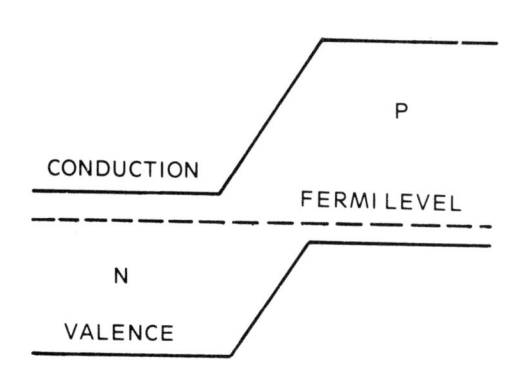

Fig. 3-2. Zener diode graph and characteristics.

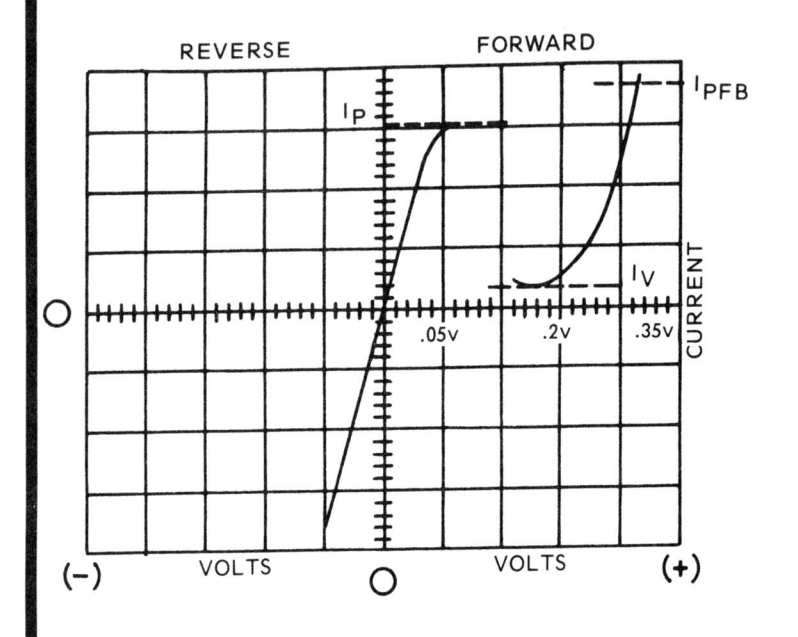

Made of Silicon

Rated in terms of I_{peak} and I_{valley} and/or a ratio of the two, approaching 10.

Normally operated in their forward region, can be operated from reverse to forward as they are in many circuits.

REMEMBER that it is the current-carrying capabilities of these devices that are different. The switching voltages are all the same.

V_{peak} equals .05 volt or 50 mv

V_{valley} equals .2 volt or 200 mv

$V_{forward\ peak}$ equals .35 volt or 350 mv

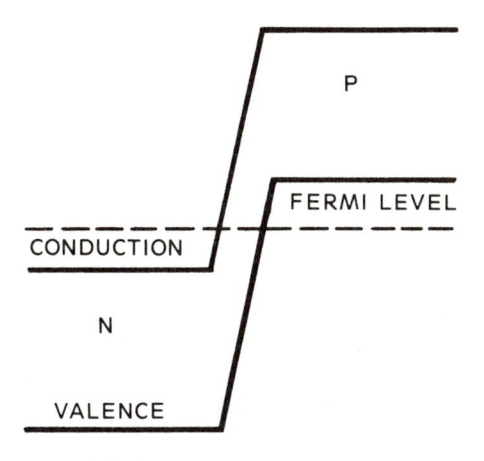

[Energy band diagram showing P and N regions with Fermi level, Conduction and Valence bands]

Can be used for their fast switching time from V_{peak} to $V_{forward\ peak}$ in the neighborhood of .15 nanosecond, or their switching time from V_{valley} to less than V_{peak} (actually switching backwards) in approximately the same time.

Once switched, they have to be reset. You can always look for something that will switch them in the opposite direction from that in which they are being used.

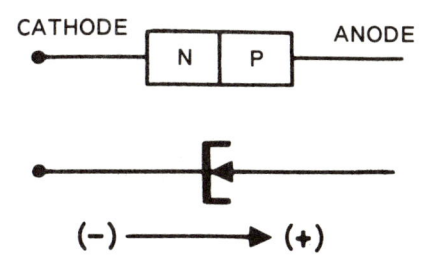

Fig. 3-3. Tunnel diode graph and characteristics.

tells us how good the individual tunnel diode is. A ratio of 10 to 1 is considered to be quite good.

There is one thing I would like to comment on here and that is that I do not believe Dr. Esaki invented tunnel diodes. He did succeed in naming the type of current flowing through the diode to give us the first peak of current flow under forward bias conditions. He said that this current appeared to tunnel around the forbidden region of the energy level diagram at the junction, and the name "tunnel" stuck. Experimental circuits describing results that I can only lay to tunnel diodes can be found as far back as 1926, long before Dr. Esaki came on the scene.

BACK DIODES

Some manufacturers dope tunnel diodes so carefully that their peak current rating is as low as 1 ma. And, this very special version of the tunnel diode is now called a back diode (Fig. 3-4). As long as we don't apply more than around 0.3V of forward bias, it won't conduct very much. But immediately upon application of reverse bias—like any other tunnel diode—they will conduct like mad. This is the reverse of the action for a normal diode (as long as we don't go to higher voltages) so this is why these devices are called BACK DIODES. They are rather special devices with rather special uses.

Refer back to Figs. 3-3 and 3-4 and note the difference between the two schematic symbols. The junction line of the tunnel diode has two little lines that tend to enclose the point of the arrow. The back diode has only one of these little lines. It is a rather subtle difference but it's easier to use than a regular diode symbol labelled with a B-D (for Back Diode), as they originally used when the device first came into being. In older schematics, you're liable to find it drawn either way.

You can't tell one diode from another just by looking at the case they are packaged in, but the graph of what each will do under applied voltages will tell you exactly what each one is.

More and more as time goes along and man's knowledge increases, the word DIODE is coming to mean simply a "two lead device." In other words, a diode doesn't have to have just one junction.

SHOCKLEY DIODE

The Shockley Diode is a good case in point. This gadget has three junctions (i.e., four alternate layers: N — P — N — P, see Fig. 3-5). This device also has a voltage-current curve all its own.

The Shockley diode doesn't conduct to speak of under either forward or reverse bias until a special limit is reached. At this special level of voltage in the forward direction, the device changes drastically from a very poor (high impedance) conductor to an excellent conductor (almost an absolute short circuit). Of course, when it switches, we must limit the current through this device with a series resistor, but as long as we do, it can be made to switch again and again. The voltage across the device will drop from its forward switching level to approximately 0.5 volt.

What occurs in the reverse bias direction is a little different. At a magnitude of voltage a bit higher than that of the forward switching level, you will get what appears to be a lousy Zener knee. In other words, in the reverse direction it will conduct, but the device maintains a high voltage across itself and thus has to dissipate a much larger amount of power than in the forward direction and is in danger of burning itself up.

Some of these Shockley diodes have forward switching voltages as high as 100 volts, and they may be stacked (hooked up in series) to get higher switching levels. They are not considered to be especially fast in their switching action—just big.

Once switched, the Shockley diode does require a minimum amount of current through itself in order to stay switched. Otherwise it will tend to oscillate back and forth a bit, and the rate is very inconsistent. Of course, each one does have a maximum amount of current it will carry in the forward direction that must not be exceeded. These devices have their own special applications too.

A variation of the Shockley diode is found in both the S.C.S. (the Silicon Controlled Switch) and the S.C.R. (the Silicon Controlled Rectifier) devices. See Fig. 3-6.

Either one or both of the inside layers of semiconductor material have outside leads connected to them and they may have a signal applied to them to switch the diode ON or OFF rather than exceeding the switching level of voltage or starving it for current as is the case with the Shockley diode.

The so called BINISTOR is another variation of the four-layer device and was manufactured by Transitron Electronic Corp. in Wakefield, Mass., but I haven't heard much about it since its introduction back around 1960.

FIELD EFFECT DIODE

In all these devices we have looked at so far, the electron current has crossed a junction, going from one type of con-

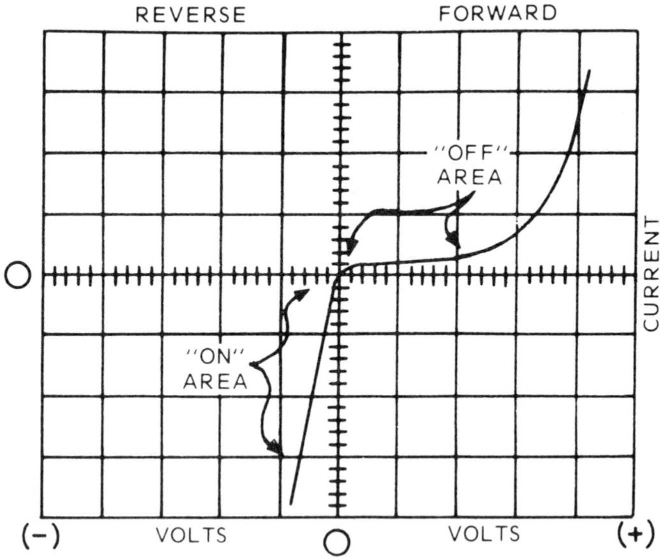
REVERSE
FORWARD
"OFF"
AREA
"ON"
AREA
O
CURRENT
(−)
VOLTS
O
VOLTS
(+)

MOST CONFUSING since the G.E. Manual draws them backwards—WE DON'T.

Usually made of Silicon

Rated by peak current just after turn on in the forward direction like a tunnel diode, but capable of carrying much higher currents in reverse.

NORMAL REGION OF OPERATION:

From slightly forward-biased to slightly reverse-biased.

A small signal device.

CATHODE — [N | P] — ANODE

CATHODE —|◁— ANODE

(+) ◀——— (−)

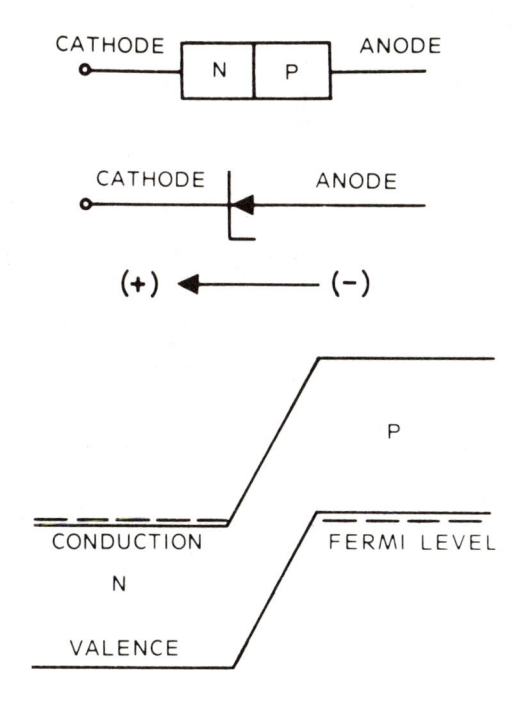

```
                    _____
                   /
                  /           P
                 /
  _ _ _ _ _ _ _/       _ _ _ _ _ _ _
  CONDUCTION    \     FERMI LEVEL
                 \
       N          \
                   \
  VALENCE          _____
```

Since these devices are used to NOT CONDUCT under SLIGHT FORWARD BIAS and TO CONDUCT under SLIGHT REVERSE BIAS they are just the opposite of a standard diode and thus are called BACK DIODES.

Fig. 3-4. Back diode graph and characteristics.

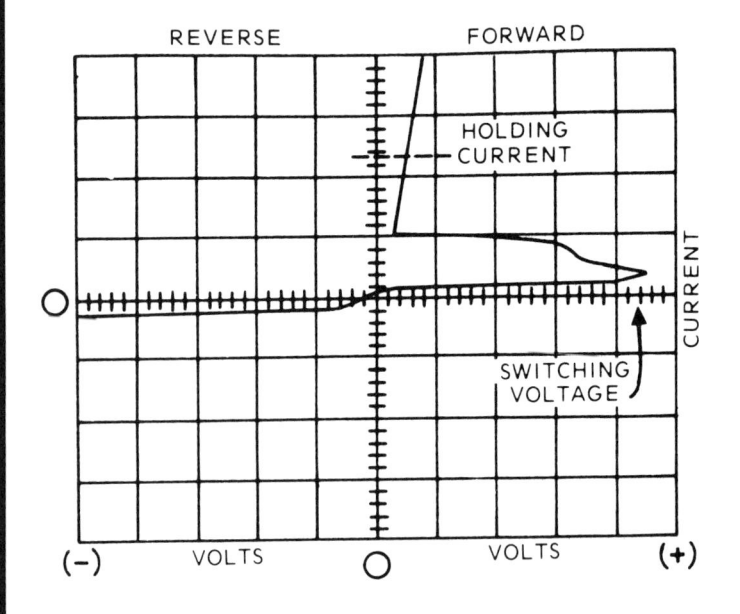

This is a 4 layer (3 junction) device.

It is rated by its switching voltage and holding current. It will also have a peak steady forward current rating that must not be exceeded unless the duty cycle is limited.

CATHODE ─── N | P | N | P ─── ANODE

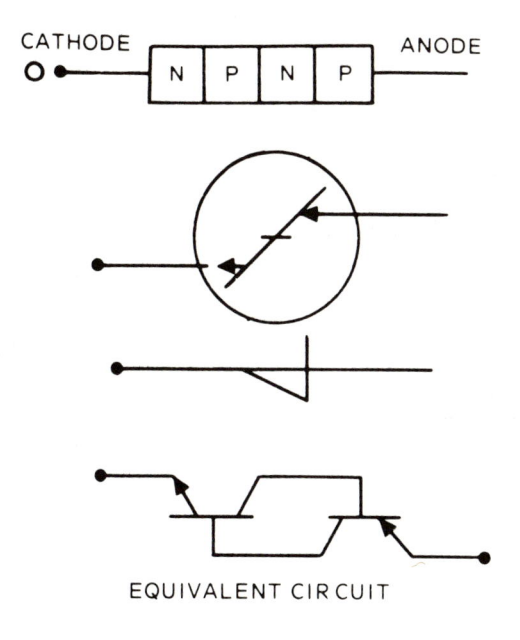

EQUIVALENT CIRCUIT

Once the diode is switched from high-impedance state to low-impedance state, the holding current must be maintained through the device or it will return to its high impedance state (it can oscillate at inconsistent rate in this mode if there is not a large controlling capacitance in conjunction with the device).

Switching time is not fast. Switching voltages range from 15 volts to 200 volts (the devices may be stacked).

Fig. 3-5. Shockley diode graph and characteristics.

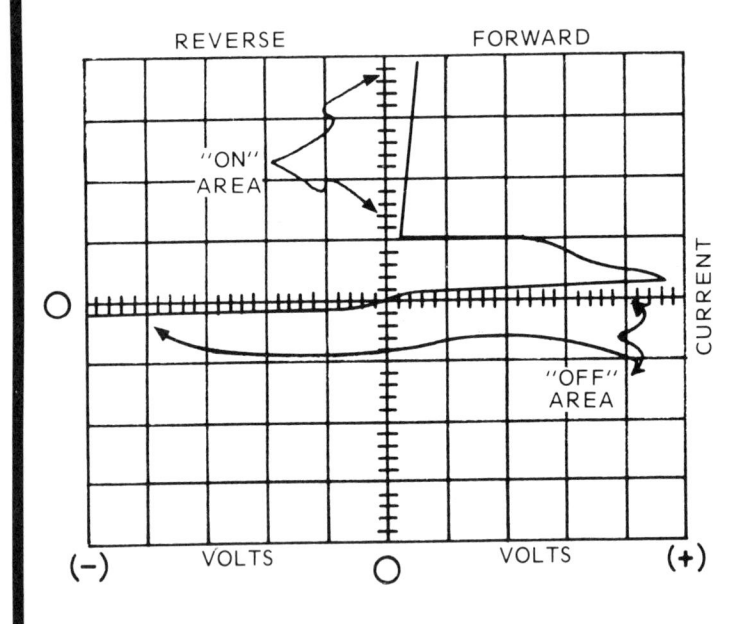

Basically the same thing as a Shockley Diode, except the two inside layers also have connections brought out for purposes of drive (turn on or turn off).

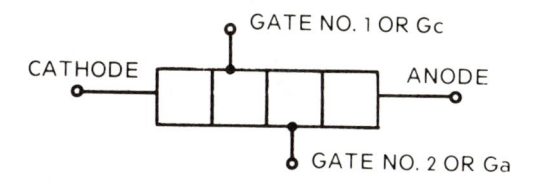

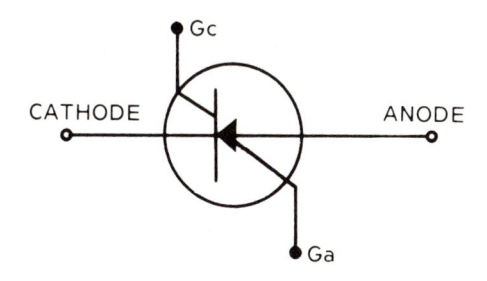

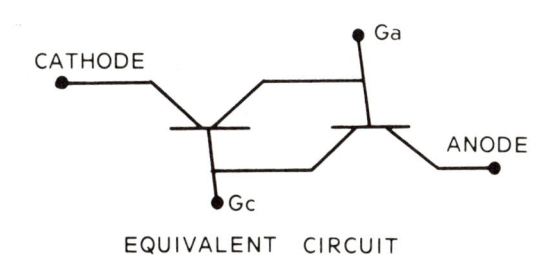

Fig. 3-6. SCS-SCR diode graph and characteristics.

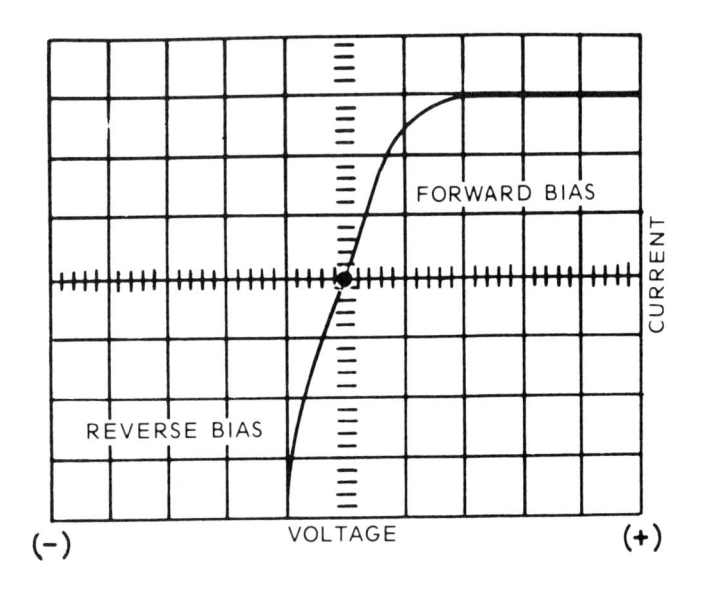

The Field Effect Diode is the counterpart to the Zener Diode. Used for its constant-current capability under wide voltage change. In reality it is a zero-biased Field Effect Transistor.

ducting material to another. Here is one where the electron current stays in the same type of material all the way from one lead to the other (Fig. 3-7). It is called the FED (Field Effect Diode). In reality it is a zero-biased Field Effect Transistor (the Source and the Gate are tied together). Its diode graph, however, shows it to be the counterpart of the Zener diode. Remember the Zener diode gave a constant voltage over a rather wide range of current levels? In other words, current could change through the device, but voltage wouldn't. Well, the FED is the opposite of a Zener. This very special diode has a constant-current level in its graph over a wide range of voltage levels. Voltage can be changed all over the place across this device and it will maintain the same

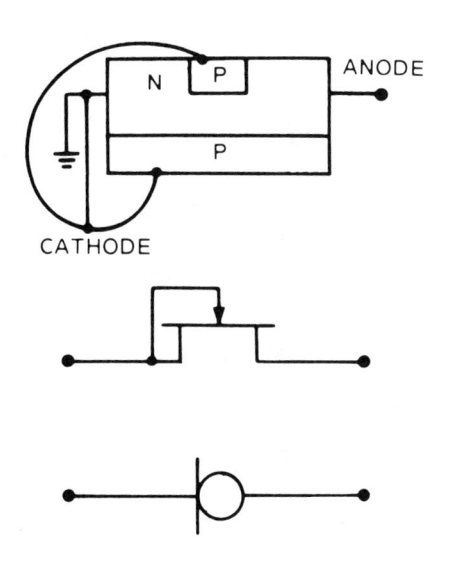

Fig. 3-7. Field effect diode graph and characteristics.

amount of current through itself. Once more, a rather special device, but you sure wouldn't know it by just looking at it. You have to see its curve to know what it is.

As I mentioned in the beginning, diodes have been specialized far beyond any implication contained in the original definition. They are no longer just a two-lead device defined by their maximum reverse breakdown voltage and maximum forward-current-carrying capability. There have been new wrinkles added to the basic curve. These new wrinkles define new uses to the imaginative engineer in new fields for almost the same old device. Each new type of diode with its own special wrinkle in the same old basic curve defines its use to new and different limitations. Each in-

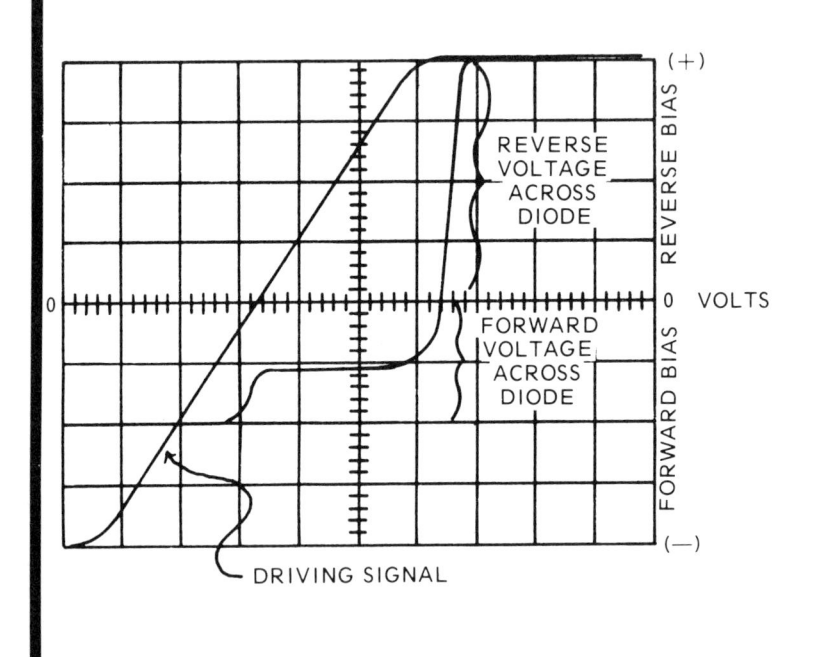

Fig. 3-8. "Snap" diode graph and characteristics.

dividual diode story as told in its own graph becomes the key to new ideas.

And, believe me, science is measuring them closer and holding them to tighter tolerances every day.

For instance, the speed with which a diode can be turned on or turned off has become quite critical to some circuitry. And, its especially interesting to note the theory behind the so called "snap" diode. See Fig. 3-8.

"SNAP" DIODE

The "Snap" diode has what might well be called a sloppy junction, with "N" and "P" majority carriers overlapping

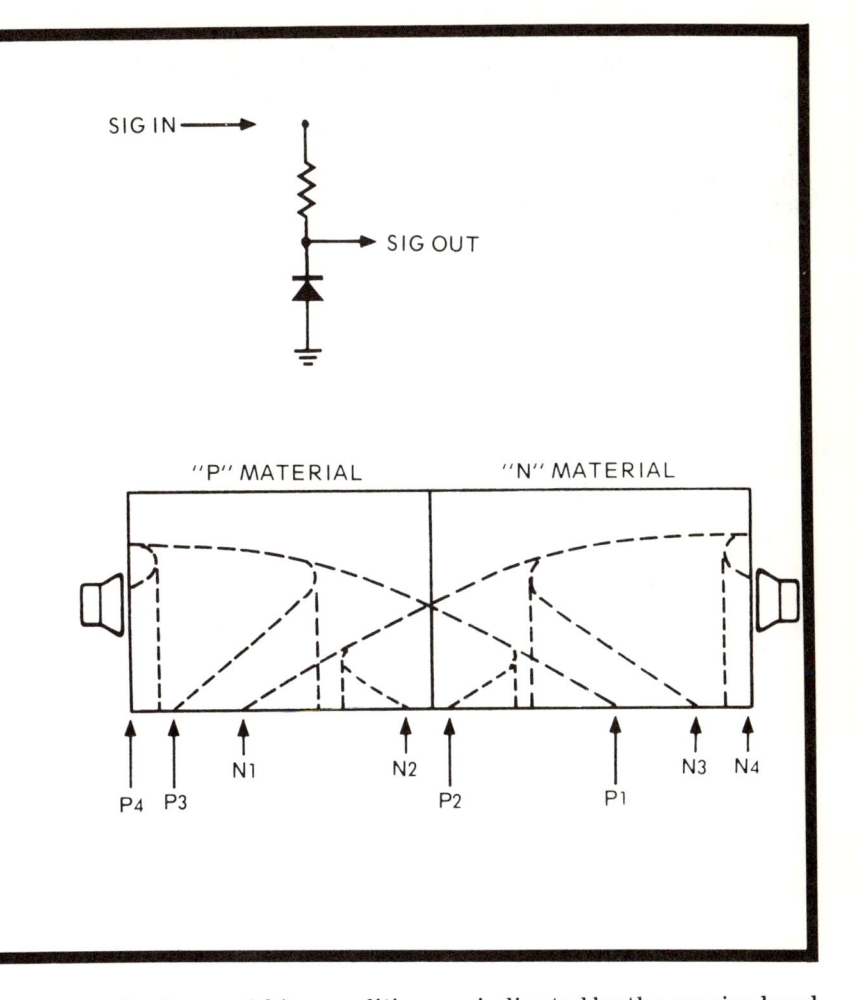

under forward-bias conditions as indicated by the carrier level graph lines N1 and P1. At the instant the circuit is reverse-biased, the current through the diode reverses direction. (The device now is almost like a capacitor with, like we used to say, a charge stored in it.) The majority carriers that are farthest away from the end of the semiconductor material feel the greatest space charge, and they are pulled out. The graph depicting majority carrier level appears to fold back on itself as time progresses and moves from P1 to P2 to P3 to P4 as the "N" material carriers shift from N1 to N2 to N3 to N4, with the result that the device runs out of both types of carriers all at once. This shows up as an extremely fast change of current in an exceptionally short period of time and is the action which gives this device its name (i.e., "SNAP" diode). But this device does not discharge its stored current like a capacitor. It releases its stored charge at a fairly stable rate, thus maintaining a stable voltage across itself until the current carriers run out. Then, SNAP!—the voltage changes. Yes, there is a slight delay between the triggering voltage change and the actual change of voltage across this diode. But some things you just have to live with.

Analyzing Time Constants of Diode Circuits

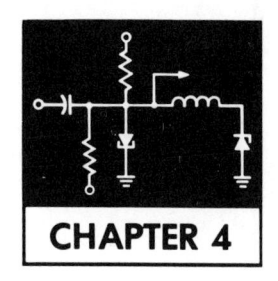

CHAPTER 4

We have covered a lot of different ideas in this book so far, but each one has more or less been worked out on an individual basis. And I think it's about time to bring things down to a more practical nature by looking at some circuits more like those you are liable to run into in present day schematics. Not that you don't find the simpler circuits, but it does take a bit more stretching of the imagination to understand some of the more complicated ones that you will run into. We now have the tools to do such work. (Remember, each idea taught so far is a tool added to your kit for analyzing new circuits.)

SIMPLE HIGH PASS FILTER (A REVIEW)

For a starter, let's refer to Fig. 4-1. Here is a high-pass circuit with three different resistors to three different voltages, which we must reduce to one resistor and one voltage before we can start trying to see what happens relative to the time-constant of the circuit. In other words, we need a Thevenin Equivalent (a voltage source with a series resistor). And, since there are more than two resistors and two supplies, let's give the Millman Equation a try (this ought to be the right tool here to start with):

$$V_{out} = \frac{\frac{V_1}{R_1} + \frac{V_2}{R_2} + \frac{V_3}{R_3}}{\frac{1}{R_1} + \frac{1}{R_2} + \frac{1}{R_3}} = \frac{\frac{100V}{12K} + \frac{0V}{60K} + \frac{(-150)}{15K}}{\frac{1}{12K} + \frac{1}{60K} + \frac{1}{15K}}$$

then change signs and

$$= \frac{\frac{100}{12} + \frac{0}{60} - \frac{150}{15}}{\frac{1}{12} + \frac{1}{60} + \frac{1}{15}} = \frac{+8.34 - 10.}{0.1667} = \frac{-1.66}{0.1667} = -9.97$$

or we might just as well round it off and say V_{out} equals $-10V$.

Draw the voltage waveform at V_{out}.
(Label time/div. & volts/div.)
100uS/DIV.   50V/DIV.

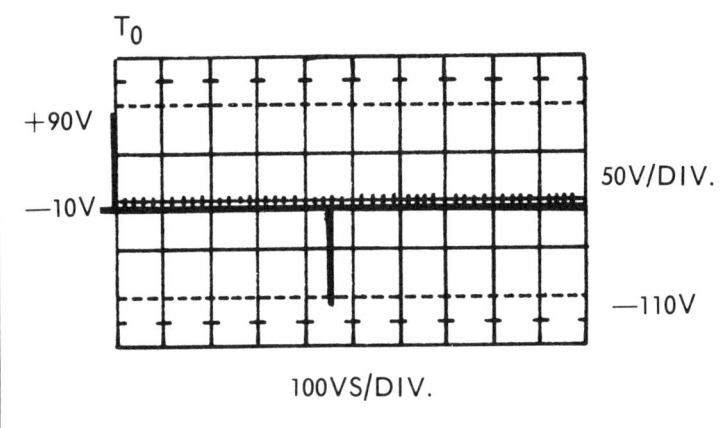

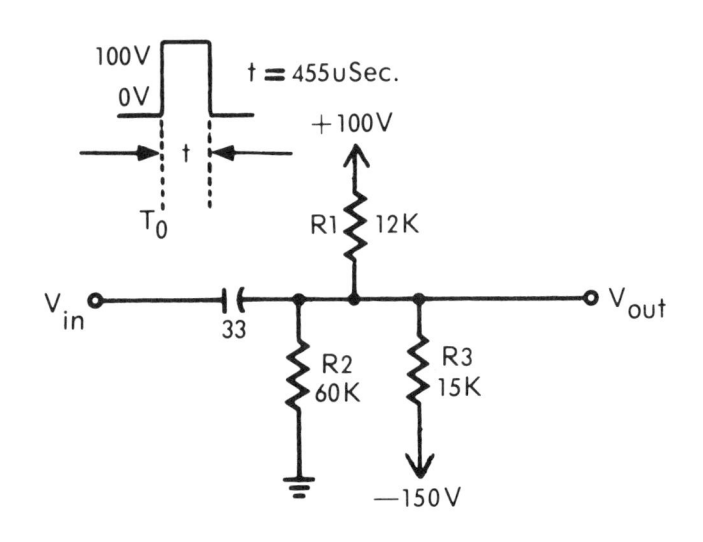

T_0

+90V

50V/DIV.

−10V

−110V

100VS/DIV.

100V / 0V   t = 455uSec.

+100V

R1 12K

V_{in} —||— 33 — V_{out}

R2 60K

R3 15K

−150V

THEVENIN EQUIVALENT:

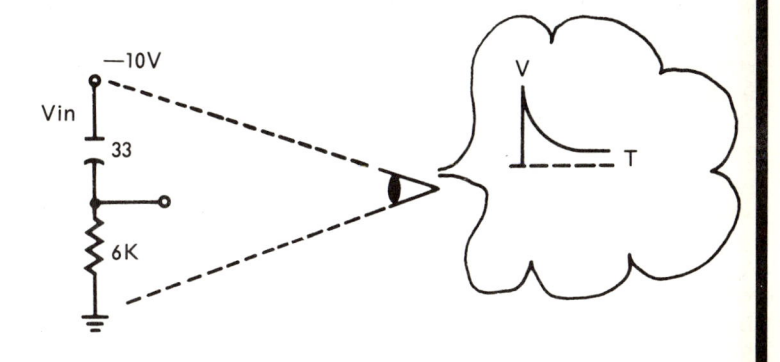

Fig. 4-1. High pass filter circuit, Thevenin equivalent, and waveform.

To find the resistor, look at it this way: The 12, 60, and 15 are all "K" ohms. The reciprocal of 0.1667 (1/0.1667 equals 6) is 6 and the reciprocal of "milli" is back to "K" again and the equivalent resistor is a 6K ohm resistor.

Now we know that the circuit in question looks like a 33 picofarad capacitor (remember—a capacitor labelled greater than 1 is in pico terms) and a 6K resistor fed by a —10-volt source. V_{out} will be starting at a —10-volt level while V_{in} is being held at 0 volts and the capacitor has a quiescent 10-volt charge across it.

Let's now consider the terms we are going to use to draw our graph. We should show what happens to both the rising edge of our incoming pulse and the falling edge, so I would like to suggest we use 100 microseconds per division for the horizontal graduations of our graticule. (Yes, we could use 50 microsec per division, but this stretches things out a bit too far to suit me.) This then says that our output waveform will be a bit over four and a half divisions wide. Then, since we should show the full amplitude of the output pulse (or pulses) let's say we are going to use 50 volts per division vertical sensitivity on the graph. (Before proceeding, review Fig. 2-17.)

Now then, if the horizontal line across the middle of our graph is considered zero volts (ground), then our V_{out} waveform will enter the graph one fifth of a division below center (at —10 volts). At T_0 the input pulse steps positive 100 volts (assuming N, Fig. 2-17, to be much greater than 100) and V_{out} does the same thing, changing from —10 volts to +90 volts since the capacitor can't change voltage instantaneously, and we draw a vertical line two divisions long showing this change in V_{out}.

Now what happens? It's going to take V_{out} five time-constants to return to —10 volts. Do we have that long, or does the driving waveform change first? Let's see. One RC equals $33 \times 10^{-12} \times (6 \times 10^{+3})$ equals 198×10^{-9} or approximately 200 nanoseconds, or, better yet (using up three of those negative powers of ten), 0.2 microseconds. Five RC will equal five times this or 1 microsecond and the pulse is 455 times this wide. We won't even see the curve of the falling part of this waveform at this horizontal speed. We will see a very fast positive spike at T_0 100 volts high (2 divisions), and four and a half divisions later in the horizontal direction we will have a 100 volt negative spike from —10 volts to —110 volts, with the output returning to its quiescent —10-volt level for the rest of the graph.

Actually, this 100-volt pulse becomes a positive and negative spike of 200 volts peak to peak existing between +90 and —110 volts.

Of course, if our driving waveform comes from a generator whose time-constant is only 10 times faster than the RC time of our circuit, the graph in Fig. 2-17 shows us that our spike will only be 77 volts high and go from −10 volts to +67 volts, and then from −10 volts to −87 volts and return to −10 volts. This would make the V_{out} waveform only 154 volts peak to peak rather than 200 volts as we figured it. But, it would still give us a pair of positive and negative differentiated spikes.

Now that we have combined Thevenin (via Millman) to the time-constant situation, let's add a diode to the time-constant picture for the next step.

ADDING A DIODE

Refer to Fig. 4-2. Here we have a diode (D1) quiescently conducting a current from the −150 volt supply through the 470K resistor. Since we do not know if the diode is made of silicon or germanium, let's assume a rough 0.5 volt drop across it (part way between 0.2 and 0.7 taken from Fig. 3-1). This tells us that the quiescent V_{out} voltage level will be about +11.5V. In the meantime (according to the input waveform), V_{in} is being held at +50V. This tells us there is a quiescent 38.5 volt charge on the 82 pf capacitor.

So far, so good. Now, what kind of a time-constant are we working with? Let's see: if the incoming pulse snaps more positive and the capacitor can't change voltage immediately, the diode will become reverse-biased and act like an open switch. It just won't be in the circuit until it becomes forward-biased again. So, one time-constant is:

$$R \times C = (82 \times 10^{-12})(470 \times 10^{+3})$$

Let's use our powers of ten to simplify the situation, by calling the 470K a 0.47 megohm resistor:

$$R \times C = 82 \times 10^{-12} \times .47 \times 10^{+6} = 82 \times .47 \times 10^{-6}$$

or $R \times C = 38.5$ microsec, and round off to 40 microsec,

and $5 \times RC = 5 \times 40$ microsec = 200 microseconds.

Now we are ready to see what happens. T_0 comes along and the input jumps 150 volts in the positive direction, reverse-

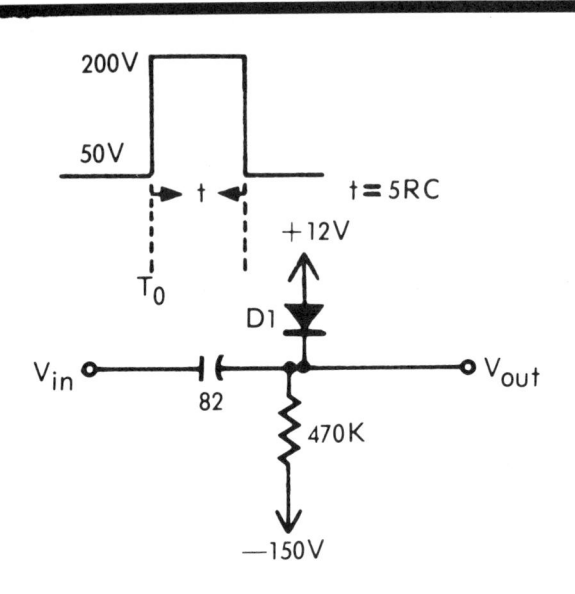

Find the time after T₀ when D1 conducts.
Draw the voltage waveform at V_{out}.
(Label time-div. & volts-div.)
20usec/div.   50V/div.

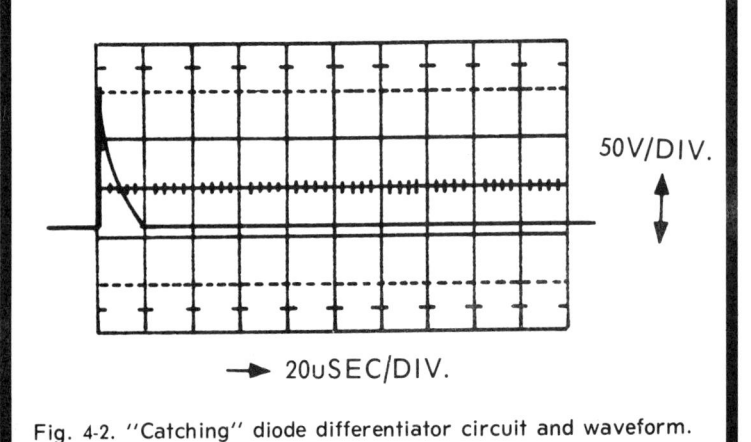

Fig. 4-2. "Catching" diode differentiator circuit and waveform.

biasing the diode. (Of course, the diode must have a reverse breakdown rating in excess of this 150 volts, and we will assume that it does.) The right side of the capacitor will move 150 volts in the positive direction too, from its quiescent value of +11.5V to +161.5V and then start to recover a five RC curve which now extends all the way down to −150V. Remember, the diode is off and is not affecting the circuit until such time as the diode can turn back on. The diode will interrupt the curve when it reaches the +11.5-volt level and then V_{out} stops changing. (This is the action of what we call a "catching" diode.) The main question is: "Just how soon after T_0 does this happen?" We can use the curve in Fig. 2-13 to tell us what portion of or how many time-constants is required for this to happen.

First, we want to find out just what percentage of the total curve does the voltage change represent? In other words, what does 150V divided by 311.5V give us?

$$\frac{150}{311.5} = 48\%$$

Now we can go to the curve in Fig. 2-13 and find 48 on the vertical axis, and move horizontally to the right to where we encounter the curve. Now, looking down at the bottom of the graph, we find that 0.65 time-constants have gone by for this percentage of change to take place.

And finally, since we know that one RC is about 40 microseconds, we can multiply 40 by 0.65 to get 26 microseconds, the amount of time after T_0 when the diode turns back on and stops all voltage change.

The voltage will remain stable and the current through the diode will also be constant for another 174 microseconds before the driving waveform (lasting 5 RC or 200 microseconds) snaps back negative to its quiescent 50-volt level. This negative surge will only send a sharp pulse of electrons through the diode, which is on during this change, giving us only a small fraction of a volt change in the negative direction (it will be the actual increase in voltage across the diode required by the extra current, with the diode representing practically a short). This small pulse disappears in an extremely short 5 RC time since the R is the resistance of the "ON DIODE." Consequently, out of this circuit we get a pulse shortened in time (the base is only 27 usec long instead of 5 RC or 200 usec long) which occurs at the same time as the rising front edge of the driving waveform, with the negative portion almost completely shorted out 200 usec later.

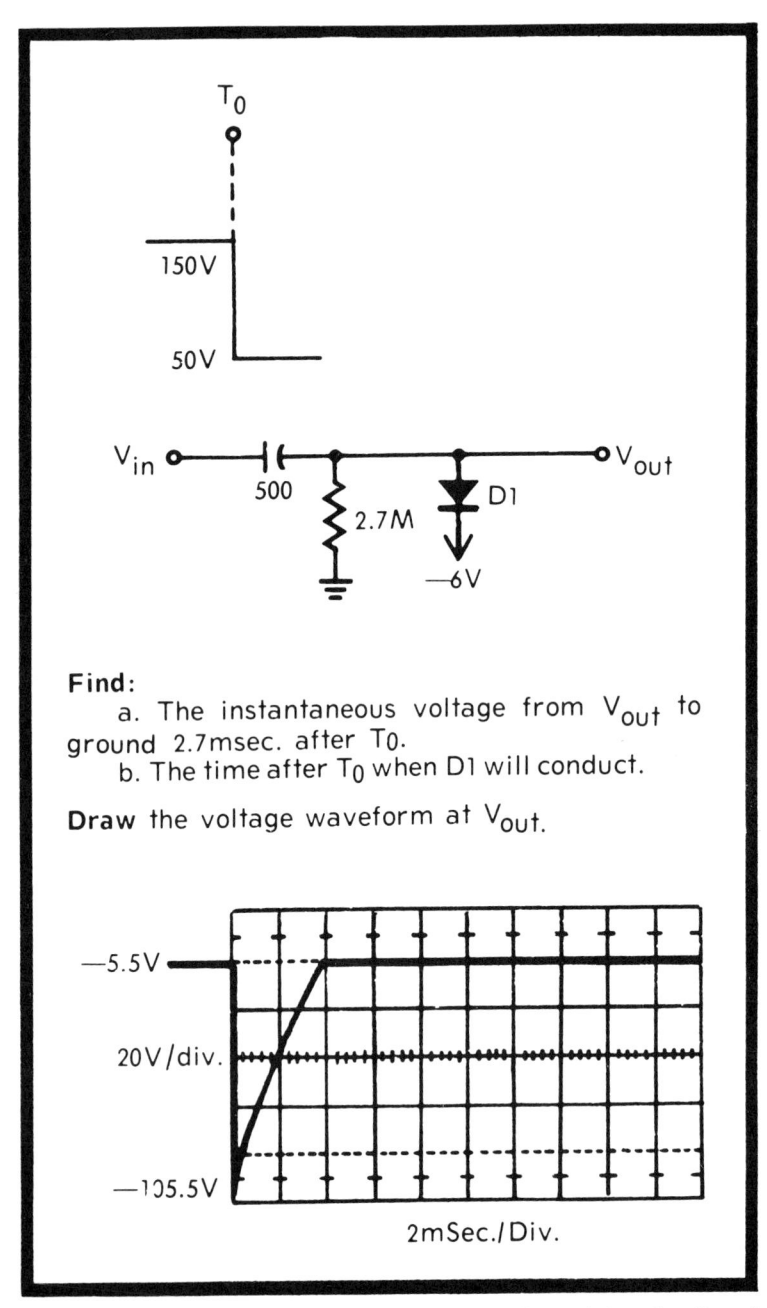

Find:
   a. The instantaneous voltage from V_{out} to ground 2.7 msec. after T_0.
   b. The time after T_0 when D1 will conduct.

Draw the voltage waveform at V_{out}.

Fig. 4-3. Reversed-polarity "catching" diode differentiator circuit and waveform.

REVERSING POLARITIES

For our third example, let's change the polarity of things and use different parts values and see how this works. See Fig. 4-3.

The diode is on again quiescently with V_{out} 0.5 volt above −6V, or at −5.5 volts. The driving waveform tells us that V_{in} is at +150 volts. Now T_0 comes along and the driving waveform drops sharply 100 volts to a +50 volt level and since the 500 pf capacitor can't change voltage instantaneously, V_{out} drops 100 volts also and the diode cuts off. This leaves the right hand plate of the capacitor at −105.5 volts but rising toward the only other reference it can see, which is ground. After the capacitor changes 100 volts of the 105 volt curve it is trying to follow, the diode comes back on and catches it, holding the voltage stable once more at −5.5 volts. In other words, it recovers over 100 volts of the 105 volt curve which says that it recovers over about 95 percent of the curve which according to the RC graph, Fig. 2-9, happens at the end of the third time-constant. So, the answer to question b in Fig. 4-3 is;

$$3 \times RC = 3 \times 500 \times 10^{-12} \times 2.7 \times 10^{+6}$$
$$= 4.05 \times 10^{-3}$$

Or, I think it would be safe to say that the diode comes back on in approximately 4 milliseonds after T_0.

Now we can concentrate on our answer to question (a). Isn't 2.7 msec equal to two time-constants? Yes, and the chart in Fig. 2-9 shows us that at the end of two time-constants we have 13.7 percent of the total amplitude of the attempted change (105.5 volts) still to go at this time and thus the V_{out} at the end of 2.7 msec after T_0 is:

13.7% of -105.5 volts = .137 x (-105.5) = -14.48 volts or say -15V.

(NOTE) The rather general percentages at the end of each time-constant are a good thing to remember. After all, you only have to remember four numbers: 63, 86, 95, and 98 percent. These are plenty close enough to give you a pretty good idea of what's going on in a circuit.

GETTING A BIT MORE COMPLEX

In the problem shown in Fig. 4-4, we find ourselves in a situation where it looks like it's going to be difficult as all heck to read the curve with any degree of accuracy at all.

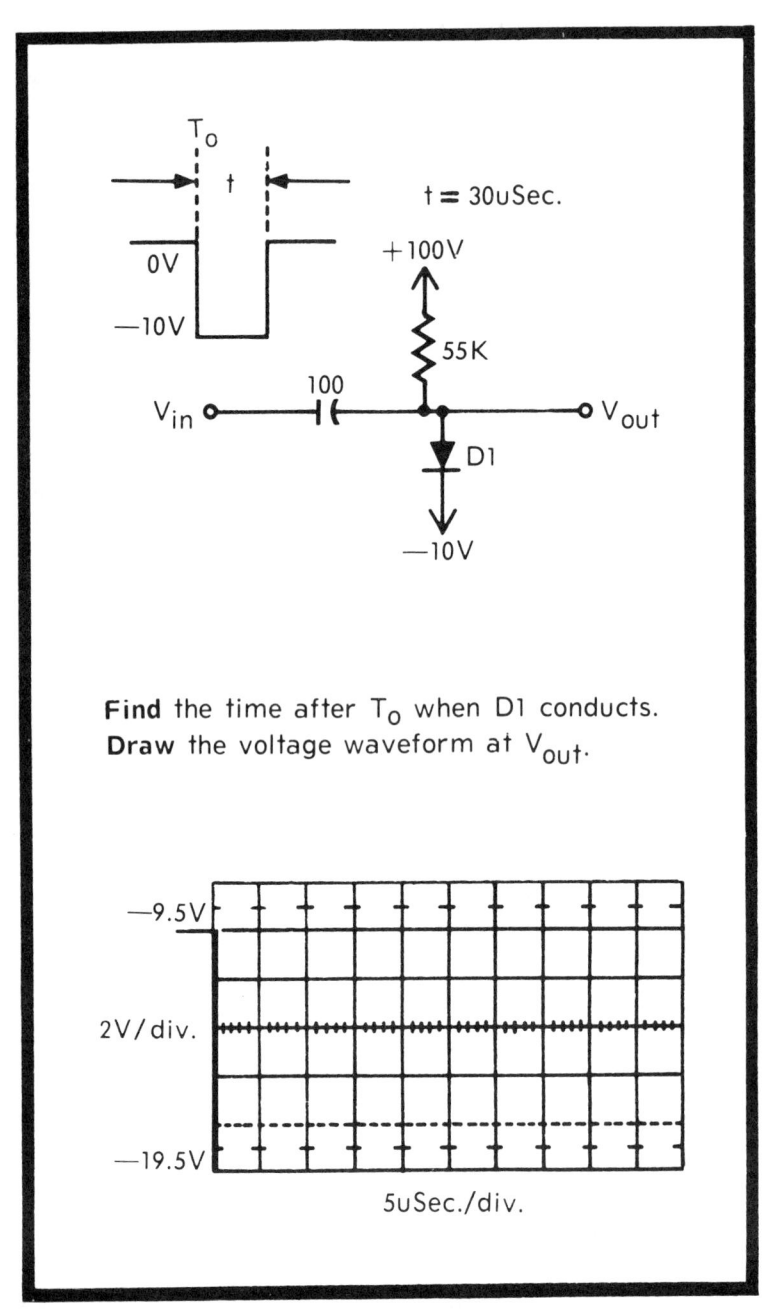

Find the time after T_o when D1 conducts.
Draw the voltage waveform at V_{out}.

Fig. 4-4. Complex differentiator circuit with "catching" diode, and waveform.

Let me show you what I mean. Quiescently, D1 is on, supplying current to the 55K resistor and holding V_{out} at approximately —9.5 volts. T_0 comes along and the 10-volt pulse hits the 100 pf capacitor, cutting off the diode and driving V_{out} to —19.5 volts. The capacitor starts to take on the change in voltage; it recovers over the 10 volts headed for the +100 volts but is caught by the diode at —9.5 volts again. In other words, the capacitor only gets to take on 10 volts out of the 119.5 volts of the total curve it started to follow. This is less than 10 percent and puts us in a region of the curve below 0.1 RC (less than one tenth of a time-constant). Remember what I suggested you do when you found yourself working in this region of the curve back in the second chapter? I said to forget the curve now; ignore epsilon and ignore the **change** of current through the resistor and treat it as though it didn't change at all. This allows us to use the basic formula:

$$\frac{dV}{dT} = \frac{I}{C}$$

Let's see how it works. We want to solve for d T. If the two fractions are equal to each other, their reciprocals are equal to each other also, so let's turn both sides of our formula upside-down.

$$\frac{dT}{dV} = \frac{C}{I}$$

and now let's transpose "d V" leaving "d T" all alone on one side of the equation:

$$dT = \frac{dV \times C}{I}$$

Now we can substitute the values for change of voltage

$$(dV = 10 \text{ v}),$$

the value of the capacitor $(C = 100 \times 10^{-12})$,

and current $(I = \frac{119.5v}{55 \times 10^3}) = 2.17$ milliamps

Thus, $dT = \frac{10 \times 100 \times 10^{-12}}{2.17 \times 10^{-3}} = \frac{1 \times 10^{-6}}{2.17}$

$$= 0.46 \text{ microseconds.}$$

119

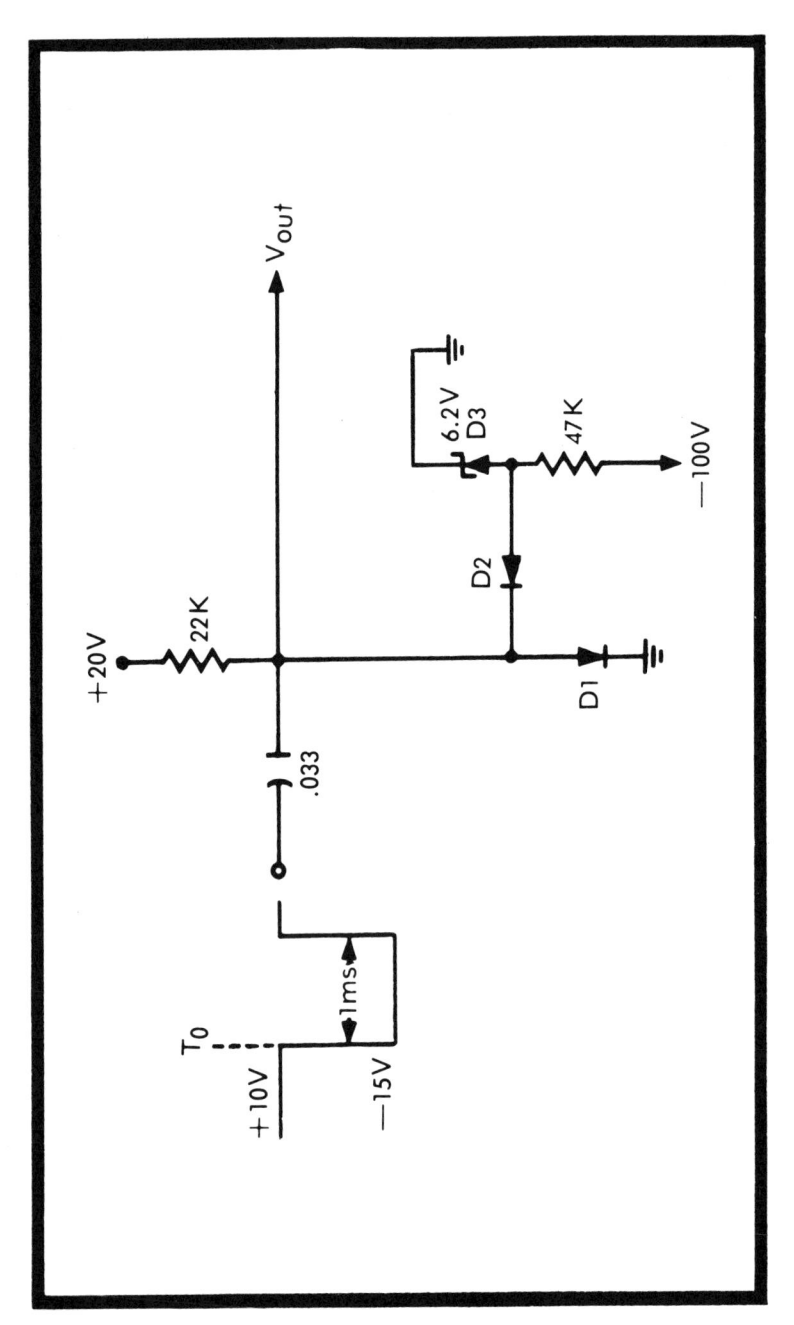

Fig. 4-5. High pass circuit with "catching" diodes in both directions, with zener clamp.

The diode then comes back on slightly less than half a microsecond after T_0. Then, 30 microseconds later, when the driving pulse snaps back positive, the ON DIODE controls the time-constant and supplies the current to the capacitor so that we hardly see anything of the positive pulse based on the "t" over "Tau" curves in Fig. 2-17. The time-constant with the diode in control of the RC product will be so short that it will approximate the RC of the generator from which the pulse comes. This is speculation, of course, but I think it is legitimate speculation. You see, if the diode represented 10 ohms resistance, our circuit time-constant would be 1 nanosecond and the RC-Tau curve shows that if you got a 2-volt positive pulse at this time, the generator would have to have an output RC product of .2 nanoseconds which is highly doubtful. So, we probably wouldn't get anything near that.

Now then, what did we use the RC curve for in this problem? We only used it to tell us that we should use the basic formula rather than try to interpret from the curve. The rest was arithmetic, and legitimate speculation.

"CATCHING" DIODES USED IN BOTH DIRECTIONS

We have used catching diodes in one direction; this time let's use catching diodes in both directions. See Fig. 4-5. Don't read my discussion yet; see how you fare as analyst first.

You're back! OK, here goes. D1 will hold the quiescent V_{out} at approximately 0.5V above ground and will function in a manner similar to the diodes in the previous problems.

D2 is quiescently off since it is back-biased because its cathode is at +.5V and its anodes at —6.2V. Why at —6.2V? Read on.

Fig. 4-5 says that D3 is a Zener diode, and the schematic tells us D3 will conduct at 6.2V of reverse bias. It is referenced to a voltage much greater than this (—100 volts) through the 47K resistor. D3, therefore, is on and operating on its Zener Knee. Thus the anode of D2 is anchored at —6.2 volts. (D2 will only conduct when its cathode drops .5 volt below —6.2 volts, that is at —6.7V.)

T_0 now comes along and our driving waveform takes a 25-volt negative step and the right side of the .033 capacitor tries to follow, heading for —24.5 volts.

The .033 value of the capacitor is .033 microfarads—remember? And the time-constant of the circuit is .033 x 10^{-6} x 22 x 10^{+3}, or about .7 milliseconds (.726 milliseconds, to be more accurate). This isn't very fast and thus the right side will follow. The only thing is, it gets interrupted by D2 turning on at

121

7mA T.D. & REVERSED BACK DIODE

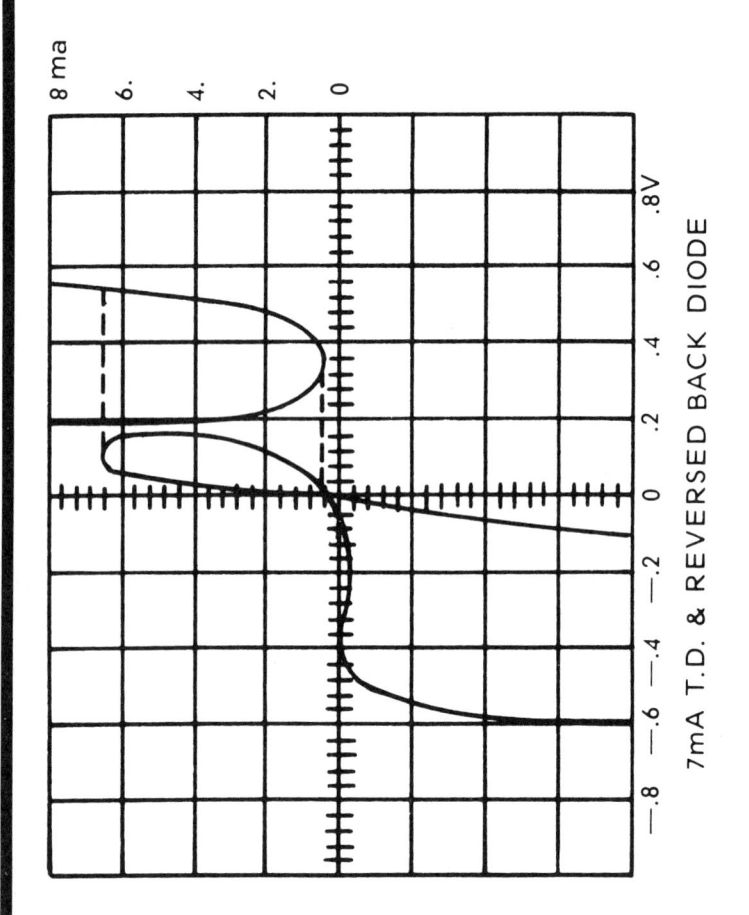

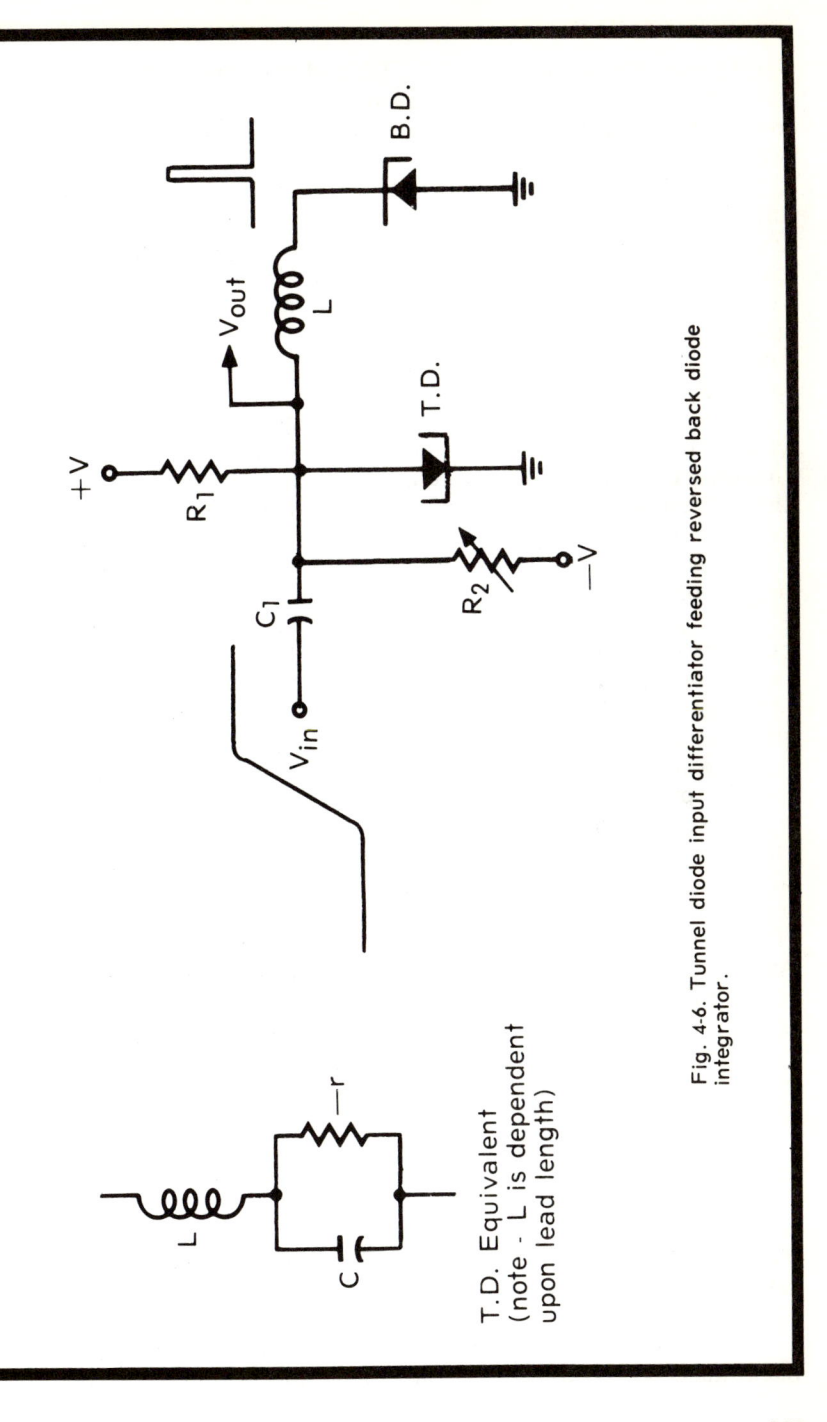

Fig. 4-6. Tunnel diode input differentiator feeding reversed back diode integrator.

T.D. Equivalent
(note - L is dependent upon lead length)

—6.7 volts (6.2 + .5 equals 6.7). D2 and the Zener (D3) both being turned on hard and representing very little resistance, reduce the acting time-constant to practically nothing and the waveform stops right there!

The fact that the driving waveform remains down does not mean that V_{out} does. The capacitor has gotten rid of its full charge of current and the waveform at V_{out} starts back positive from the —6.7 volt level headed for the +20 volt reference of the 22K resistor (D2 turns off almost immediately after it turns on). Thus V_{out} will go from —6.7 volts to +0.5 volt before D1 comes back on again for a total change of 7.2 volts out of a curve that is 26.7 volts high.

The question then is: "What percentage of 26.7 volts is 7.2 volts?" We divide 7.2 by 26.7 and get 27 percent. Now we can go to the curve to get a solution. Recovery of 27 percent as read on the vertical axis of our curve (use Fig. 2-13) gives something very close to .33 time-constants, or divide 0.726 milliseconds by 3 to obtain 0.242 milliseconds (approximately one quarter of a millisecond).

The pulse lasts for a full millisecond, so about three quarters of a millisecond after D1 has turned back on we might see a little glitch as D1 does the honors of shorting out the positive-going portion of the waveform, keeping it from showing up at V_{out} to any appreciable degree.

To summarize this point, I suppose you could say (if you have to put a fancy handle on it) that we've been studying time-related current flow, determined, in part at least, by the RC curve. In other words, we've been studying circuits in which different amounts of current flow at different times and predicting the resultant voltage waveshapes as this changing current goes through a resistor or comes from a capacitor. And, of course, the RC curve isn't the only thing that determines the waveshape of a resultant voltage even though we can learn a lot from it (more than we have learned so far).

We have looked into a number of other items that are used in changing voltage waveshapes such as Tunnel Diodes, Back Diodes, Shockley Diodes, Field Effect Diodes and Snap Diodes. And I would like to have you go through at least one application of each of these devices with me, so let's start with Fig. 4-6.

TUNNEL-DIODE DIFFERENTIATOR FEEDING BACK-DIODE INTEGRATOR

In Fig. 4-6 we have a circuit using a tunnel diode and a back diode (B, D) as well as the conduction graphs of these two devices on a common applied voltage grid, showing what each is trying to do. The back diode is up-side-down in the circuit, so its graph is up-side-down also, with its forward bias region being shown in the lower left-hand quadrant and its reverse-bias conduction curve shown in the upper right-hand quadrant. For the tunnel diode, we have the forward-biased conduction curve shown in the upper right-hand quadrant and reverse-bias conduction curve in the lower left-hand quadrant.

Just from inspection, I think it is fair to say that V_{out} has to be slightly positive. If it were negative, the driving signal would have to be excessively big in order to overcome this condition and get the tunnel diode into its switching region or larger change-of-voltage region. So, we shall assume that R_2 is adjusted so that V_{out} is about +0.05 volt (or +50 millivolts). The T.D. curve tells us that it must be conducting something close to 6 milliamps at this voltage. Now with this same voltage (50 mv) across the series combination of the inductance and back diode, the B.D. curve shows that it isn't quite conducting much of anything in the way of current yet. So, V_{out} at quiescence may be considered to be at +50 mv, with R_1 carrying 6 ma from both R_2 and the T.D., and with just a dribble of current from the L - B.D. leg of the circuit.

We now have the conditions set to receive a signal at V_{in} and predict what will happen. The positive signal appears on the left of C_1 pulling electrons out as extra electrons flow in on the right. Where do these extra electrons come from? The lowest impedance (greatest source of electrons) for the right hand plate of C_1 is the T.D. and as soon as anything more than 7 ma total current is demanded from this guy, he insists on operating at a new voltage. The only place in his forward-bias curve where more than 7 ma can flow through him is around .55 volts, and he switches to this voltage level. A small voltage change—yes—but an extremely fast one. (We'll go into how fast it is soon.) This new voltage level appears across the L - B.D. combination also, but the inductor opposes current change. It is trying to operate on the L-R time-constant curve with the R being the ON resistance of the reverse-biased B.D. As the inductor reduces its opposition to current flow, so does the B.D. and V_{out} drops below the .35 volt level, where the T.D. takes control of the output voltage change, pulling it back down to the 0.05-volt level where the T.D. can supply all the current to the circuit necessary to meet the existing conditions.

When the input waveform drops back to its quiescent level, trying to reverse-bias V_{out}, the T.D. shorts everything out, not allowing V_{out} to go below −0.1 volt. And, of course,

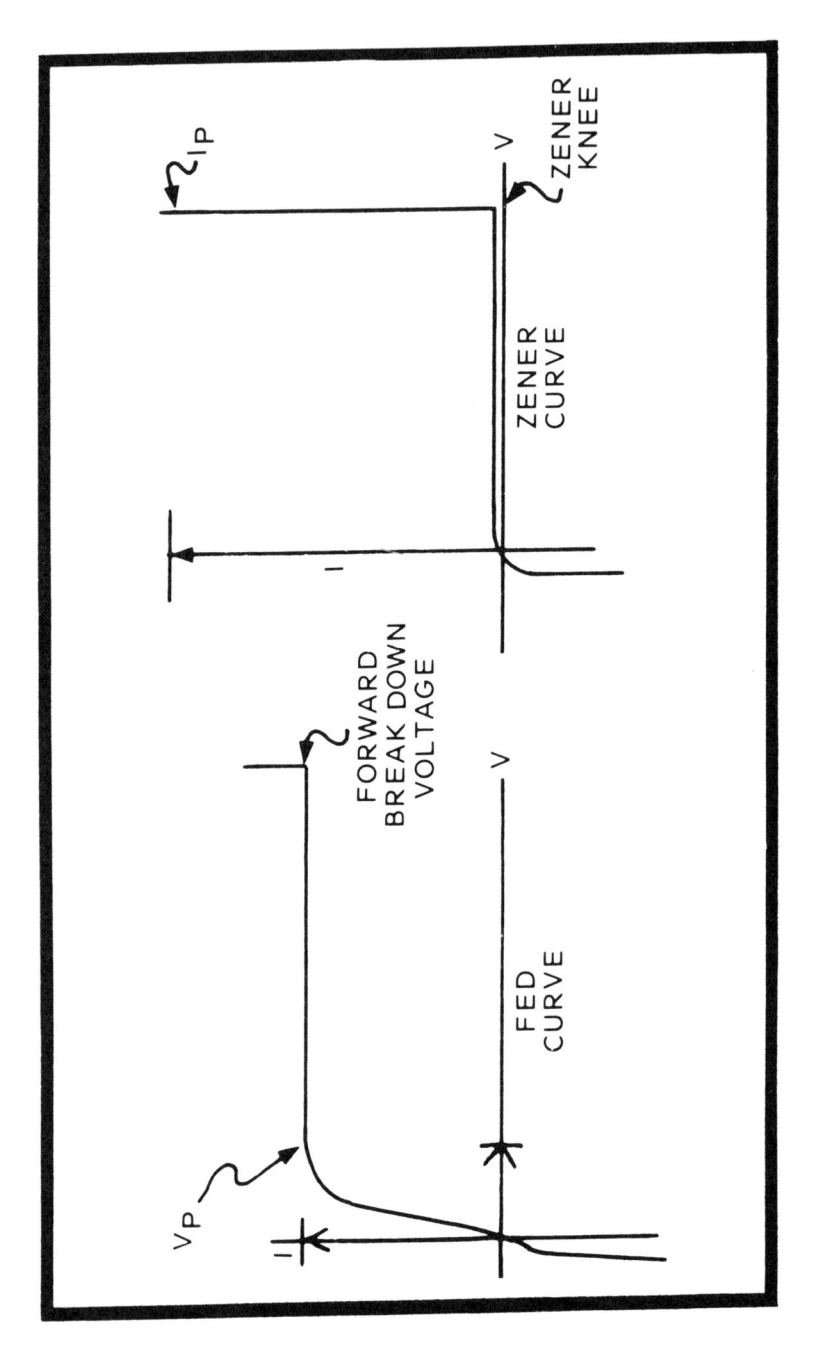

Fig. 4-7. Field effect diode characteristics.

everything returns to normal in an extremely short RC time since the T.D. reverse bias resistance is extremely low (less than 1 ohm).

The V_{out} change of voltage is 0.5 volt in magnitude and its time duration is largely controlled by the size of L (its size is probably found by the old "cut and try" method, depending on the purpose for which the pulse is to be used).

Now then: just how fast is this positive-going 0.5-volt change at V_{out} going to be? Well, our old favorite formula dV/dT equals I/C can give us a fairly accurate estimate of this change time. If you will refer to the characteristics of a tunnel diode, you will find that it does have a capacitance characteristic in the neighborhood of 7 picofarads (this value does depend on the T.D. type—some are lower—some higher). And when this device changed voltage from 0.05 volt to 0.55 volt, 6 of that 7 ma of current had to go somewhere. It did. It changes the voltage of this inherent capacitance. So, C (in our formula) is 7×10^{-12} and I is about 6×10^{-3} while the dV (change in voltage) is 0.5 volt. Let's do it and figure:

$$dT = \frac{dV \times C}{I} = \frac{.5 \times 7 \times 10^{-12}}{6 \times 10^{-3}}$$

Then,
$$dT = \frac{3.5}{6} \times 10^{-9}$$

or about 0.6 nanoseconds—less than a billionth of a second. Yes, that's pretty fast. The reciprocal of this time represents a frequency in excess of 1 GHs (greater than 1,000 megacycles per second).

FIELD EFFECT DIODE CIRCUITS

One of the more interesting solid-state diodes is the so-called FED or Field Effect Diode. And, if you can consider a Zener diode to be a Thevenin Equivalent (a constant source of voltage with very low series resistance), then the FED becomes a Norton Equivalent (a source of constant current with a very high shunt impedance. See Fig. 4-7. Yes, the FED curve is a forward-bias curve for the device and the Zener diode curve in the first quadrant is a reversed-bias curve for the actual junction. I drew them both in the same quadrant for comparison purposes.

Let's look at the circuit shown in Fig. 4-8 to see how such a device might be used. FED_1 and FED_2 are matched Field Effect Diodes. They carry the same amount of constant current in the forward direction and have similar low values of

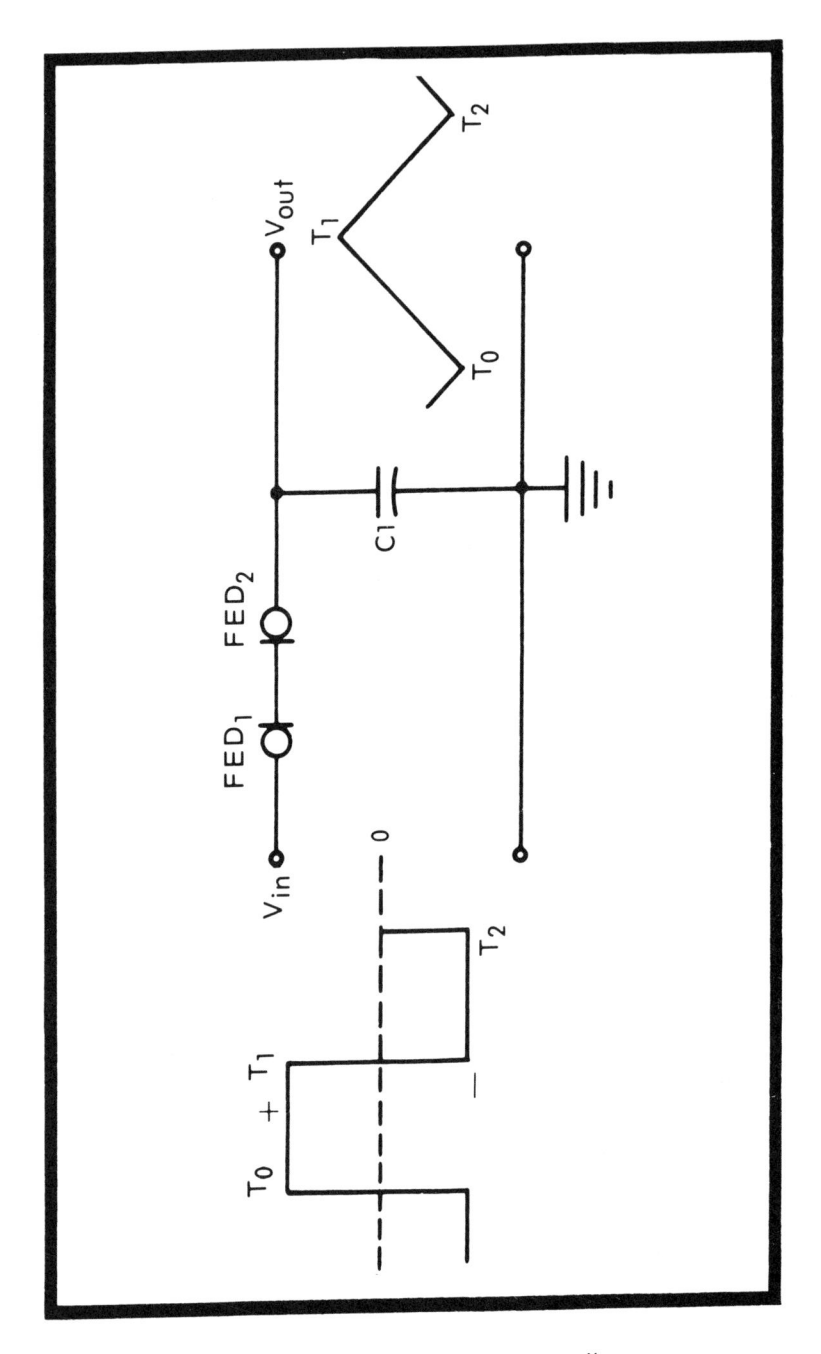

Fig. 4-8. FED circuit with shunt capacitor.

Pinch-Off-Voltage ratings (V_p). The size of C_1 will go along with the magnitude of current for the FEDs via the dV/dT equals I/C formula so that the amplitude of V_{out} will still leave the input square wave at least V_p above the V_{out} waveform as it approaches time T_1. Otherwise the peak in the output waveform at T_1 will be rounded or flattened.

During the time T_0 to T_1, FED_1 will be forward-biased and controlling the amount of electron current being taken out of C_1. Thus the voltage across the capacitor will move toward positive at a steady rate. During the time T_1 to T_2, the driving waveform has switched negative and electron current, controlled by FED_2, is being forced back into C_1 and the output waveform goes in the negative direction at a steady rate.

This circuit then is more of a true integrator than was our low-pass R-C circuit. It changes a square wave into a more linear ramp waveform. And, yes, we do have to watch out for one other parameter of the FEDs here. The change of voltage from + to — at T_1 must not exceed the breakdown voltage of the FEDs. They're pretty good in this respect but it does pay to check, however.

Another circuit that changes one waveshape into another that uses FEDs back to back is shown in Fig. 4-9. This circuit substitutes a Zener diode in place of C_1 in the previous circuit.

The input sine wave will have to reach a level of voltage of 10 volts plus Vp before the output signal will snap positive. The Zener will hold V_{out} at +10 volts while FED_1 controls the current (and FED_1 will have to have a breakdown voltage in excess of 30 volts to withstand the input voltage peak). As soon as the input waveform drops below 10 volts plus V_p for FED_1, the output voltage will snap negative to ground, and there will be a slight glitch here (crossover distortion) before FED_2 and the Zener turn on in the other direction. When this happens V_{out} will go negative 0.6 volt just like a normal silicon diode junction under forward bias does. (FED_2, of course, will have to have a breakdown voltage in excess of the negative 40 volt peak of the input waveform.) The 10 volt Zener diode has to be compatible with the FEDs in that its Ip (peak current) rating must exceed the constant-current level of the FEDs. But, this is not hard to find, since FEDs come capable of all kinds of different current levels.

If you wanted a square wave output that goes as far negative as it does positive in voltage, all you have to do is add another Zener diode to the picture. Just hook it up in series with the top of the first one but make sure its arrow will point in the other direction. Note that the higher the forward breakdown voltage of the FEDs, the larger the sine wave input

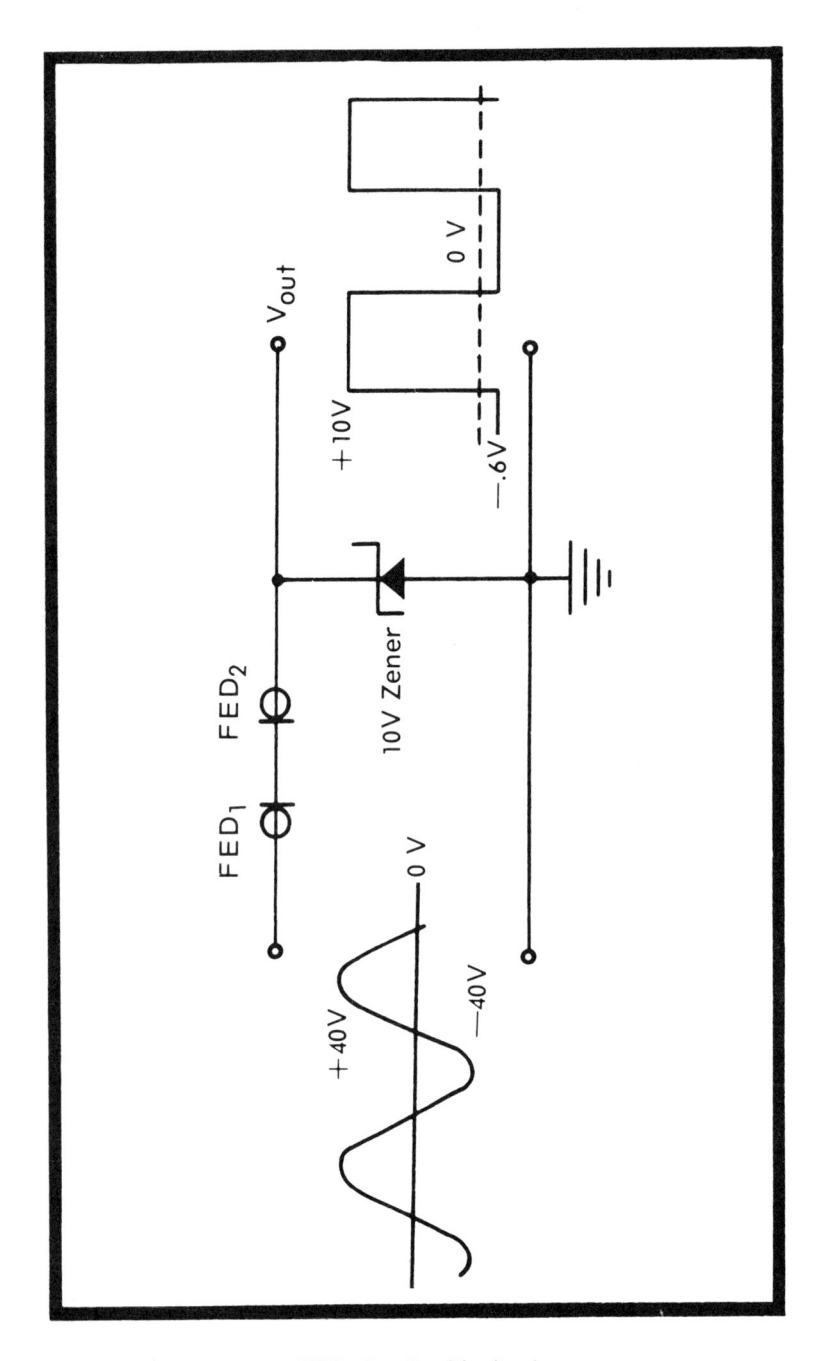

Fig. 4-9. FED circuit with shunt zener.

voltage can be and the less crossover distortion near zero volts you will have.

SAWTOOTH GENERATORS

There are some simple circuits that only require the proper level of DC voltage be applied to them to make them do their thing. The circuits shown in Fig. 4-10 and 4-11 are two such examples. Let's take Fig. 4-10 first. It is what we can call a negative-going sawtooth generator. In this circuit, the Shockley diode must switch from its low resistance state to its high resistance state. This says that the constant current from the FED must be less than holding current for the Shockley Diode (see Fig. 3-5). For, you see, once the Shockley is conducting, if it isn't handling more than holding current, it will turn itself off or go into its high resistance state. Also, we must realize that the DC voltage we apply at the input cannot exceed the breakdown voltage of the FED. C_1 will be a large filter capacitor holding the DC input voltage stable regardless of the current surges caused by the Shockley turning on and off.

With these limitations in mind, let's see if we can describe what happens. At T_1, the voltage across the Shockley has reached the switching level of the device and it turns on. The voltage across it goes to just about 1 volt as it slurps just about all the electrons from the upper plate of C_2. After this, it only has the FED as a source of current and this current is not enough to keep it turned on, so the Shockley turns off. However, now the FED has maximum voltage across it and supplies its steady current to the upper plate of C_2, causing it to go negative in a linear fashion until time T_2 is reached and switching voltage across the Shockley is once more achieved, which turns the Shockley on once more, causing the output voltage to snap positive once more. This goes on and on and on and on until you disconnect the DC voltage. The slope of this negative-going sawtooth waveform is determined by the I over C ratio, while its amplitude is determined by the Shockley diode switching voltage. The level of DC voltage at the input must at least equal the Shockley switching voltage plus V_P for the FED. Actually, a little bit more than this is advisable but do not go beyond the forward breakdown voltage for the FED.

Just about the same things can be said for the circuit in Fig. 4-11. Note that the Shockley and FED have traded places and there is no need for a filter capacitor (C_1 in Fig. 4-10) since the surge current caused by the Shockley turning on goes into the timing capacitor (C_1 in Fig. 4-11) this time.

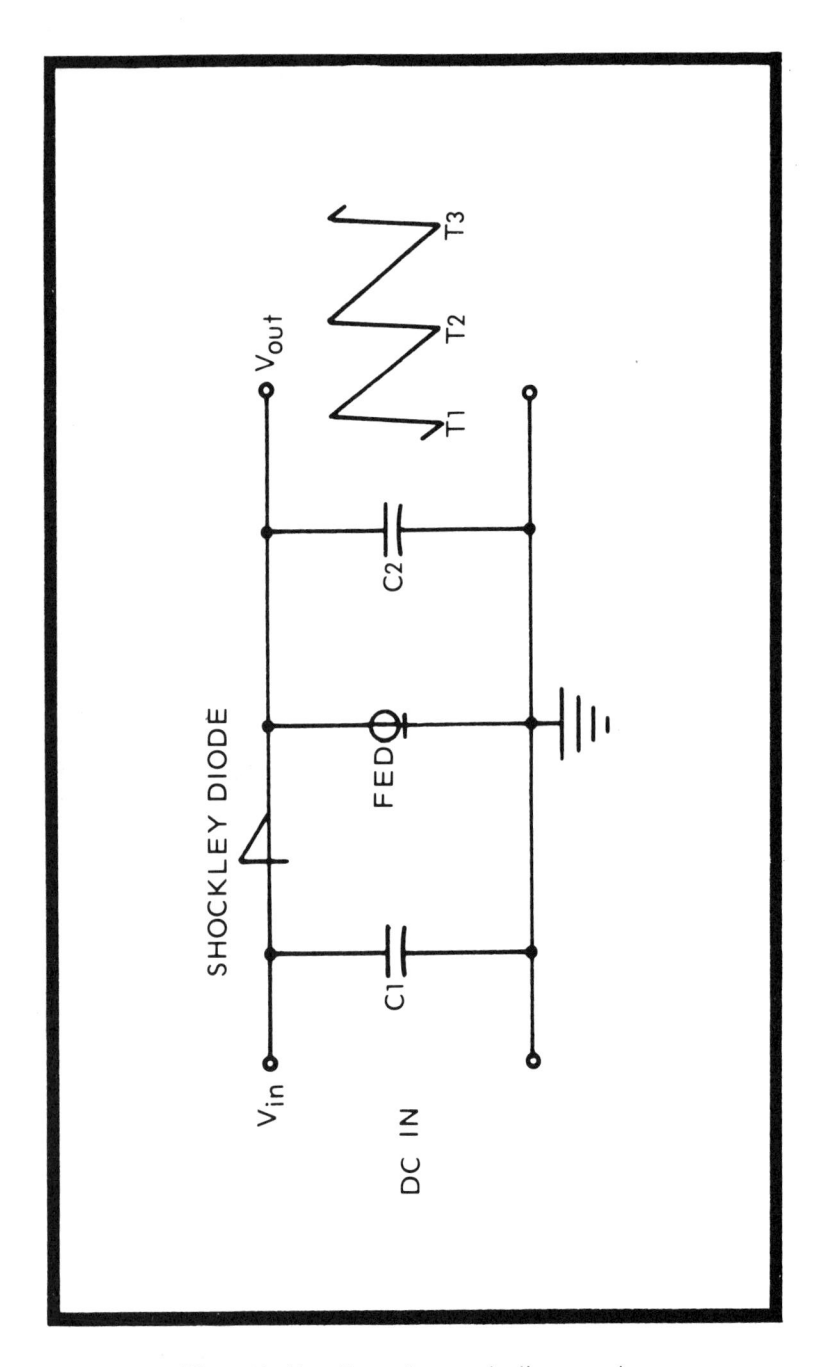

Fig. 4-10. Negative-going sawtooth generator.

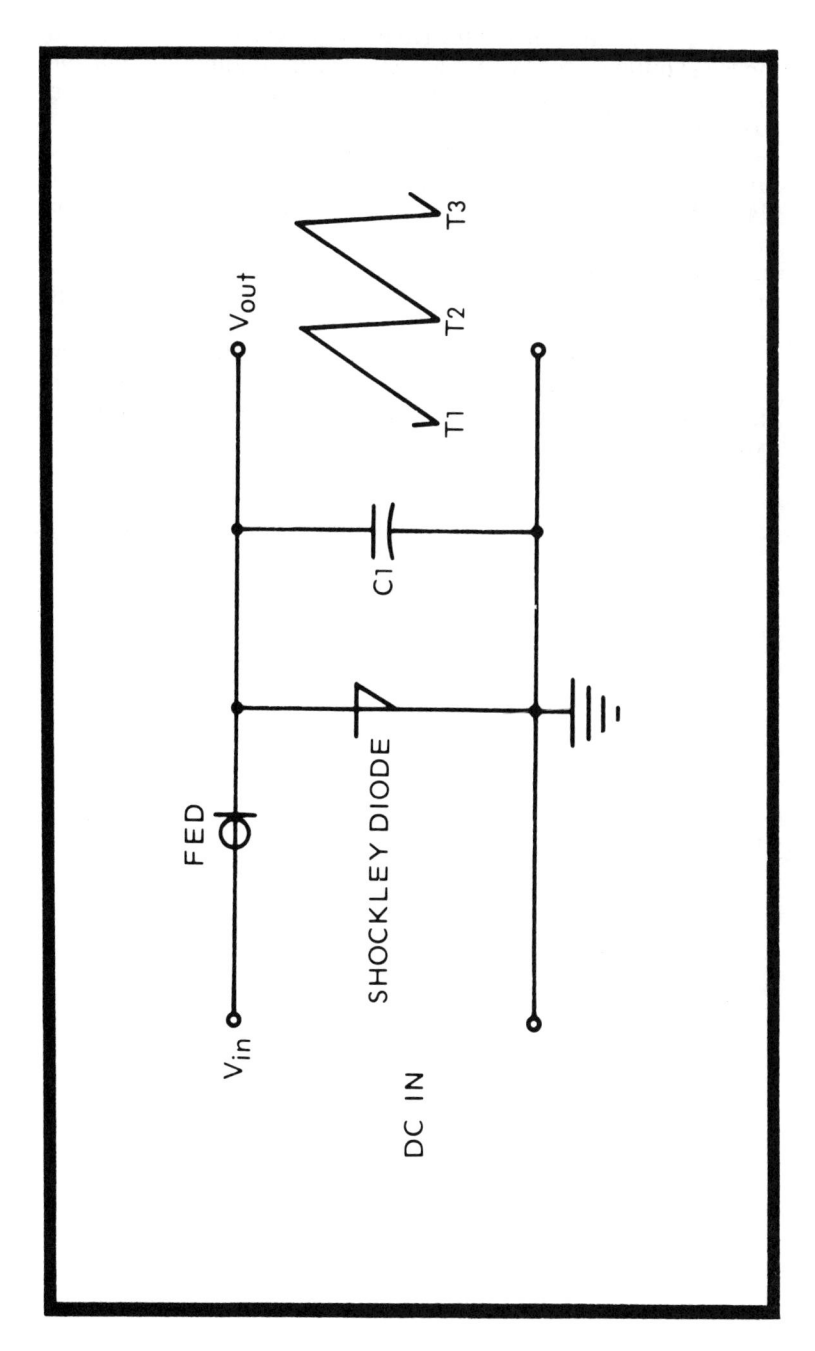

Fig. 4-11. Positive-going sawtooth generator.

At T_1, the Shockley is off (in its high resistance state) and the FED pulls a steady current (less than holding current for the Shockley) out of C_1, causing the output voltage to move positive in a linear fashion. This continues until the level of switching voltage for the Shockley is reached. The Shockley turns on, dumping a sudden supply of electrons into the upper plate of C_1, causing its voltage to go to the 1-volt ON level of the Shockley. At this time the Shockley cannot carry holding current since the FED will not allow it to, and the Shockley turns off again, so the whole process repeats itself. This circuit thus generates a positive-going sawtooth waveform.

Why don't you get the parts, and have the fun of building these circuits? Before you look at the output waveforms with an oscilloscope, you ought to be able to predict them almost exactly.

Amplifier Devices

CHAPTER 5

I suppose you might say that the history of the material this chapter is all about began with a feud between two of the pioneers in electronics. The two I speak of are J.A. Fleming and Lee DeForest. Fleming claimed to his dying day that DeForest merely added a piece of bent wire to "his valve." However, it was early in 1907 that DeForest applied for an American Patent on a wireless telegraph receiver that disclosed the very first three-electrode vacuum tube. It used a zigzag piece of wire as a control element located between the filament and what was called the plate of the tube.

The next big step came about close to 1927 with the general use of the tetrode or screen-grid tube as a result of the work by such men as Austin and Starke, and Hull and Shottky.

The end of World War II came along with the Bell Telephone Co. engineers development of the NPN and PNP transistors (the first solid-state "triodes"). This development was followed by rather wide experimentation in France and other parts of the world as well as America with the Field Effect Transistor (the so-called FET). Actually, as I understand it, the very first transistor that was made to work by the Bell Lab engineers was a Field Effect device, but part of it was a liquid and thus not very practical.

At any rate, the developments have come hot and heavy, with each engineer with a new idea trying to get a patent. Consequently, each new device has come along with a vocabulary all its own, and with a system of measurement all its own, with each one trying to be different and having a different set of circuit equations related to its use. This has mixed things up like mad for the electronics technician. Also, all this has been much more exacting than the average technician has to be. He just hasn't had the time, the teachers, the equipment (to measure these devices within an inch of their lives as well as the parts in the circuit) nor the need to predict them so closely. He'd just like to be able to study the circuit diagram and get an idea within 10 percent of what the

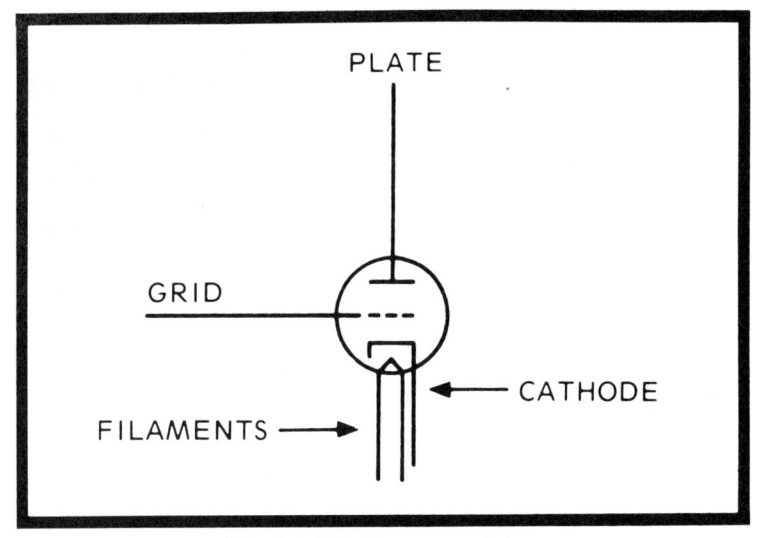

Fig. 5-1. Triode vacuum tube.

thing ought to do. With this information he can begin to recognize troubles where they exist and correct them.

But have heart, you students of electronics. Engineers from all over the world have come up with a simpler (almost universal) way of thinking about these active devices (tubes, transistors and FETs—amplifying devices) which lumps them all into the same line of thought.

TRIODE TUBE ANALYSIS

So, let's start with the triode vacuum tube to get the methods of measurement firmly in mind, find their interrelationships, and get their uses firmly established. First, let's name the parts so we know what we are talking about (see Fig. 5-1). The part which releases the carrier (or whatever it is that goes through the tube) is called the cathode. Since the cathode only releases these carriers when it is hot, early vacuum tubes passed a current through the cathode to make it hot, and the heater and cathode were the same thing. However, modern tubes separated these two functions and each cathode has a heater known as the filament of the tube. The carriers go to what was first known as the anode, but common usage has changed the name of this part to the plate of the tube. DeForest introduced the third element of our triode between the cathode and the plate to control the flow of carriers and this was called the grid of the tube.

Up until just a few short years before the start of the 20th Century, it was thought that charged molecules were the carriers that went from cathode to plate in the vacuum tube; but, in 1897, J. J. Thomson demonstrated that it was the electron that made the trip. So, for vacuum tubes, NPN transistors, and N-Channel FETs, the carrier that moves from the cathode (or similar element) to the plate (or that part which takes its place) is the electron. In PNP transistors and P-Channel FETs, the carrier is the absence of an electron (or hole) in the atomic structure which moves from the effective cathode to the effective plate of the device and is thought of as being a positive (+) charge. In any case, we always have something that is a carrier which moves from the cathode to the plate (or whatever takes their places).

To understand vacuum tubes we must go back to the very basics of electronics and remember that Like Charges Repel each other and that Unlike Charges Attract each other. Thus, if it is the electron (a negative charge) that leaves the cathode, the plate must be positive (in voltage) with respect to the cathode in order to attract the electron. So, with DC voltage applied to our tube, we will have a steady flow of electron carriers from the cathode to the plate. If we are going to control this steady flow of negative charges, we had better let like charges repel each other and make the grid a bit more negative in voltage than the source of the electroncs—the cathode. This difference in voltage between grid and cathode is spoken of as bias.

The thermionic vacuum diode only had one curve of current through the device as opposed to voltage applied to it. The triode, though, has as many curves as you may wish to draw—one curve for each different bias voltage applied to the grid (see Fig. 5-2). This set of curves for a triode tube is called the family of curves or its plate characteristics. Note what this group of curves is telling you. It says that as a more and more negative voltage is applied to the grid with respect to the cathode (which in this picture remains at zero volts), it takes a higher and higher plate voltage to get the same amount of current (electron flow) through the device. For instance, when the grid is 9 volts negative with respect to the cathode, it takes 300 volts positive on the plate with respect to the cathode to get 10 milliamps to flow through this tube. It only took about +35 volts on the plate to get this same amount of current to flow through this tube when the grid was at the same voltage as the cathode (the grid at zero bias).

This family of curves can be made to show us many things. Pay attention to this one. It says that we have three

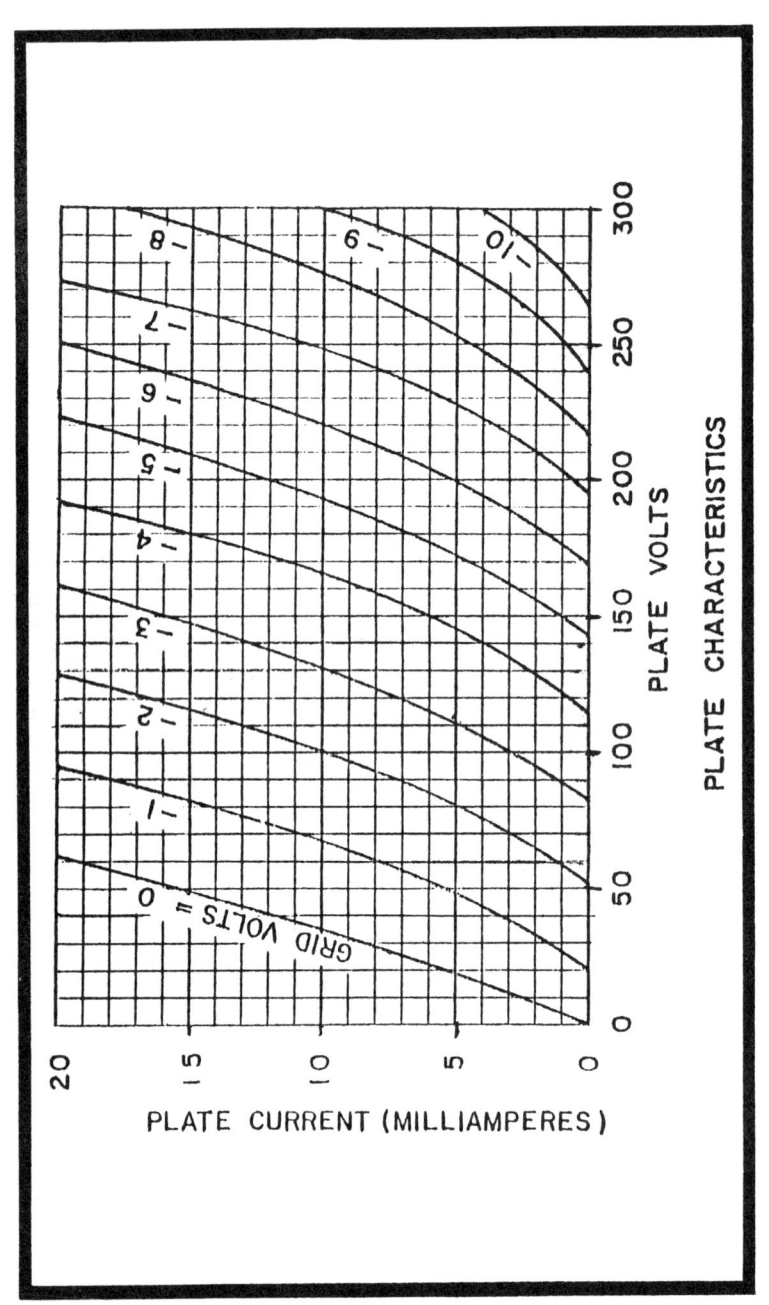

Fig. 5-2. Plate characteristics, 6DJ8-ECC88 triode vacuum tube. (Courtesy Amperex.)

variables (i.e., plate voltage, grid bias, and the current that flows in the plate—plate current) **any two of which will determine the third**:

(1) Plate Voltage (V_p) and Plate Current (I_p) determine the amount of bias this tube must have to operate at this point. (For example, for V_p equals 150V and I_p equals 6 ma, grid bias must be —4V.)

(2) V_p and Bias (grid voltage with respect to the cathode) determine I_p even if you ask the tube to burn itself up by using too little bias for the amount of plate voltage you supply. (For example, for V_p equals 250V and bias of —2V instead of proper bias of —9V, the plate current will be off the chart and the tube will be destroyed.)

(3) Bias and I_p determine the amount of V_p necessary to achieve their levels of operation. (For example, knowing that the bias should be —3V and wanting a plate current of 13 ma, read down to Vp equals +140V.)

Remember, bias is not the instantaneous voltage on the grid as measured to ground; it is the voltage on the grid measured with respect to the cathode.

With these three variables (V_p, I_p, and bias) we must have three ratios of change (deliberately change one and observe the change in the other, with the unused variable held stable) that we can establish.

For instance, if we change the plate voltage from +150 volts to +100 volts and observe the change in I_p along the —3-volt bias line, we will find that it shifts from 16 ma to 3 ma due to this process. We have a 50-volt change on the plate and a 13 ma change of current that goes with it. Using Ohm's Law, this can determine a resistance for us and since it is observed from the plate, it is called Plate Resistance (r_p).

$$r_p = \frac{\text{change in } V_p}{\text{change in } I_p} = \frac{50}{13 \times 10^{-3}} = 3.84K.$$

Let's hold I_p stable at 10 ma this time and change the bias from —2 volts to —4 volts (a change of 2 volts) and note the change of V_p that has to go with it if I_p is to remain at 10 ma (a change of 65 volts — from +100V to +165V). This gives us a maximum "voltage out" change for a given "voltage in" change which is called the Amplifications Factor or mu (u).

$$u = \frac{d\,V_p}{d\,V_{bias}} = \frac{65V}{2V} = 32.5.$$

Now for the third ratio, let's hold V_p stable at about +130 volts. Starting with a bias of —3 volts, you'll note that we will have almost 10 ma of I_p. Change the bias of —2 volts (a change of 1 volt) and note that we will have just over 20 ma of I_p (a change of about 10 ma). And we will relate this ratio to the conductance of the cathode. Since resistance is voltage over current, then conductance G (the opposite of resistance) must be the change of current divided by the change of voltage that caused it.

$$G = \frac{\text{change in } I_p}{\text{change in bias}}$$

Since Ohms is the term for resistance they turned it around backwards for the units of conductance (corny—isn't it) and called them Mhos. So, 10 ma divided by 1 volt is 10 millimhos or, more commonly, 10,000 micromhos. The name of this ratio is not just the conductance of the device, it is the Transconductance of the tube and it is not just labelled "G," it is called "Gm."

Yes, we could anchor the grid at —3 volts with respect to ground and apply +130 volts with respect to ground to the plate and then move the cathode from zero volts to +1 volt with respect to ground, measuring the change in I_p that resulted. This could be set up to be called cathode resistance, but the measurements we made for Gm would be more accurate since they change bias only and the plate-to-cathode voltage is held absolutely stable. And, we can use a resistance ratio of these two variables by dividing the 1-volt bias change by the 10-ma change in I_p to get 100 ohms of resistance (cathode resistance, to be exact).

So now we have:

Plate resistance
$$r_p = \frac{\text{change in } V_p}{\text{change in } I_p} \text{ (with bias held stable)} \quad \text{(Eq. 5-1)}$$

Amplification factor
$$\text{Mu } \mu = \frac{\text{change in } V_p}{\text{change in bias}} \text{ (with } I_p \text{ held stable)} \quad \text{(Eq. 5-2)}$$

Transconductance
$$Gm = \frac{\text{change in } I_p}{\text{change in bias}} \text{ (with } V_p \text{ held stable)} \quad \text{(Eq. 5-3)}$$

Cathode resistance
$$r_K = \frac{1}{G_m} \quad . \quad \text{(Eq. 5-4)}$$

Note: The lower case "r" has been used for the internal resistance of an active device since capital R will be used for resistance external to the active device. In the case of r_K, the K was used for Cathode since C is used with transistors for the collector (plate) of the device and we don't want to mix our terms up now or later on.

It is interesting to note what we get when we take the product of G_m and r_p in the form of their basic equations:

$$\frac{d\,I_p}{d\,V_{bias}} \times \frac{d\,V_p}{d\,I_p} = \frac{d\,V_p}{d\,V_{bias}} = u \quad \text{(Eq. 5-5)}$$

This of course states that:

$$G_m \times r_p = u \quad \text{(Eq. 5-6)}$$

And, substituting:

$$u = \frac{r_p}{r_K} \quad \text{(Eq. 5-7)}$$

Transposing:

$$r_K = \frac{r_p}{u} \quad \text{(Eq. 5-8)}$$

and

$$r_p = u\,r_K \quad \text{(Eq. 5-9)}$$

This development is not exact, but it is plenty close enough to use as a rule of thumb approach. Note what it tells us. It says that r_p is u times r_K, or that r_K is the same as r_p divided by u. It says that our vacuum tube is kind of like a pair of binoculars; you look through it in one direction and it magnifies the resistance on the other end (looking into the plate). And have you ever looked through a pair of binoculars backwards? It makes everything look smaller, doesn't it? Well, our vacuum tube is going to do the same thing because if we look into the cathode we will see the plate resistance made smaller by the factor u. See Fig. 5-3. And, these two ideas can

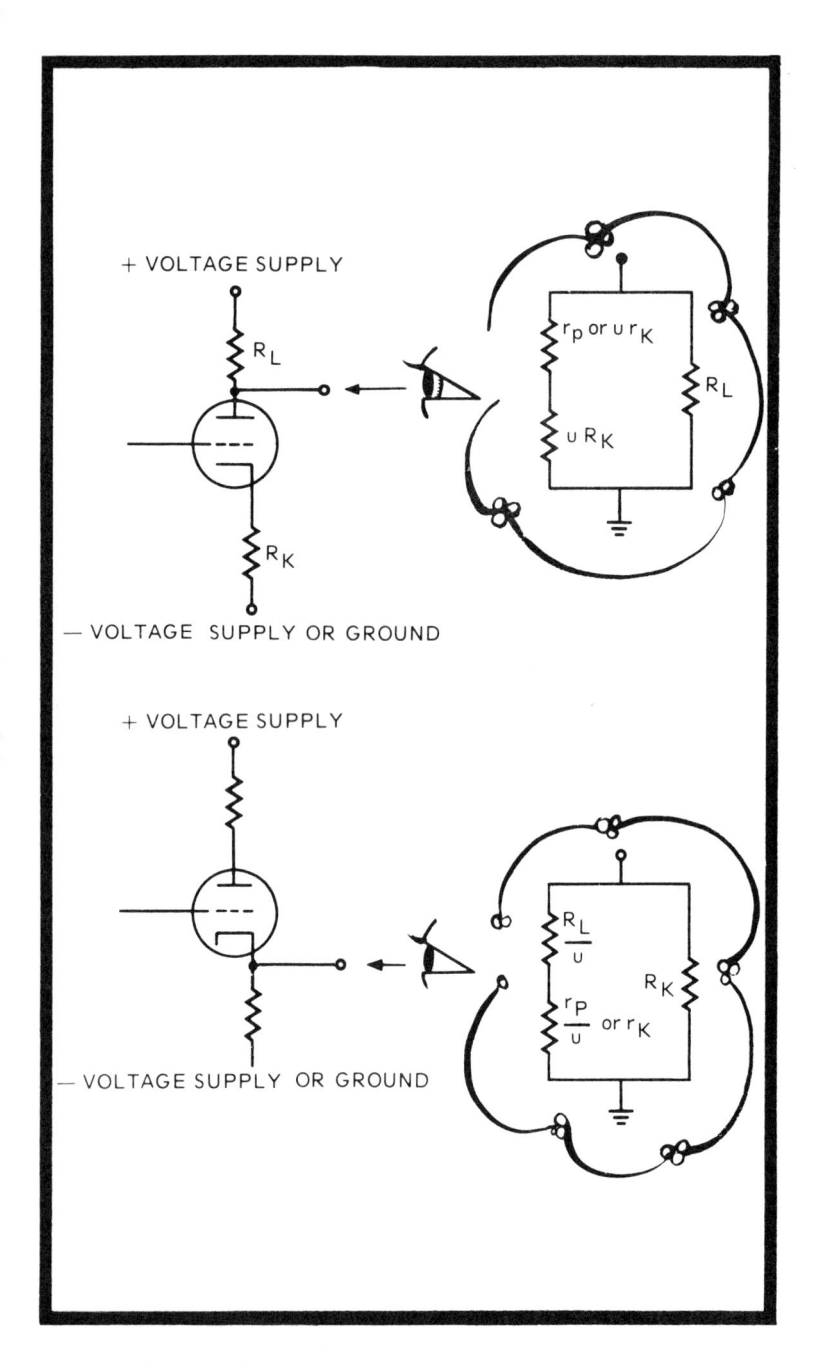

Fig. 5-3. Thevenin-type look at a triode vacuum tube.

be proven. The first (multiplication by u) looking into the plate circuit of the tube can be proved by use of the plate family of curves. The second (division by u) looking into the cathode of the tube requires experimentation and measurement.

What do you say we prove this first bit about u multiplying r_K when looking into the plate circuit. Fig. 5-4 will be our work sheet. The two circuits are the same except for the voltage applied to the plate. First, we'll determine the grid voltage. It's a simple resistive divider, so let's go: 47K divided by 100K plus 47K is about 32 percent and 32 percent of 150 volts is 48 volts, so that's where the grid is (+48 volts). Now then, if the cathode were at the exact same voltage as the grid, there would be 43 volts over the cathode resistor of 5K. This would give the cathode 9.6 ma of current (locate this point on the zero bias curve in Fig. 5-2). But what if the cathode had 4 volts of bias (was 4V above the grid voltage, which is anchored)? Well then we would have 52 volts over the same 5K cathode resistor and 10.4 ma of current (locate this point on the —4-volt bias curve). Now use a ruler to connect these two points. Since this tube will have (within a volt or so) about 100 volts from cathode to plate, draw a line vertically up the graph from the +100 volts indicated on the base line. Where your two lines cross is the operating point of this tube. It's almost exactly on the —2-volt bias line and 10 ma of current level, isn't it? Well, this is our first condition.

Go to the second diagram in Fig. 5-4 now and note that the grid is at the same old +48 volts as before, but the plate is now at +300 volts. This demands more current and thus the cathode must have moved more positive, giving the tube a higher amount of bias. What if we now had 8 volts of bias (the cathode at +56V)? This would give us a current available to the cathode equal to 56 volts over 5K or 11.2 ma of current for the plate (locate this point on the 8V bias line). Now connect this point with a straight line to the point previously found on the 4-volt bias line. Now then, what plate-to-cathode voltage does the tube see? Must be close to 245 volts, the way I see it. Let's draw our next vertical line just to the left of 250 volts on the base line of our tube curves. Well, we must have just under 7 volts of bias and about 11 ma of plate current.

Comparing the two problems now, you can see that increasing the plate voltage by 150 volts got us a 1 ma increase in plate current and (since R equals V/I) the tube and its cathode resistor must look like about 150K ohms of resistance. If I now take away the r_p value we calculated before (3.8K) we'll have about 146.2K ohms left, and if I divide this by the value of the

143

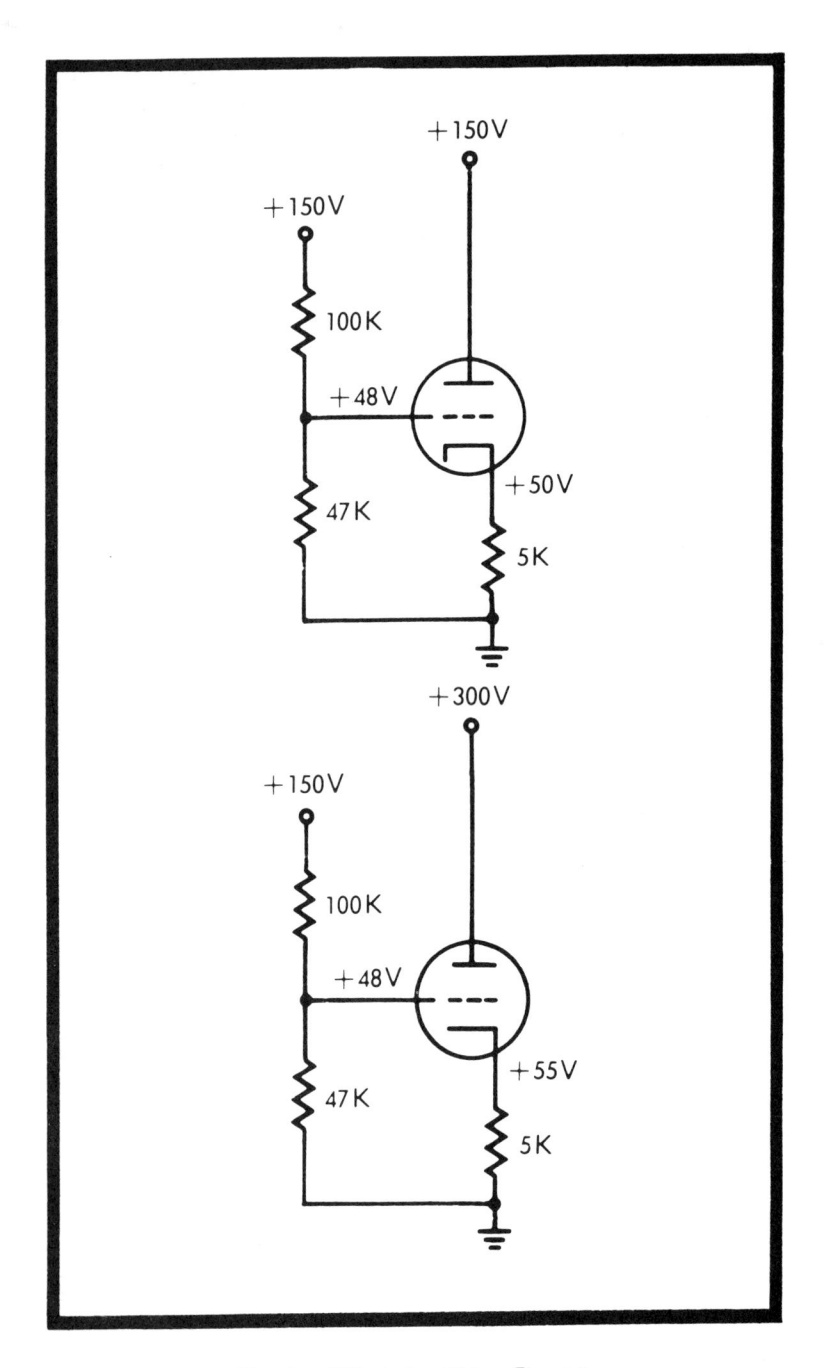

Fig. 5-4. "Worksheet" for Fig. 5-3.

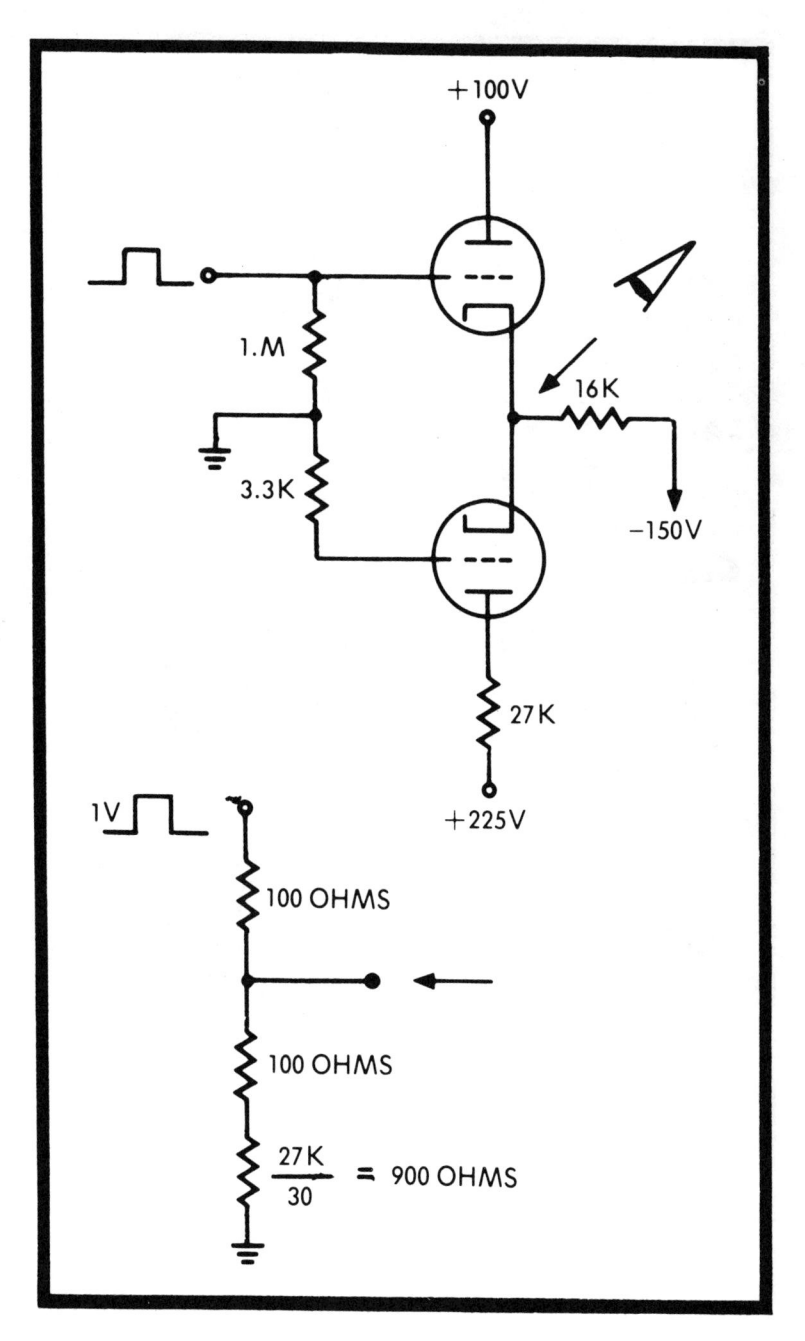

Fig. 5-5. R_L is viewed from the cathode.

AMPEREX TUBE TYPE 6DJ8/ECC88

TENTATIVE DATA

The 6DJ8/ECC88 is a frame grid sharp cut-off twin triode with separate cathodes designed for use in cascode circuits, RF and IF amplifiers, mixer and phase inverter stages. The tube features high transconductance, low noise properties, as well as extreme reproducibility of characteristics as a result of the frame grid construction. The heater is designed for parallel operation from a 6.3 volt supply.

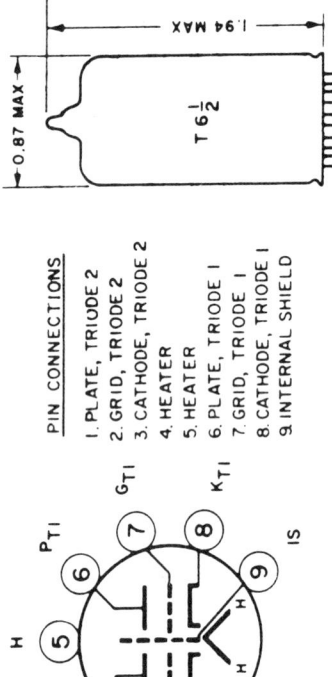

PIN CONNECTIONS

1. PLATE, TRIODE 2
2. GRID, TRIODE 2
3. CATHODE, TRIODE 2
4. HEATER
5. HEATER
6. PLATE, TRIODE 1
7. GRID, TRIODE 1
8. CATHODE, TRIODE 1
9. INTERNAL SHIELD

GENERAL CHARACTERISTICS

MECHANICAL

Cathode	coated, unipotential
Mounting Position	any
Overall Length	2.19 inches max
Seated Height	1.94 inches max
Diameter	7/8 inch max
Bulb	T6½
Outline	6-2
Base	E9-1
Base Connection	9DE

ELECTRICAL

Heater Characteristics

Heater Arrangement	series supply
Heater Voltage (ac or dc)	6.3 volts
Heater Current	365 mA

Fig. 5-6A. Complete tube data, 6DJ8-ECC88 dual triode (part 1 of 6 parts) (Courtesy Amperex).

6DJ8/ECC88

Interelectrode Capacitances (Without External Shield)

Grounded Cathode Input Section [1]

Plate to Grid	1.4 µµf
→ Input Capacitance	3.3 µµf
Output Capacitance	1.8 µµf
Grid to Heater	0.15 µµf

Grounded Grid Output Section [1]

Plate to Cathode	0.2 µµf
Input Capacitance	6 µµf
Output Capacitance	2.9 µµf
Cathode to Heater	2.7 µµf
Plate to Grid	1.4 µµf

Between Input and Output Section

Plate of Input Section to Plate of Output Section	0.045 µµf
Grid of Input Section to Plate of Output Section	0.005 µµf

Ratings (Design Center Values: Each Section)

Plate Supply Voltage	550 volts max
Plate Voltage [2]	130 volts max
Plate Dissipation	2 watts max
Cathode Current	25 mA max
Grid Bias	50 volts max
External Grid Resistance	1 megohm max
External Resistance Between Cathode and Heater	20,000 ohms max
Voltage Between Cathode of Output Section and Heater (Cathode Positive with Respect to Heater)	130 Vdc + 50 Vrms max
Voltage Between Cathode of Input Section and Heater	50 Vrms max

Characteristics (Each Section)

Plate Voltage	90 volts
Negative Grid Bias	1.2 volts
Plate Current	15 mA
Transconductance	12,500 μmhos
Amplification Factor	33
Equivalent Noise Resistance	275 ohms

[1] Triode No. 1 should be used as the grounded-cathode input section of the cascode amplifier and Triode No. 2 as the grounded-grid output section.

[2] In order not to exceed the maximum permissible plate voltage when the cascode amplifier is controlled, it is necessary to use a voltage divider for the grid of the grounded-grid section. With grid current biasing for the grounded cathode section the plate voltage across this section should not be more than 75 V in the not-controlled condition.

Fig. 5-6B. Complete tube data, 6DJ8-ECC88 dual triode (part 2 of 6 parts) (Courtesy Amperex).

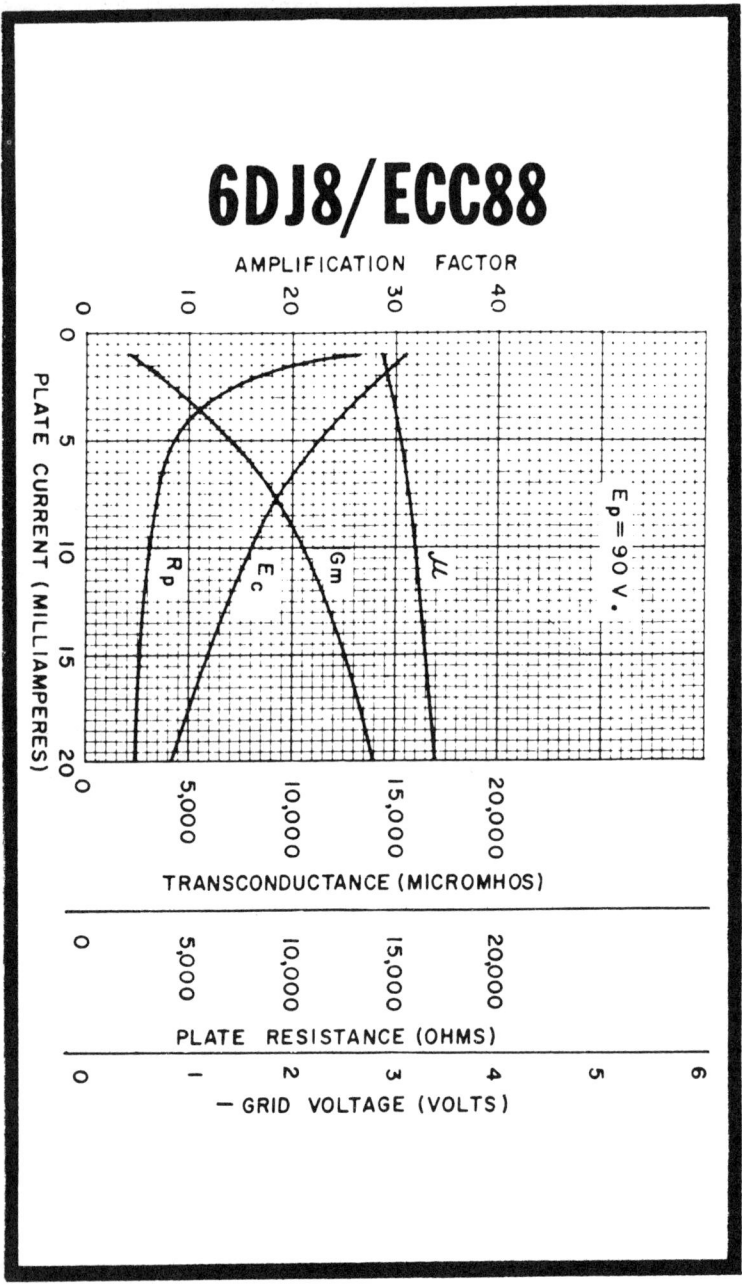

Fig. 5-6C. Complete tube data, 6DJ8-ECC88 dual triode (part 3 of 6 parts) (Courtesy Amperex).

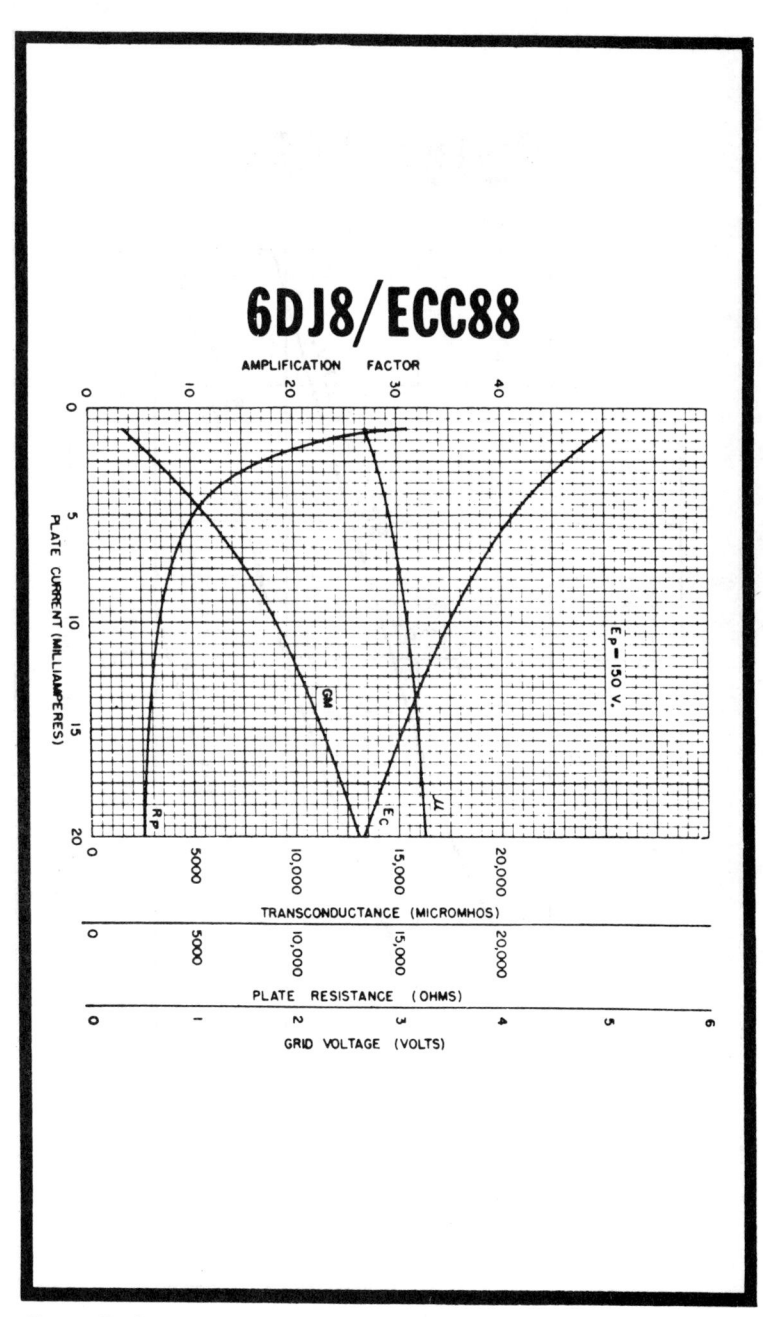

Fig. 5-6D. Complete tube data, 6DJ8-ECC88 dual triode (part 4 of 6 parts) (Courtesy Amperex).

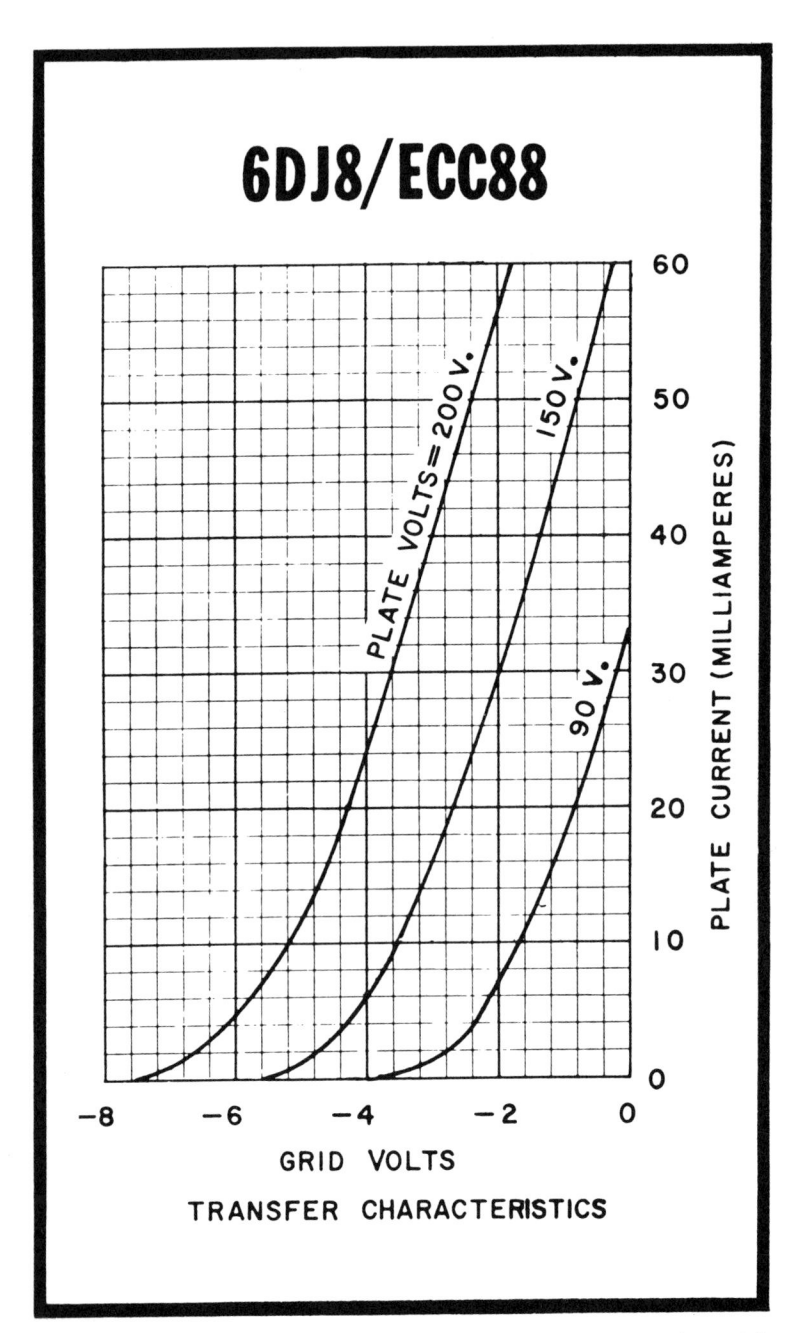

Fig. 5-6E. Complete tube data, 6DJ8-ECC88 dual triode (part 5 of 6 parts) (Courtesy Amperex).

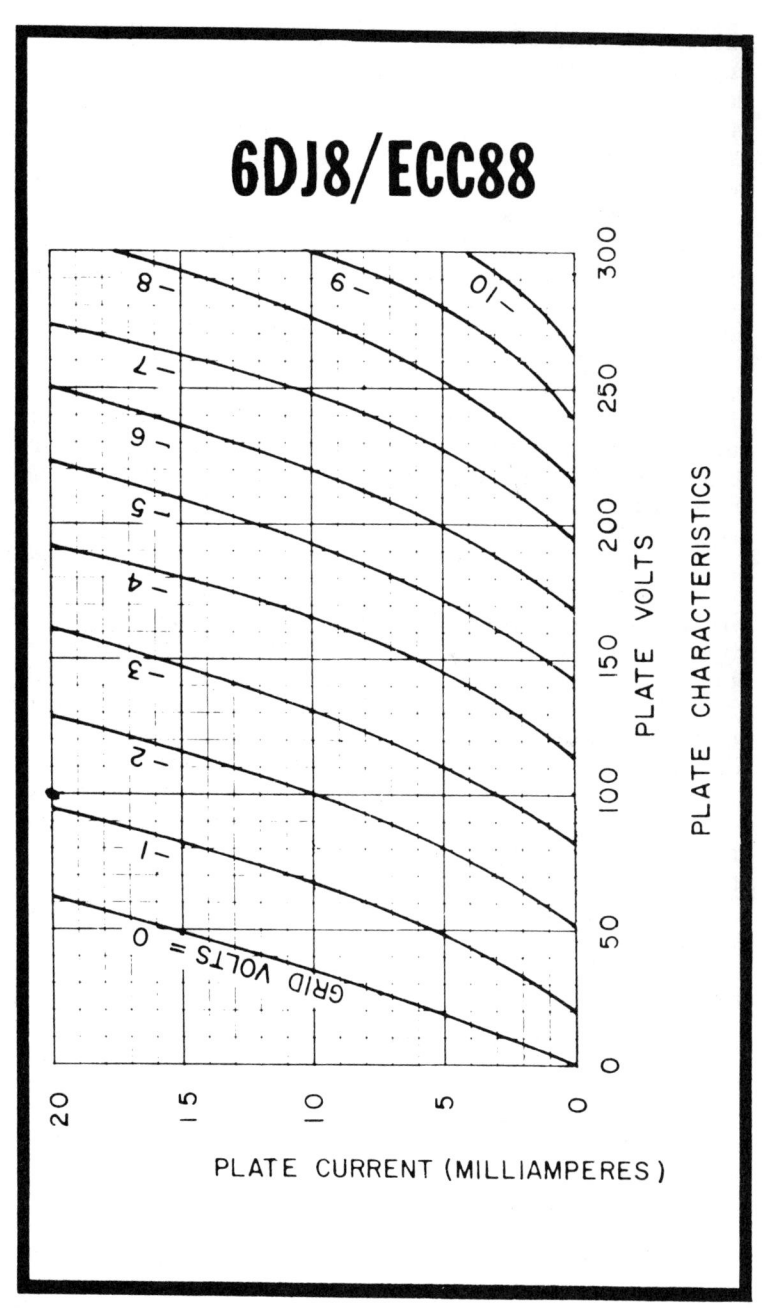

Fig. 5-6F. Complete tube data, 6DJ8-ECC88 dual triode (part 6 of 6 parts) (Courtesy Amperex).

153

cathode resistor (5K ohms) we get 146.2/5 equals 29.2 This ought to be almost exactly the same as u for the tube. What u does the tube have? Well, we changed the plate voltage 150 volts and the bias changed from 2 to 7 volts, didn't it? What's 150/5? It's 30, isn't it? There you go! Positive proof that a resistor in the cathode of a vacuum tube will look u times greater when viewed from the plate.

I agree. The answers aren't exactly the same, but they are close enough to prove our point. I frankly don't think anyone can draw curves accurately enough to have the answers come out exactly the same, but I haven't stopped hoping.

To prove that R_L is divided by u when viewed from the cathode, you'll have to construct the circuit shown in Fig. 5-5. Use the 6DJ8 (ECC88) again—after all it is two triodes in the same envelope. The plate, grid, and cathode of one tube are pins 1, 2 and 3, while the plate, grid, and cathode of the second triode are pins 6, 7, and 8. Pins 4 and 5 are the filaments requiring 6 volts AC.

If what we say is correct, you should be able to measure (with an oscilloscope) the same signal you would get from the circuit at the right in Fig. 5-5. What might that be? Let's see: R_L over u is 27K/30 or 900 ohms. And r_K in each case is 100 ohms. What kind of division of signal would this give us? Well, 1K over 1.1K is about 91 percent. That is, if you put in a 1-volt signal, you should see a signal on the cathodes about 0.9 volts high. Note that if the plate resistor did not reflect into the cathode circuit, you would be applying the 1 volt signal to just two 100 ohm resistors and thus would see only half of it on the common cathodes. Try it out; I think you'll find that u divides R_L and that this much resistance does reflect into the cathode.

If you want to try other experiments, use the 6DJ8-ECC88 data that follow (Fig. 5-6).

TRIODES AND PENTODES

As I mentioned in the first part of this chapter, 1927 saw the general use of the tetrode or screen-grid tube come into being. In other words, what we call the screen grid was added to the triode. See Fig. 5-7. This gave the plate curves a new wrinkle, very similar to the tunnel diode curve but over a much bigger swing in voltage. Another way to say it is that it gave a region of the curves where the plate current went down as the plate voltage continued to go up. What actually happened was that the electron current from the cathode was hitting the plate so hard that it was knocking off more electrons than were actually hitting it, and these secondary

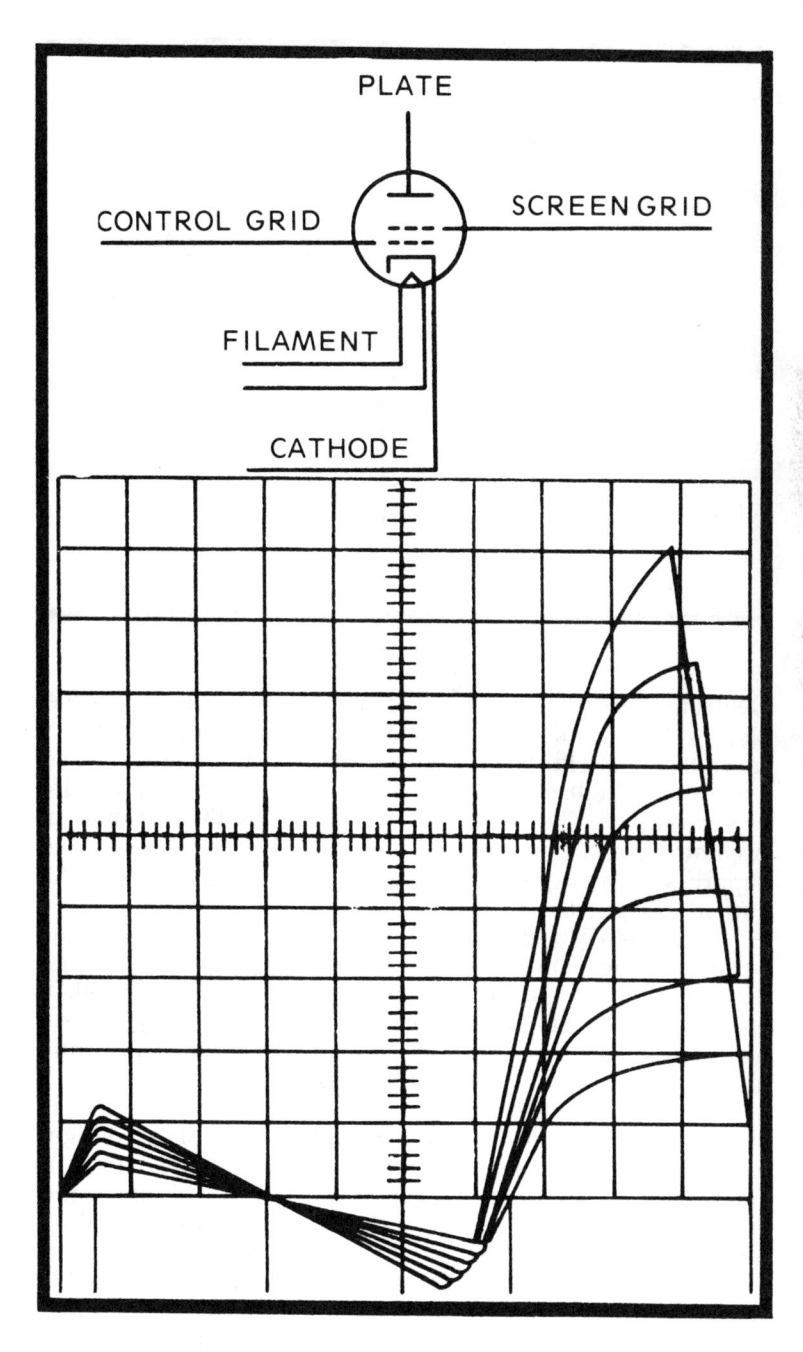

Fig. 5-7. Tetrode tube and I_P — E_P curves.

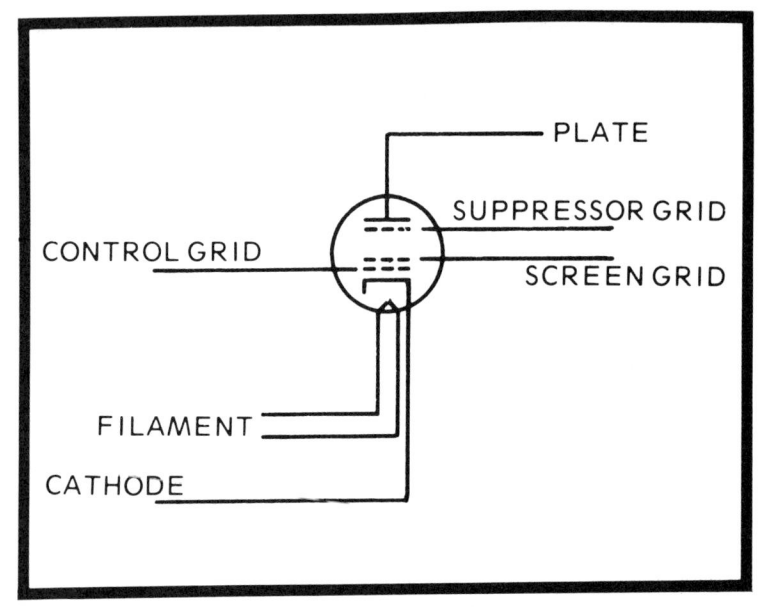

Fig. 5-8. Pentode tube.

electrons were all going to the new (screen) grid. Of course, plate voltage could be made high enough to recapture this secondary current, but this left us with a tube that required excessively high plate voltage to achieve even a moderate power output. The tube did lend itself to the building of oscillators, and then somebody came along and added the suppressor (3rd) grid which is usually connected back to the cathode (Fig. 5-8).

The addition of the suppressor grid eliminated the secondary emission region of the curves, gave the tube a high plate resistance characteristic and a resultant very high u characteristic (remember, u equals r_p divided by r_K). Also, we find that the input capacity was reduced quite a bit and higher frequencies became more of a reality.

Whenever we have a screen grid (2nd grid) tube, since it is operated at a high positive voltage similar to the plate voltage, we find that we have a new measurement to take into consideration. Since this second grid is positive, it will take some of the current from the cathode away from the plate. This includes a similar percentage of signal current and (different than the triode) the ratio of plate current to cathode current is considerably less than 1. In the case of the 6AU6, we can go to the curves to illustrate what I mean. Fig. 5-9C shows the plate current with respect to No. 1 grid bias and a different screen grid (grid No. 2) voltage for each curve. Check the plate current determined by −1 volt of bias and 125 volts on the screen grid. It's about 7.6 ma. Now go to Fig. 5-9D, a graph of screen grid current as determined by control grid bias and screen voltage. Again use −1 volt of bias and 125 volts on the screen to get approximately 3 ma of current. The sum of these two currents is the cathode current (7.6 + 3 equals 10.6 ma). To get this new measurement, divide 7.6 by 10.6 to get something close to 0.72 (or 72 percent). This ratio or its result has been known by several names: The name "plate efficiency" actually being a "signal power out" divided by the product of "signal voltage in" times "cathode signal current," but this does not exactly match it so I'd like to call this "ratio of currents"—ALPHA. You see, we make a very similar measurement on transistors and it is named alpha. So, rather than actually misname this current ratio "plate efficiency," let's just call it "alpha" in both cases and have only one name to remember. Any one familiar with solid-state devices will immediately recognize what you mean anyhow and it will simplify things for you.

What we do with these measurements and how we combine them with previously introduced theory is subject matter for the next chapter, so I'll not go into it here. However, this chapter is a long way from being completed. We have yet to go into the solid-state devices and their similar measurements.

Transistors came into being at a time when there were a lot more scientists around than there were when vacuum tubes came into being. Science itself had made some rather giant steps in its development during the time between these two events. The scientific method of analysis had evolved to a much higher degree and practically no one was willing to let the transistor become public without complete scientific analysis. Every little reaction had to be taken into account, and it took technicians a long time to discover that they didn't have to split hairs so finely in order to find out what was going on. The surprising thing about it was that most textbooks mentioned the key to the situation, but then didn't develop Transistor Theory along these lines (which would have paralleled vacuum tube analysis). Instead, each group developed Transistor Theory along its own specialty line and used their own method of scientific notation. They even changed the mathematical language in which they spoke about the device. (i.e., transistors don't have characteristics, they have parameters), and this was a strike against the novice before he even got started studying transistors because he didn't even know what a parameter was. If he tried to look

SYLVANIA engineering data service

SYLVANIA 6AU6
6AU6A, 3AU6
4AU6, 12AU6

QUICK REFERENCE DATA

The Sylvania Types 3AU6, 4AU6, 6AU6, 6AU6A and 12AU6 have a sharp cutoff pentode contained in a miniature T-5½ envelope. They are designed for use in RF or IF amplifier applications.

Types 3AU6, 4AU6 and 6AU6A have controlled heater warm-up time for series string operation.

MECHANICAL DATA

Bulb	T-5½
Base	E7-1, Miniature Button 7-Pin
Outline	5-2
Basing	7BK
Cathode	Coated Unipotential
Mounting Position	Any

ELECTRICAL DATA

HEATER CHARACTERISTICS

	3AU6	4AU6	6AU6	6AU6A	12AU6	
Heater Voltage	3.15	4.2	6.3	6.3	12.6	Volts
Heater Current	600	450	300	300	150	Ma
Heater Warm-up Time[1]	11	11			11	Seconds
Heater-Cathode Voltage (Design Maximum Values)						
Heater Negative with Respect to Cathode						
Total DC and Peak	200	200	200	200	200	Volts Max.
Heater Positive with Respect to Cathode						
DC	100	100	100	100	100	Volts Max.
Total DC and Peak	200	200	200	200	200	Volts Max.

DIRECT INTERELECTRODE CAPACITANCES

	Shielded[2]	Unshielded	
Grid No. 1 to Plate	.0035	.0035	µµf Max.
Input: g1 to (h+k+g2+g3+I.S.)	5.5	5.5	µµf
Output: p to (h+k+g2+g3+I.S.)	5.0	5.0	µµf

RATINGS (Design Maximum Values)

Plate Voltage	330 Volts	Max.
Grid No. 2 Supply Voltage	330 Volts	Max.
Grid No. 2 Voltage	See Rating Chart	
Plate Dissipation	3.5 Watts	Max.
Grid No. 2 Dissipation	0.75 Watts	Max.
Positive Grid No. 1 Voltage	0 Volts	Max.

CHARACTERISTICS AND TYPICAL OPERATION

Plate Voltage	100	250	250	Volts
Grid No. 3 Voltage	Connected to Cathode at Socket			
Grid No. 2 Voltage	100	125	150	Volts
Cathode Bias Resistor	150	100	68	Ohms
Plate Current	5.0	7.6	10.6	Ma
Grid No. 2 Current	2.1	3.0	4.3	Ma
Transconductance	3900	4500	5200	μmhos
Plate Resistance (approx.)	0.5	1.5	1.0	Megohms
Grid No. 1 Voltage for Ib = 10 μa	−4.2	−5.5	−6.5	Volts

NOTES:

1. *Heater warm-up time is defined as the time required for the voltage across the heater to reach 80% of the rated heater voltage after applying four (4) times rated heater voltage to a circuit consisting of the tube heater in series with a resistance equal to three (3) times the rated heater voltage divided by the rated heater current.*

2. *Shield No. 316 connected to Cathode Pin No. 7.*

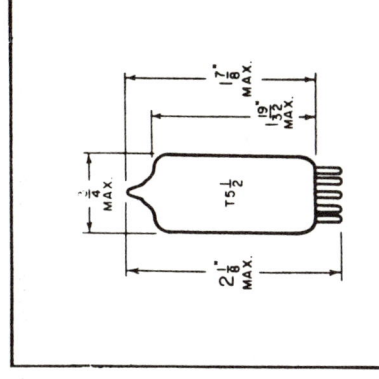

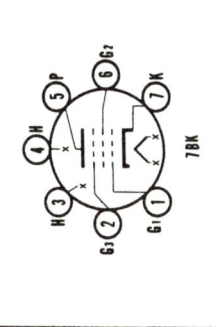

Fig. 5-9A. 6AU6 pentode data (part 1 of 5 parts). (Courtesy Sylvania Electronic Tubes Div.)

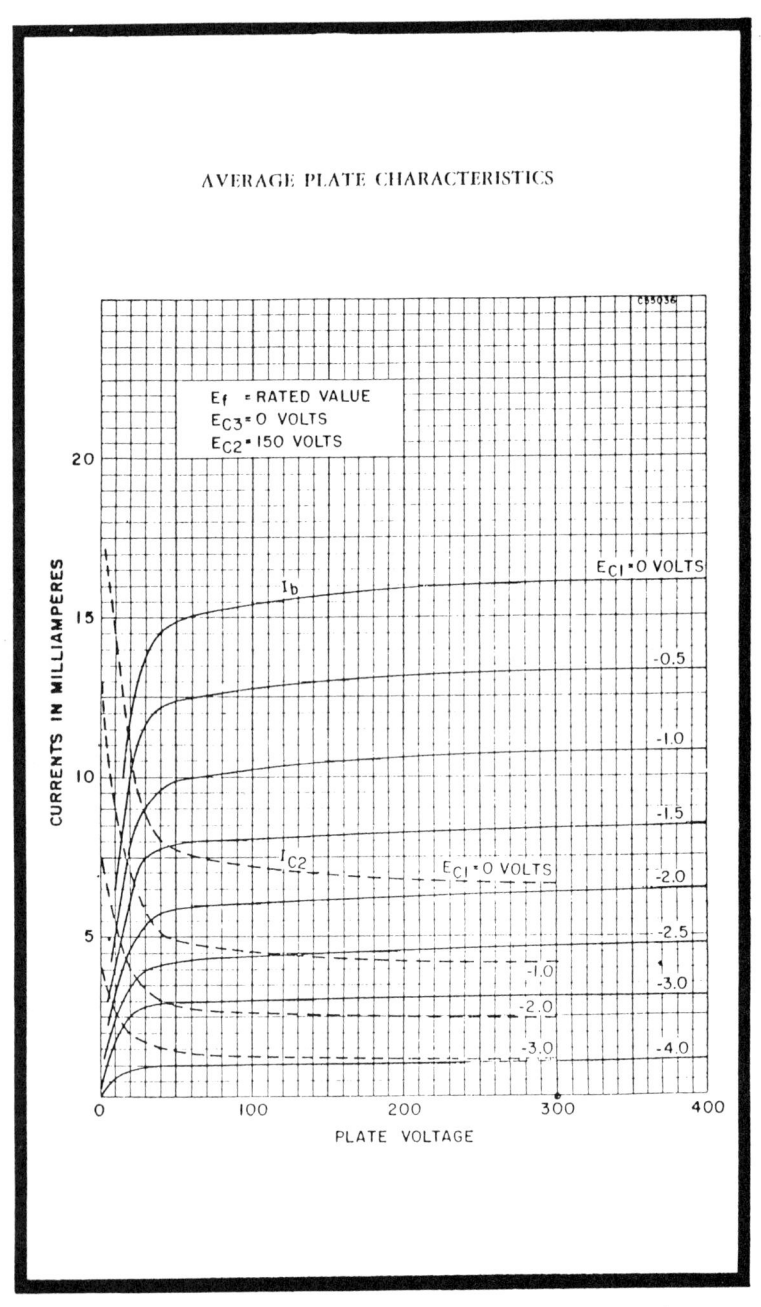

Fig. 5-9B. 6AU6 pentode data (part 2 of 5 parts). (Courtesy Sylvania Electronic Tubes Div.)

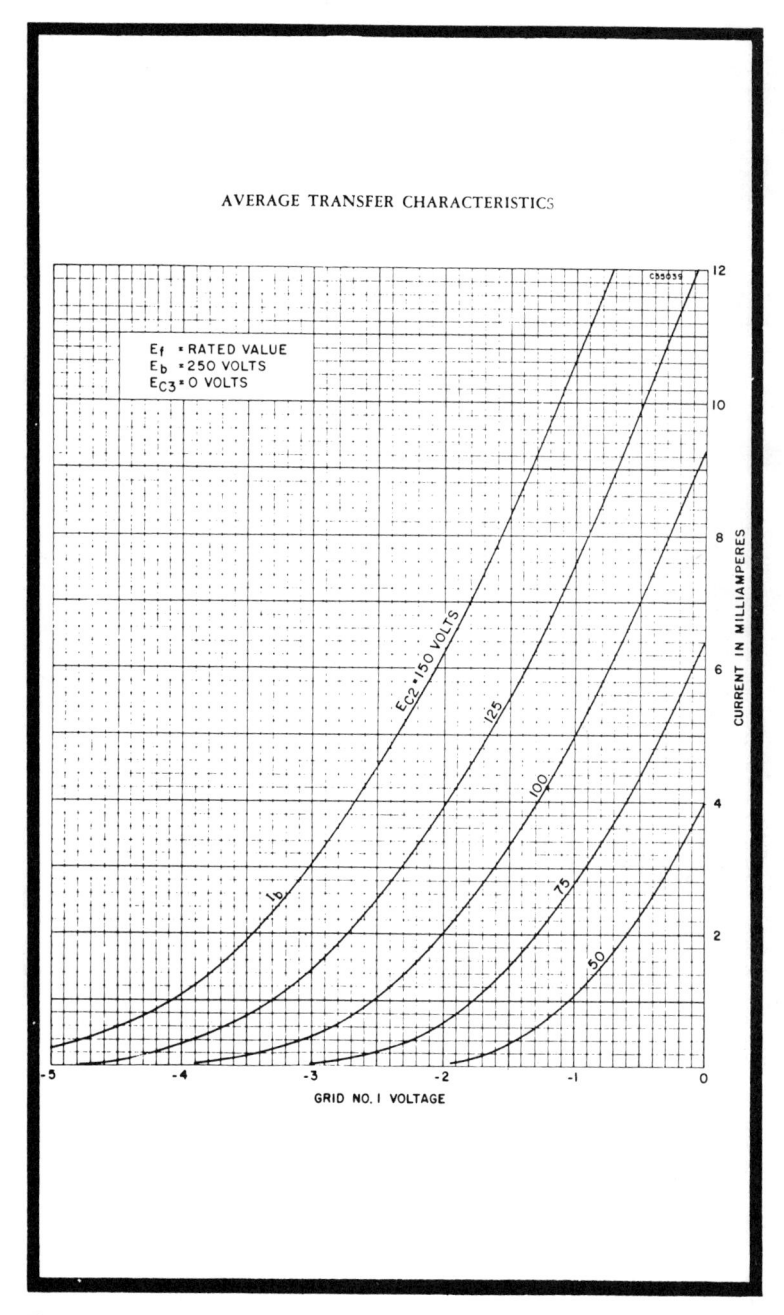

Fig. 5-9C. 6AU6 pentode data (part 3 of 5 parts). (Courtesy Sylvania Electronic Tubes Div.)

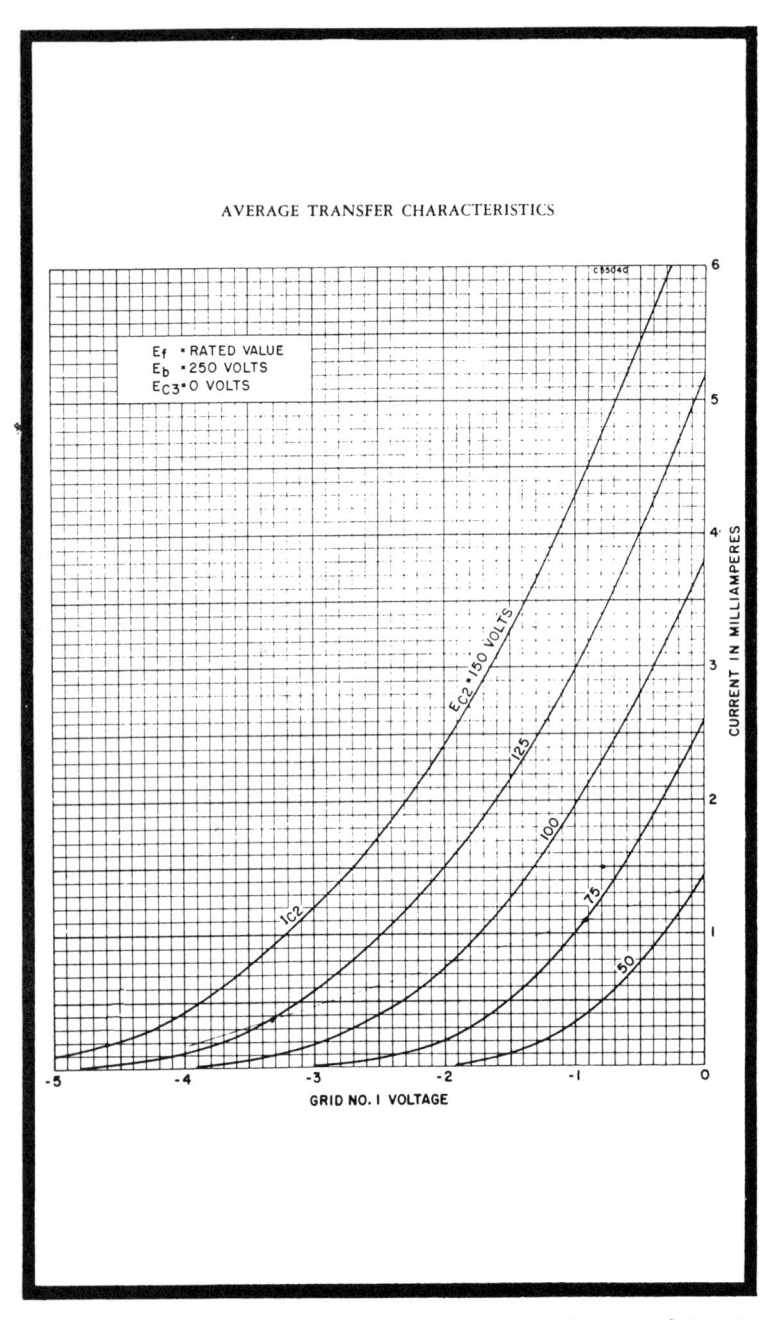

Fig. 5-9D. 6AU6 pentode data (part 4 of 5 parts). (Courtesy Sylvania Electronic Tubes Div.)

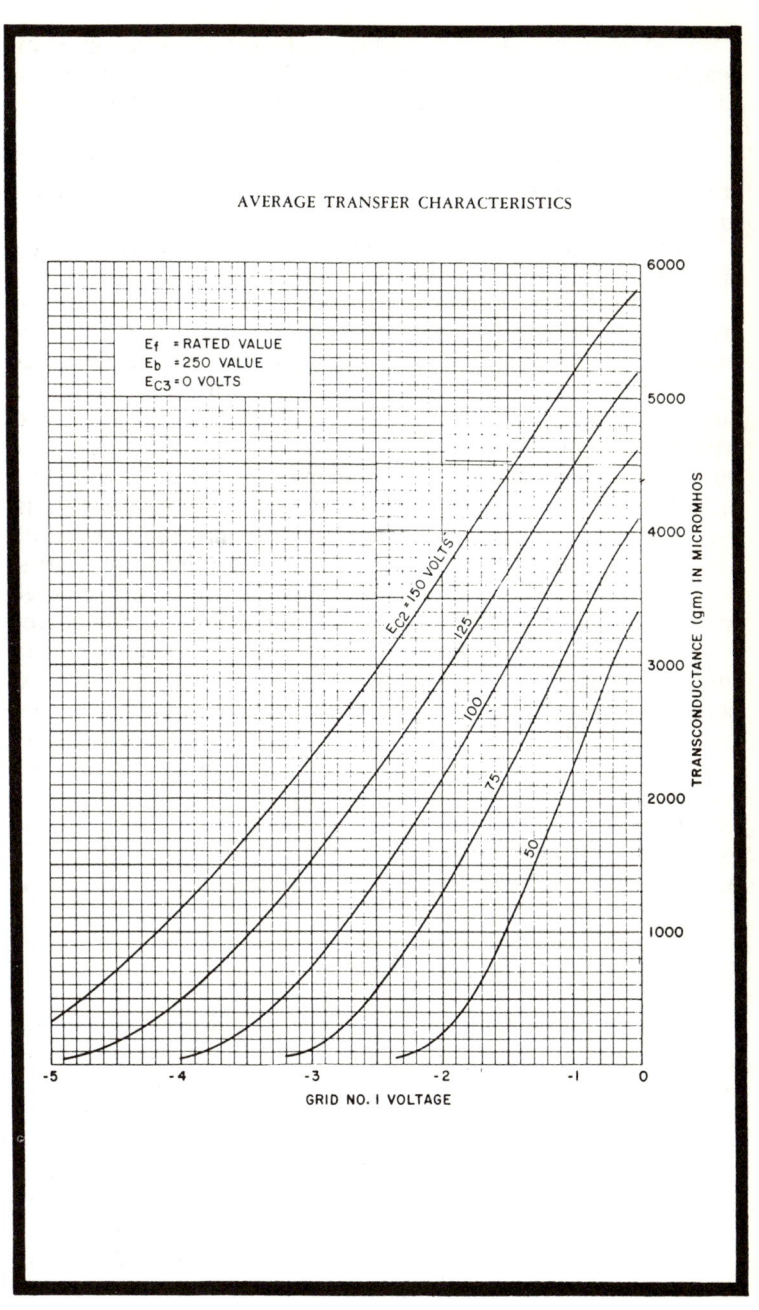

Fig. 5-9E. 6AU6 pentode data (part 5 of 5 parts). (Courtesy Sylvania Electronic Tubes Div.)

it up, he probably failed to find it. All he could do was go to his teacher and ask. So, he more than likely came out of this discussion with the idea that Characteristics and Parameters were the same thing. This is not quite true.

A Parameter is a ratio of two **measurements** (one measurement change directly caused by the other) on a transistor.

A Characteristic of a Vacuum Tube is an **estimate** of how a tube should react to a certain situation.

Read the last two sentences again, please. Vacuum-tube technology, remember, is good enough so that we can use the plate curves of the US-made 6DJ8 for the European-made ECC88 with full expectations that the two will react in a similar fashion (the differences will be quite small). However, with transistors, the situation is quite different. Just because you have a 2N169A in your hand doesn't mean that it will have exactly the parameters as those you find in the manual for it. For that matter, those you find in the manual won't be exact. They will probably have maximum and minimum limits, and—on top of that—you are told that the limits will change with the magnitude of DC current you force through the device (the published parameters were probably taken at 1 ma through the device and 5 volts across it). Then, you are expected to use these parameters in equations that mathematicians split hairs to get. I suppose this is OK for the engineer who wants to predict what changes in parameters will do to his circuit, but for the average technician who wants to know how the circuit operates in general, they are much too exacting. For this reason, I am not going into the "H" parameters. They are covered quite adequately in just about every other text that I've seen.

EQUATING TUBES AND TRANSISTORS

Roughly, NPN and PNP transistors are triodes (three-lead devices) with a set of output curves that look like they belong to a pentode. Compared to a triode vacuum tube, they have a very high r_p, a very high u (voltage gain characteristic) and a high G_m (or a very low cathode resistance).

Since the transistor is a solid-state device, its cathode does not need to be heated up to make it work. All you have to do is apply the proper voltages. Consequently, there are no heater leads to worry about and no warm-up time required. When you turn it on, it's on, immediately.

Of course, the three leads that make this thing operate were given different names than they had in the triode vacuum tube, but they do serve similar functions. The equivalent Cathode was called the Emitter; the equivalent Grid, the Base, and the equivalent Plate, the Collector. See Figs. 5-10 and 5-11.

In the case of the NPN transistor, electron current enters the emitter and goes to both the base and collector (so they must both be positive with respect to the emitter).

In the case of the PNP transistor, electron current enters both the collector and base from the outside world, flowing through the device and out the emitter (so both collector and base must be negative with respect to the emitter).

Note: In either case, NPN or PNP, the emitter carries the sum of the base current and the collector current.

It is of special interest to note also that with either the NPN or PNP transistors, the emitter-to-base junction is an "ON" diode and the voltage across it is determined predominantly by the material the transistor is made from (i.e., .6 volt for silicon or .2 volt for germanium). The thing that determines just how hard the transistor is turned "ON" is the amount of electron current that flows in the base lead, so be careful how you apply drive voltage to a transistor. Be sure there is a resistor to limit the current in the circuit so the swing in voltage you are using doesn't overdrive the transistor. Once you get the transistor turned on, you are NOT going to change its emitter-to-base voltage very much unless you just want to turn it off or burn it up. (A few hundred ohms in series with the emitter lead will also help safeguard your transistor.) You see, maximum breakdown ratings of transistors are really more important than the parameters for r_p, u and G_m. Just as long as these parameters are better than a certain minimum (whatever the design engineer chose as "worst case"), the circuit will work as it should. And, transistor circuits are usually designed so that their parameters can change quite radically (50 percent below design center to 100 percent above design center) without the circuit changing its action by very much. This makes it easy on the circuit analyst because it allows us to make an educated guess as to what the actual values of these parameters are and go ahead just as though we had the absolutely accurate ones.

Look at Fig. 5-12A and note the difference between the "hob" parameter and the "hoe" parameter. These are both measurements of the output conductance (change of plate voltage and resultant change of current for an I/V ratio). The "hoe" parameter has a grounded emitter (cathode) and thus the slope of the output curves is greater (less flat) and the conductance is higher. (This isn't good; the higher r_p is, the

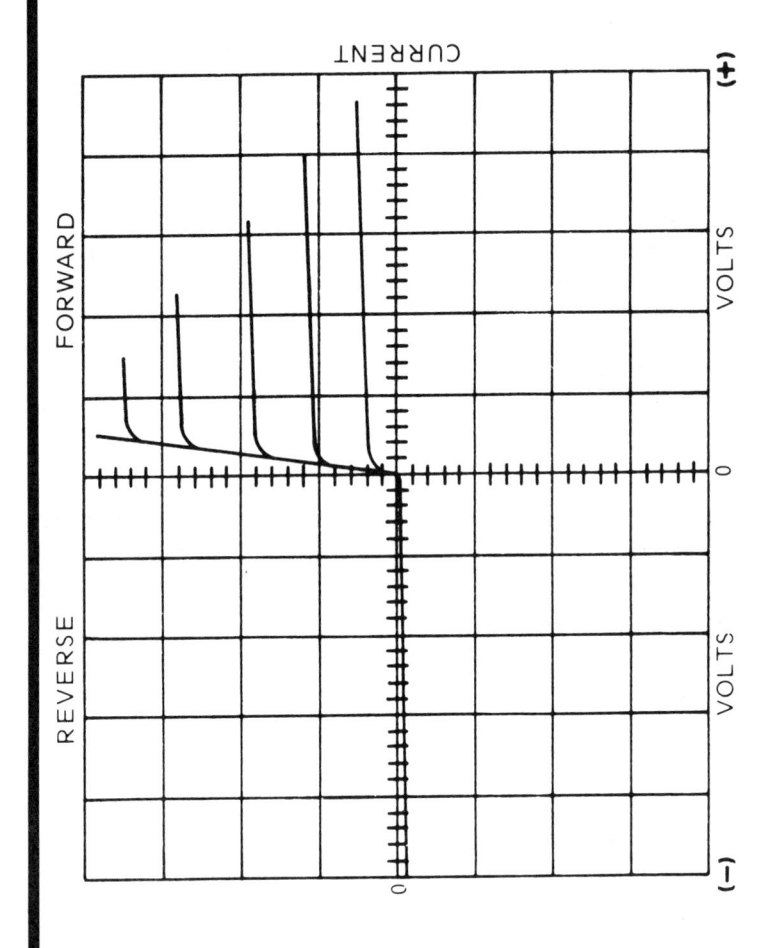

Fig. 5-10. Solid-state counterpart to pentode vacuum tube, i.e., NPN transistor.

The Solid-State Counterpart to Pentode Vacuum Tubes.

Current flows emitter to collector as controlled by bias current. It is normally "off" and has to be turned "on" by some bias current drawn from the base lead.

$r_{tr} = $ "Transresistance"

$$r_{tr} = \frac{.026}{I_e} + R_{eb} = \frac{1}{Gm}$$

$$r_p = \frac{1}{hoe} = u \times \frac{1}{Gm} = u \times r_{tr}$$

$$u = r_p \times gm = \frac{r_p}{r_{tr}} = \frac{1}{hoe \times r_{tr}}$$

RESISTANCE in the Collector circuit when seen from the Emitter is divided by $\varkappa$.
RESISTANCE in the Base circuit when seen from the Emitter is divided by β.
RESISTANCE in the Emitter circuit when seen from the base is multiplied by β.
RESISTANCE in the Emitter circuit when seen from the Collector is multiplied by $\varkappa$.

NPN TRANSISTOR

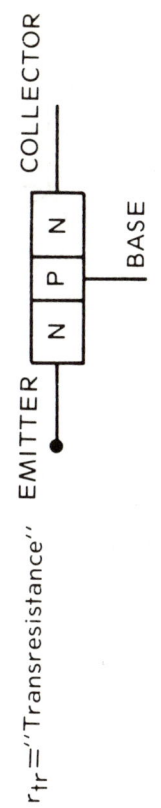

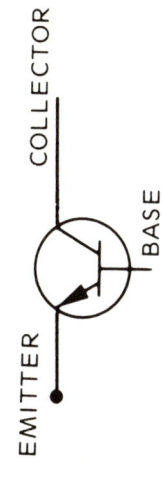

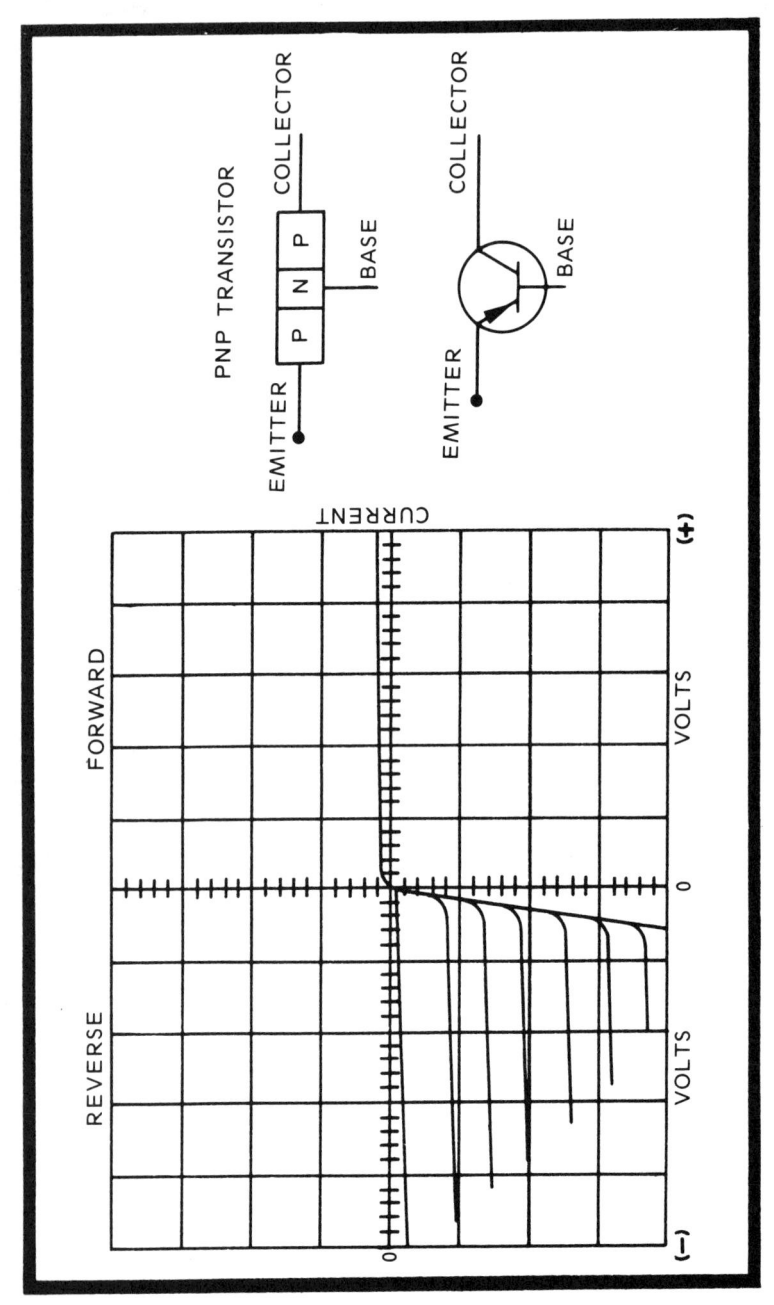

Fig. 5-11. Solid-state counterpart of a backwards pentode, i.e., PNP transistor.

better we like it.) Note though, not only does the type of circuit make a difference, but frequency also makes a big difference, as well as the operating point. According to our information sheet, when "hob" equals .2 umhos at 270 Hz and hoe equals 140 umhos at 455 kHz (r$_p$ is the reciprocal of these values), r$_p$ can equal anything between 5 Meg down to almost 7K ohms. Quite frankly, we can choose anything between 100K and 50K ohms to fit just about any case. I usually use 100K ohms, even though it might be a little on the high side of "Worst Case."

The wonderful part of this idea is that we can use this value of 100K for r$_p$ for just about any transistor in a worst-case, mid-frequency analysis situation and be quite confident that we are being reasonable.

Now, to the G$_m$ of a transistor. Remember, the reciprocal of Gm is emitter (cathode) resistance and the base-to-emitter junction is an "ON" diode. The slope of any small part of the diode curve is predictable (within reason). This is just another way of saying that all transistors operating with the same amount of current going through them will have the same G$_m$. And that is rather a wild statement, but it is predominantly true. This is a small-signal characteristic, but is extremely useful to us later on, especially in the form of a certain amount of resistance.

Since a vacuum tube G$_m$ characteristic is called "Transconductance" and is measured in "Mhos," I am going to suggest that we call this reciprocal of a transistor G$_m$, "Transresistance." (Please understand me—I did NOT invent this name. It was suggested to me by Mr. Ron Olson, an engineer who is a friend of mine.) And I will use the notation, "r$_{tr}$" to mean Transresistance for the rest of this text.

Science tells us that the equation for the slope (resistance) of any small part of the perfect germanium diode curve is:

$$r_e = \frac{K T}{q I_e m} \qquad \text{(Eq. 5-10)}$$

where K is Boltzmann's constant (1.38×10^{-23}) in watt-sec/degrees C
T is the absolute temperature in degrees Kelvin (T equals degrees C + 273)
q is the charge of an electron (1.6×10^{-19} coulomb)
I$_e$ is the current flowing through the junction
and m is a constant whose value is 1 for germanium and between 1 and 2 for silicon. All of which sounds pretty complicated until we take the trouble to put these values into the

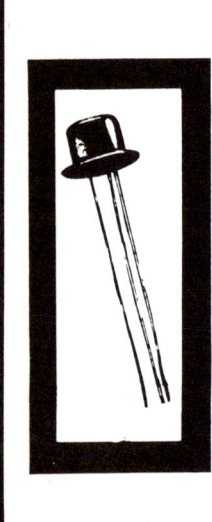

The General Electric type 2N169A is an isolated case NPN rate grown germanium transistor recommended for high gain RF and IF amplifier service and general purpose industrial applications where high beta, high voltage, low collector capacity and extremely low collector cut-off current are of prime importance.

absolute maximum ratings (25°C)

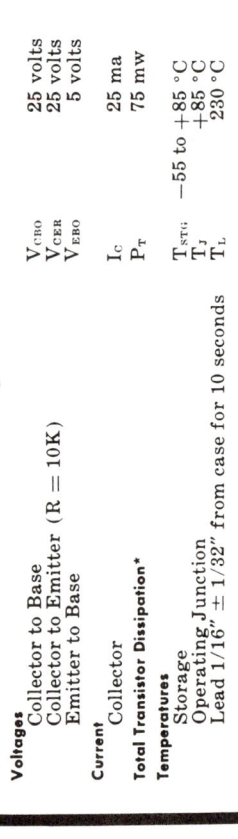

Voltages
Collector to Base	V_{CBO}	25 volts
Collector to Emitter (R = 10K)	V_{CER}	25 volts
Emitter to Base	V_{EBO}	5 volts

Current
Collector	I_C	25 ma

Total Transistor Dissipation*
	P_T	75 mw

Temperatures
Storage	T_{STG}	−55 to +85 °C
Operating Junction	T_J	+85 °C
Lead 1/16" ± 1/32" from case for 10 seconds	T_L	230 °C

*Derate 1.25 mw/°C increase in ambient temperature.

electrical characteristics (25°C)

D-C CHARACTERISTICS

		Min.	Design Center	Max.	
Collector to Emitter Breakdown Voltage ($R_{BE} = 10K$, $I_C = .3$ ma)	V_{CER}	25			volts
Reach-Through Voltage	V_{RT}	25			volts
Forward Current Transfer Ratio ($I_C = 1$ ma, $V_{CE} = 1$ V)	h_{FE}	34	72	200	
Base Input Voltage ($I_C = 1$ ma, $V_{CE} = 1$ V)	V_{BE}		.14	.2*	volts
Saturation Voltage ($I_B = .5$, $I_C = 5$ ma)	$V_{CE(SAT)}$	.1*	.23	.4*	volts
Collector Current ($I_E = 0$, $V_{CB} = 15$ V)	I_{CBO}		.9	5	µa
Emitter Current ($I_C = 0$, $V_{EB} = 5$ V)	I_{EBO}	.13*	.9	5	µa

LOW FREQUENCY CHARACTERISTICS (COMMON EMMITTER)
($V_{CE} = 5$ V, $I_C = 1$ ma, $f = 270$)

		Min.	Design Center	Max.	
Forward Current Transfer Ratio	h_{fe}		50		
Output Admittance	h_{ob}		.2		µmhos
Input Impedance	h_{ib}		55		ohms
Reverse Voltage Transfer Ratio	h_{rb}		2×10^{-4}		

HIGH FREQUENCY CHARACTERISTICS (COMMON EMMITTER)
($V_{CB} = 5$ V, $I_E = 1$ ma, $f = 455$KC)

		Min.	Design Center	Max.	
Base Spreading Resistance	r'_b		250		ohms
Output Capacity	C_{ob}		2.4		µµf
Forward Current Transfer Ratio	h_{fe}		30		
Output Admittance	h_{oe}		140		µmhos
Input Impedance	h_{ie}		700		ohms
Reverse Voltage Transfer Ratio	h_{re}		10×10^{-3}		
Noise Figure ($B_w = 1$ cycle), ($f = 1$KC, $V_{CB} = 1.5$ V, $I_E = -0.5$ ma)	NF		6	6	db
Power Gain (Typical IF Test Circuit)	G_{pe}	27	28		db
Available Power Gain	G_a		39		db
Cutoff Frequency	f_{hfb}		9		MC

*These limits are design limits within which 98% of production normally falls.

Fig. 5-12A. Data for 2N169A NPN germanium transistor (part 1 of 7 parts). (Courtesy General Electric Co.)

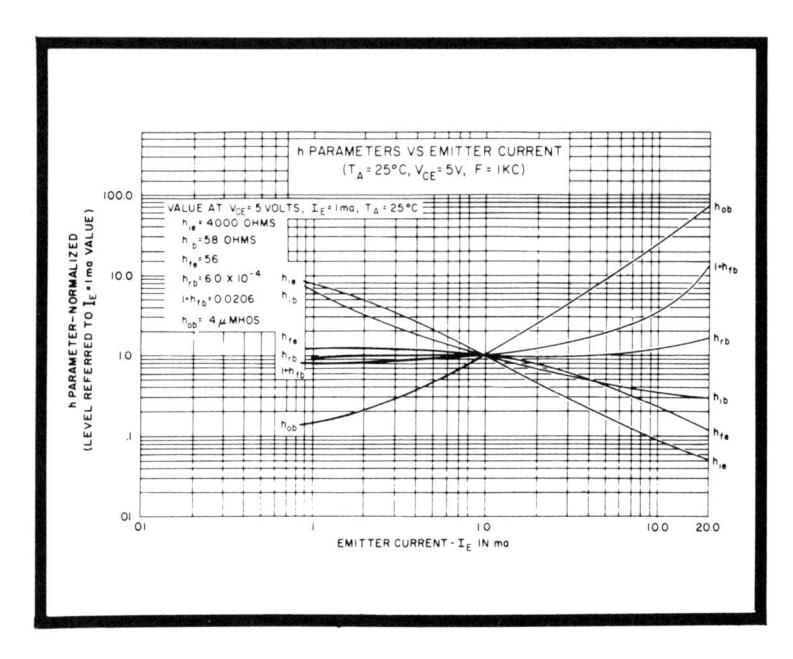

Fig. 5-12B. Data for 2N169A NPN germanium transistor (part 2 of 7 parts). (Courtesy General Electric Co.)

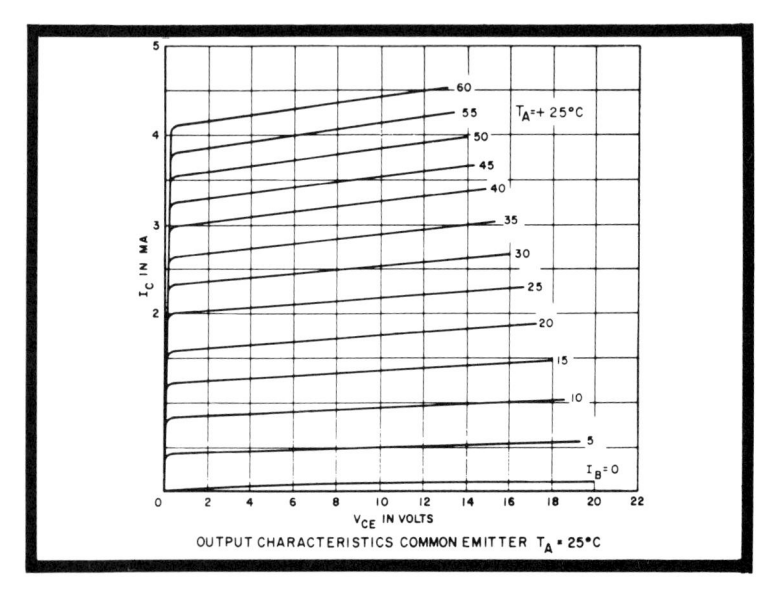

Fig. 5-12C. Data for 2N169A NPN germanium transistor (part 3 of 7 parts). (Courtesy General Electric Co.)

172

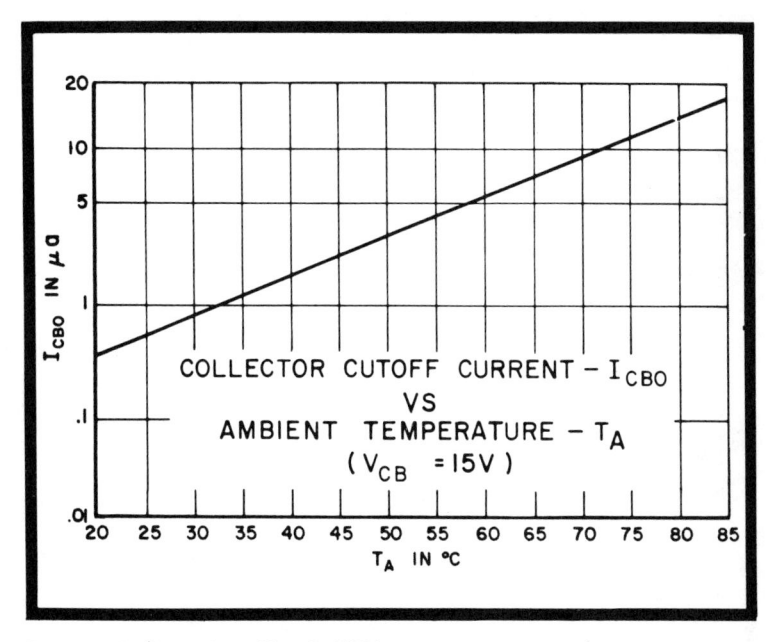

Fig. 5-12D. Data for 2N169A NPN germanium transistor (part 4 of 7 parts). (Courtesy General Electric Co.)

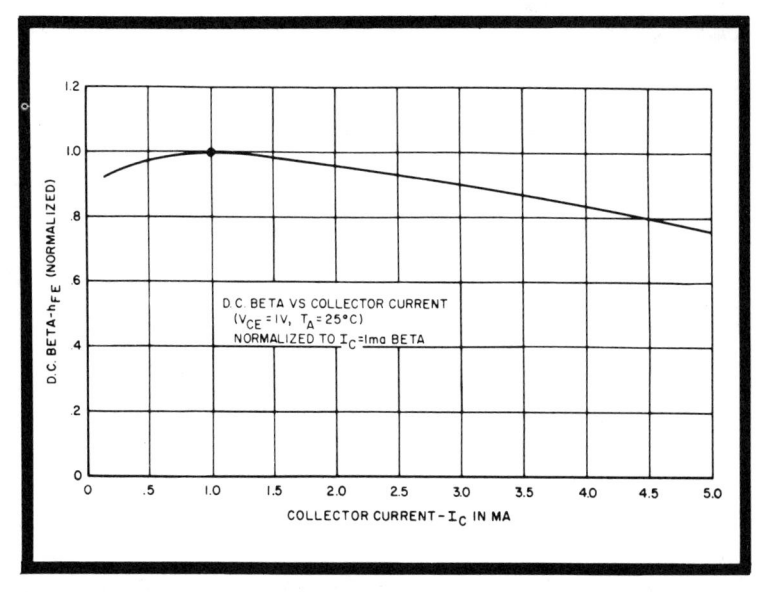

Fig. 5-12E. Data for 2N169A NPN germanium transistor (part 5 of 7 parts). (Courtesy General Electric Co.)

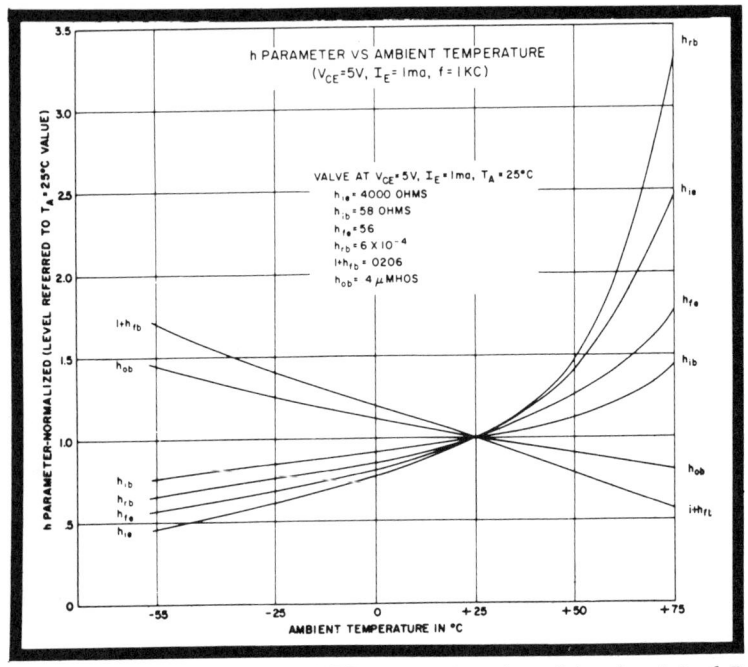

Fig. 5-12F. Data for 2N169A NPN germanium transistor (part 6 of 7 parts). (Courtesy General Electric Co.)

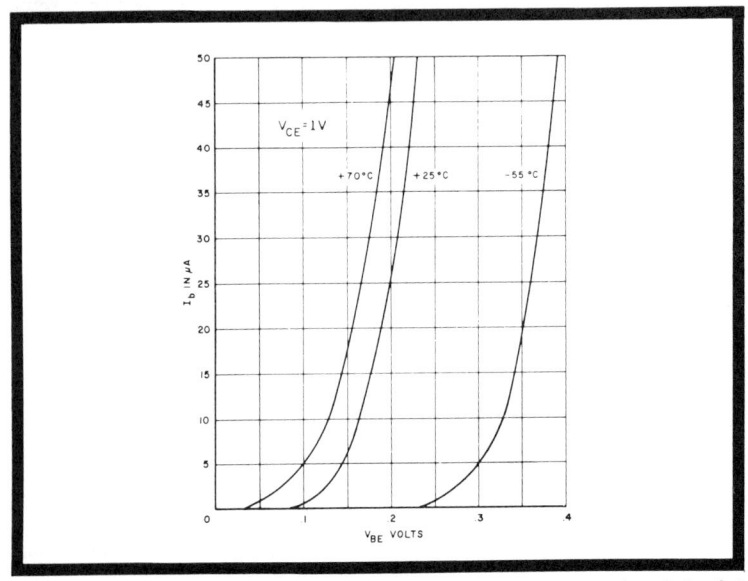

Fig. 5-12G. Data for 2N169A NPN germanium transistor (part 7 of 7 parts). (Courtesy General Electric Co.)

equation and figure out just what it is we are going to divide I_E into. K times T divided by q times m comes out equal to 0.026. And, all you have to remember is "point oh two six — .026 —." This number is already expressed in units and there is no further power of ten associated with it. But—NOTE—most diodes (and transistors) carry current in terms of the milliamp, so if we are going to divide this term by milli units, we had better write it in terms of milli units—thus:

$$r_e = \frac{26. \text{ milli units}}{I_e \text{ milli amps}}$$

The milli units will cancel and the resistance of a germanium diode carrying 1 ma of current is predicted to be 26 ohms.

Now all of this is great except for the fact that it is true only of the perfect PN Junction made of germanium and you have to put it into a practical sense. By that I mean, you will have to add a little something extra to make the answer come out closer to reality. It is my practice to add something between 3 ohms and 5 ohms (anything within this range to round out the answer to a convenient value). Also, we must remember that this is true only of regular small-signal transistors and not for power transistors. Any transistor carrying up to maybe 40 ma can be figured in this fashion. This points out that no transistor operated in this range of current will be figured to have an emitter resistance less than 5 ohms. And believe me—it pays to remember this.

For power transistors carrying less than 100 ma, you may figure the emitter resistance to be equal to 1 ohm. For power transistors carrying 100 ma or more, just figure 0.5 ohm and let it go at that.

Don't let the fact that you may be dealing with a silicon transistor throw you a curve either. This twenty six over I sub e (the quantity) plus something between 3 and 5 ohms works just as well for silicon as for germanium. After all, what you are after is a worst case approximation of what this might be so you can take its effect into account.

Now, how about the u characteristic. This (remember) is the greatest possible voltage gain we could possibly expect (working into an infinite load) between the signal-in (on the base) (grid) to the signal-out (on the collector) (plate) of this device. With the triode vacuum tube it was the ratio of the plate resistance to the cathode resistance, and this is the same. Just divide that 100K collector resistance by the value of emitter resistance (for 1 ma of emitter current it was 25 ohms) to get an estimated u equals 4,000. At higher currents, remember that (r_p), the collector resistance, goes down, and this tends to stabilize the value of u a little bit; but never let the size of this number bother you. It is realistic. However, you'll very seldom—if ever—find a transistor working into a load resistance high enough to even start to challenge its value of collector resistance. (In the next chapter you'll find out more about this subject and this situation.)

So transistors have characteristics just like a vacuum tube! By rule of thumb, they are easily predictable and you shouldn't hesitate to use these rules when you analyze a circuit. The most important thing of the bunch is to remember the equation for Transresistance:

$$r_{tr} = \frac{26}{I_e \text{ (in milliamps)}} + \text{(something between 3 and 5 ohms)}.$$

This is the resistance seen looking into the emitter of either an NPN or PNP transistor made of either germanium or silicon.

The collector resistance may be estimated at between 50K and 100K ohms, and u is your estimated value of collector resistance divided by transresistance.

Since NPN and PNP transistors are solid-state devices that have a linear output (the same change in the output for the same change of the input at successive levels of operation) for a successive series of changes of input CURRENT (not voltage, as with vacuum tubes), we have another characteristic with these devices that we didn't have with the triode vacuum tube. This is the ratio of collector current change to the base current change that caused it. It is an amplification factor similar to u, but it is a **current** ratio, not a voltage ratio. Transistors are designed so that a small change of base current will produce a much larger change in collector current. This small-signal output current change (change in I_c) divided by the input change of current (change in I_b) is commonly called "BETA." However, in the "h" parameter system, there are at least three standard forms of this measurement:

Small-Signal Beta
$$= \frac{\text{change of collector current}}{\text{small change of base current}} = hfe$$

Large-Signal Beta
$$= \frac{\text{change of collector current}}{\text{large change of base current}} = HFE$$

$$\text{DC Beta} = \frac{\text{DC Collector current}}{\text{DC Base current}} = hFE$$

It is the small-signal beta (hfe) that we are interested in as being the one of most use to us. And once again, this parameter has such a broad range in most types of transistors that we can use a standard "Worst Case Estimate of 50" for all transistors in regular use. There are special cases and special transistors that don't fit this situation, but they are usually given adequate publicity so that there is no doubt in your mind that it is a special case.

You will recall the first part of this chapter where we discussed the fact that the u of the vacuum tube was the factor which multiplied any unbypassed resistance in the cathode circuit when it was viewed from the plate through the tube, and that this same factor divided any load resistance connected to the plate when it was viewed through the tube in the other direction, from the cathode. Well, the u of a transistor does the same thing. Any unbypassed (without a capacitor connected to the emitter) resistance connected to the emitter is magnified tremendously when viewed from the collector, and any collector load resistor is reduced to practically nothing when it is viewed from the emitter (due to the very high u of a transistor).

However, beta is a more usable value (worst case estimate — near 50). The smaller this type of number (characteristic) is, the more we have to take it into consideration. And, beta operates in a similar fashion to u but relative to a little bit different viewpoint. Beta is the factor that is used between base resistors and emitter resistors. Let's see what I mean.

(1) Assume that we have a transistor with 5 ma going through it.

   (note) — 26 divided by 5 gives a bit more than 5 ohms—add almost 5 ohms more and get an r_{tr} equals 10 ohms.

(2) Assume a Beta (collector current divided by base current) of 9.

   (note) — With 9 units of current in the collector and 1 in the base, we must have 10 units of current in the emitter.

(3) Assume that r_{tr} is made up of (Beta + 1) resistor; that is, 1 resistor for each unit of current that flows in the emitter. They are all in parallel and each one is the same value of resistance as the next.

   (note) — r_{tr} must be made up of 10 resistors. One of these resistors must connect directly to the base while the other nine connect to the collector. All ten must feel any voltage that is

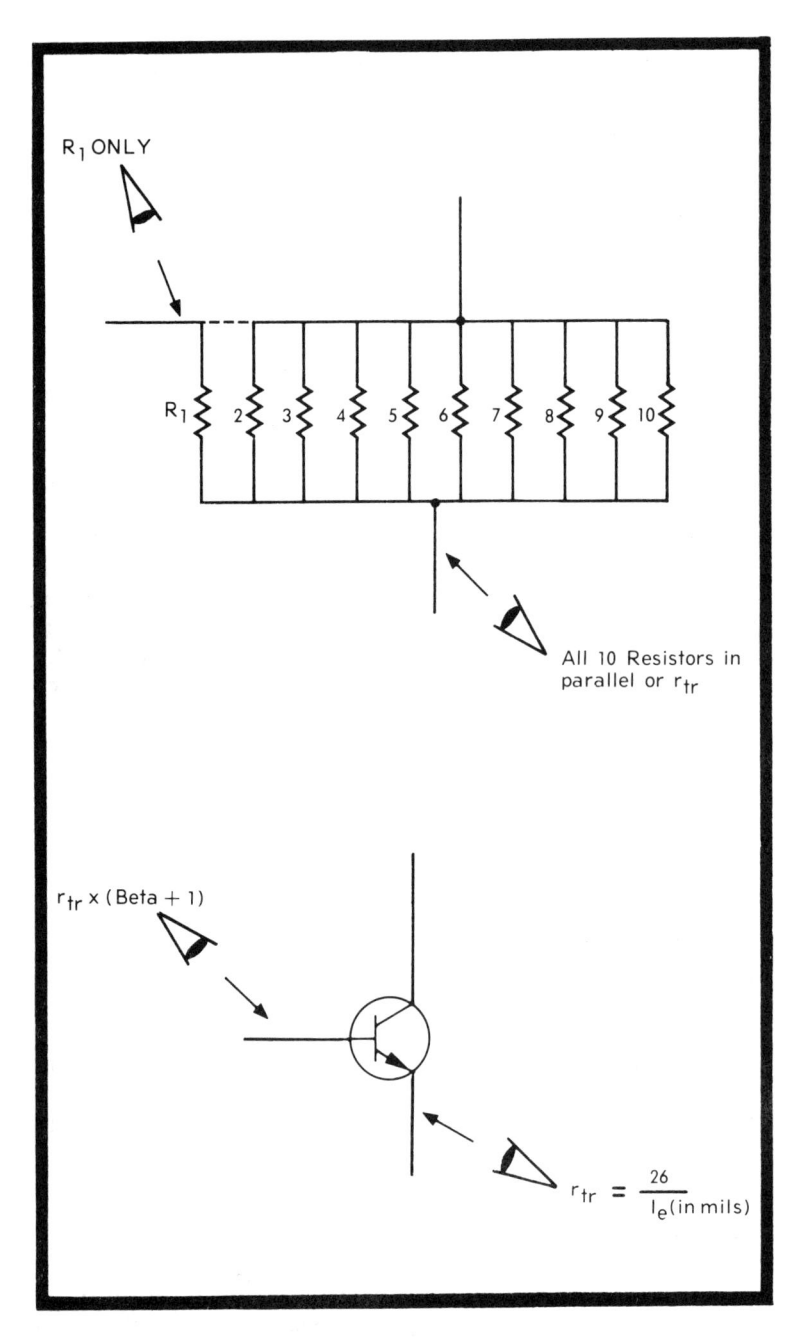

Fig. 5-13. Finding base to emitter resistance by using "transresistance."

178

applied to the base with respect to the emitter. See Fig. 5-13. Now, the question is, "How much resistance do I have between the base and the emitter?" In other words, how much resistance do I see when I look into the base?

If you need to, refer back to Chapter 1 for finding the total resistance of a parallel group of resistors. R_{total} was equal to the reciprocal of the sum of the reciprocals (remember?). OK—so let's write our equation like this:

$$r_{tr} = 10 \text{ ohms} = \frac{1}{\frac{1}{r_1} + \frac{1}{r_2} + \frac{1}{r_3} + \frac{1}{r_4} + \frac{1}{r_5} + \frac{1}{r_6} + \frac{1}{r_7} + \frac{1}{r_8} + \frac{1}{r_9} + \frac{1}{r_{10}}}$$

All of these resistors are the same, so we already have a common denominator and should be able to write:

$$r_{tr} = 10 \text{ ohms} = \frac{1}{\frac{(10)}{(r_1)}} = \frac{r_1}{10}.$$

Remember, when dividing by a fraction, we invert it and multiply with it. This then tells us that r_{tr} is equal to the base-to-emitter resistance divided by (beta + 1). Or (of more use to us yet), we can transpose the value (beta + 1) and get this:

The base-to-emitter resistance equals (beta + 1) x r_{tr}.

The answer to the problem I posed is a simple thing now. Beta was 9 and r_{tr} was 10 ohms. The base-to-emitter resistance must equal 10 x 10 or 100 ohms.

Note also that if we have an unbypassed resistor connected to the emitter of our transistor and view it from the base of this device, beta +1 will not only magnify r_{tr} but will magnify this other resistor also. And looking into the base, we will see the sum of the two magnified resistances. See Fig. 5-14.

Of course, this factor (beta + 1) operates in the opposite manner when looking into the emitter of a transistor with regard to any resistance connected to the base. That is, if we had 100 ohms connected in series with the base and looked into the emitter of our transistor with its beta of 9, we would see 10 ohms more resistance than we started with.

What you end up having to remember is that u is the factor used between plate (collector) and cathode (emitter), while

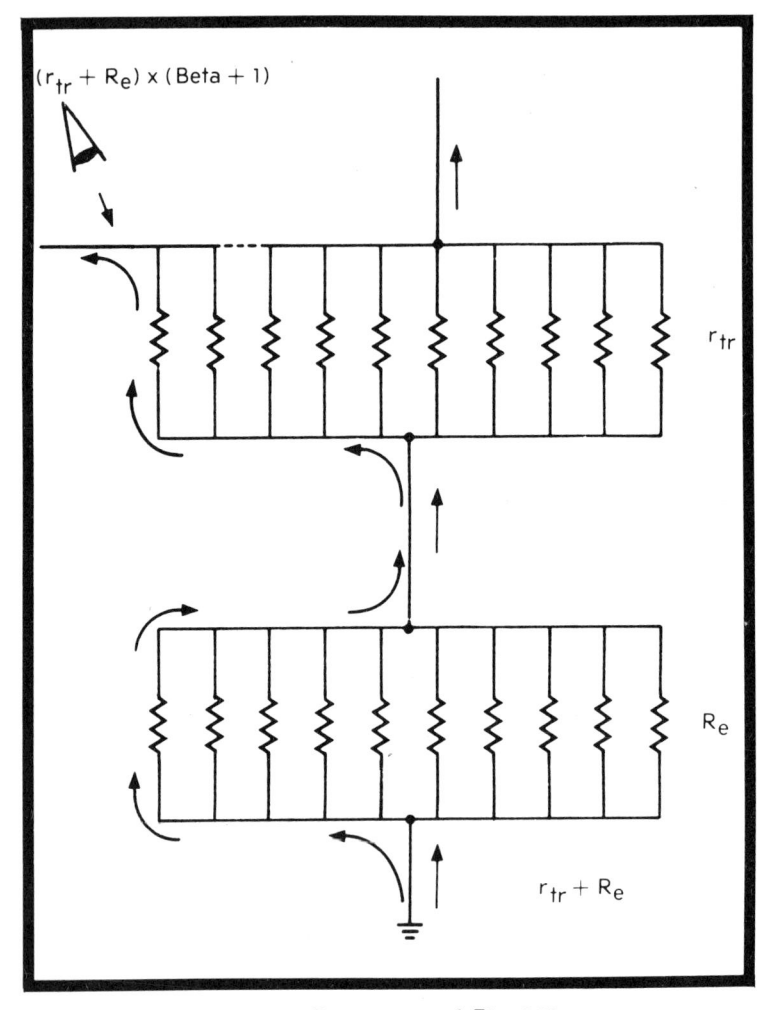

Fig. 5-14. Further use of Fig. 5-13.

beta + 1 is the factor used between base and emitter (not applicable to vacuum tubes).

It is interesting to note that although (beta + 1) is used as a divider when viewing series resistance in the base circuit from the emitter, the inverse happens when viewing capacitance in the base circuit from the emitter. Since the emitter accepts (beta + 1) units of current for each single unit of current that goes to the base, the capacitor changes voltage as though it were (beta + 1) times greater in size than it actually is. This system of magnifying the value of a capacitor can be taken advantage of sometimes where space does not allow a large size and value capacitor to be used. See Fig. 5-15 for an example. If you don't understand this paragraph and this illustration, you had better review before proceeding!

(The remainder of this chapter is paraphrased material that is reprinted by permission of Tektronix, Inc., Beaverton, Oregon. It appeared in SERVICE SCOPE, Feb. 1966.)

The Field Effect Transistor (FET) is a little bit different type of solid-state transistor; its operation more closely parallels the operation of the vacuum tube than the NPN or PNP types of transistors. You might even go so far as to say that the major current that goes from the effective cathode to the effective plate is 90 degrees out of phase with that of the more common NPN or PNP types of transistors. See Fig. 5-16 and let me show you what I mean.

Drawings "A" and "B" of Fig. 5-16 shows a PNP semiconductor structure. In "A," we have a PNP transistor in which the holes (+ charges) travel from the emitter up through the N material base region and, because there are not enough electrons (— charges) in the base region, go on to the collector as a minority current (the current that flows in a reverse-biased junction). Voltage-wise, this says that the base of this transistor will be about —0.6 volt with respect to the

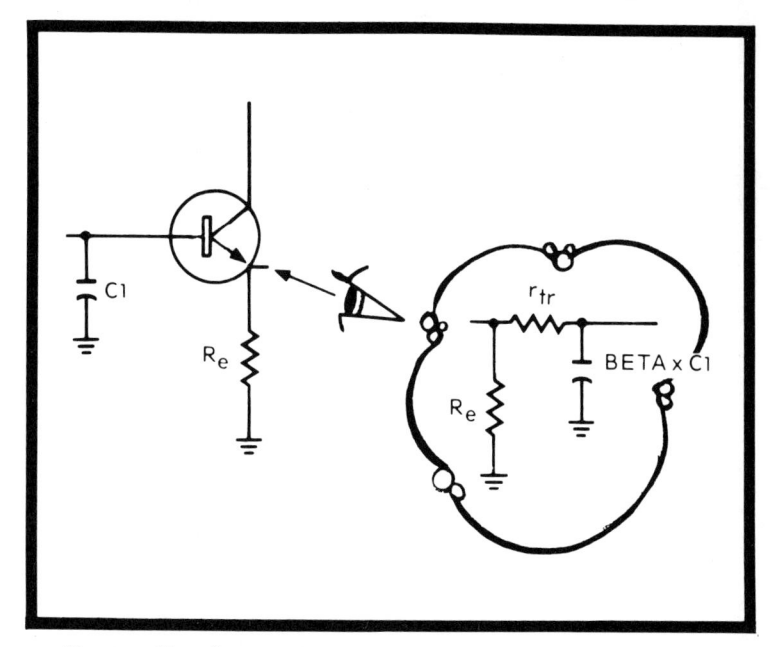

Fig. 5-15. How (beta + 1) can magnify the value of a capacitance.

"A" — A PNP TRANSISTOR

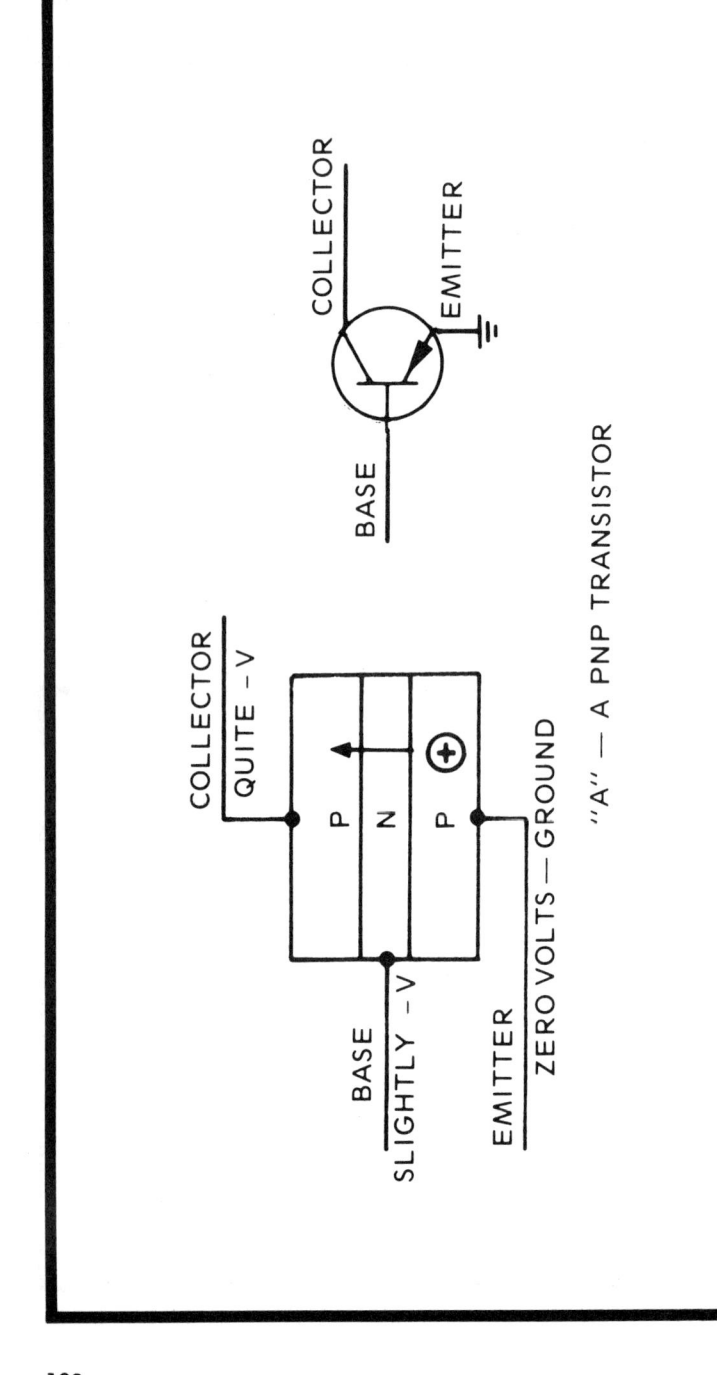

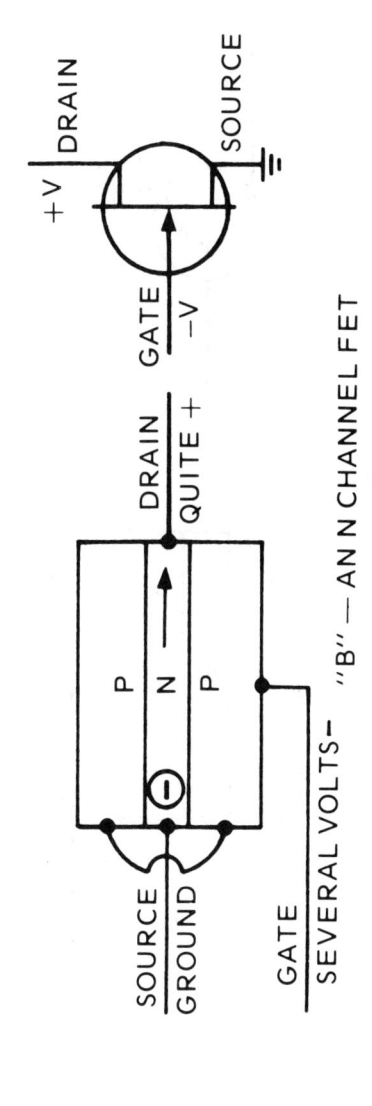

Fig. 5-16. Field effect transistors vs conventional (part 1 of 2 parts).

"C" — AN NPN TRANSISTOR

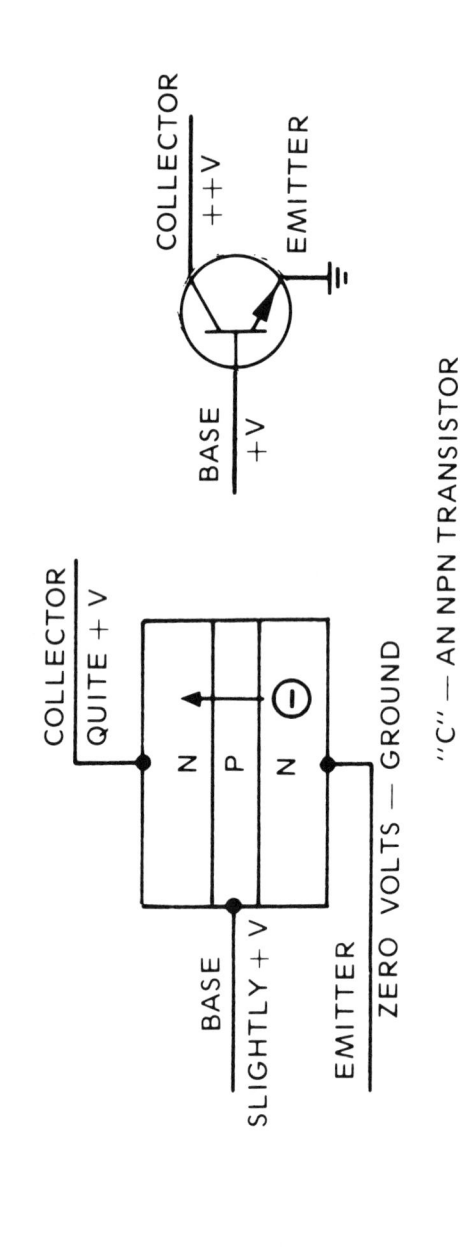

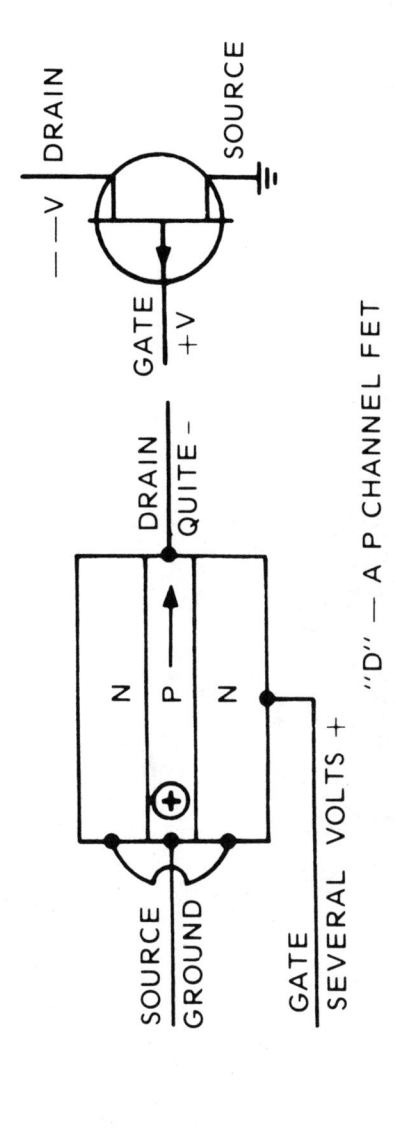

Fig. 5-16. Field effect transistors vs conventional (part 2 of 2 parts).

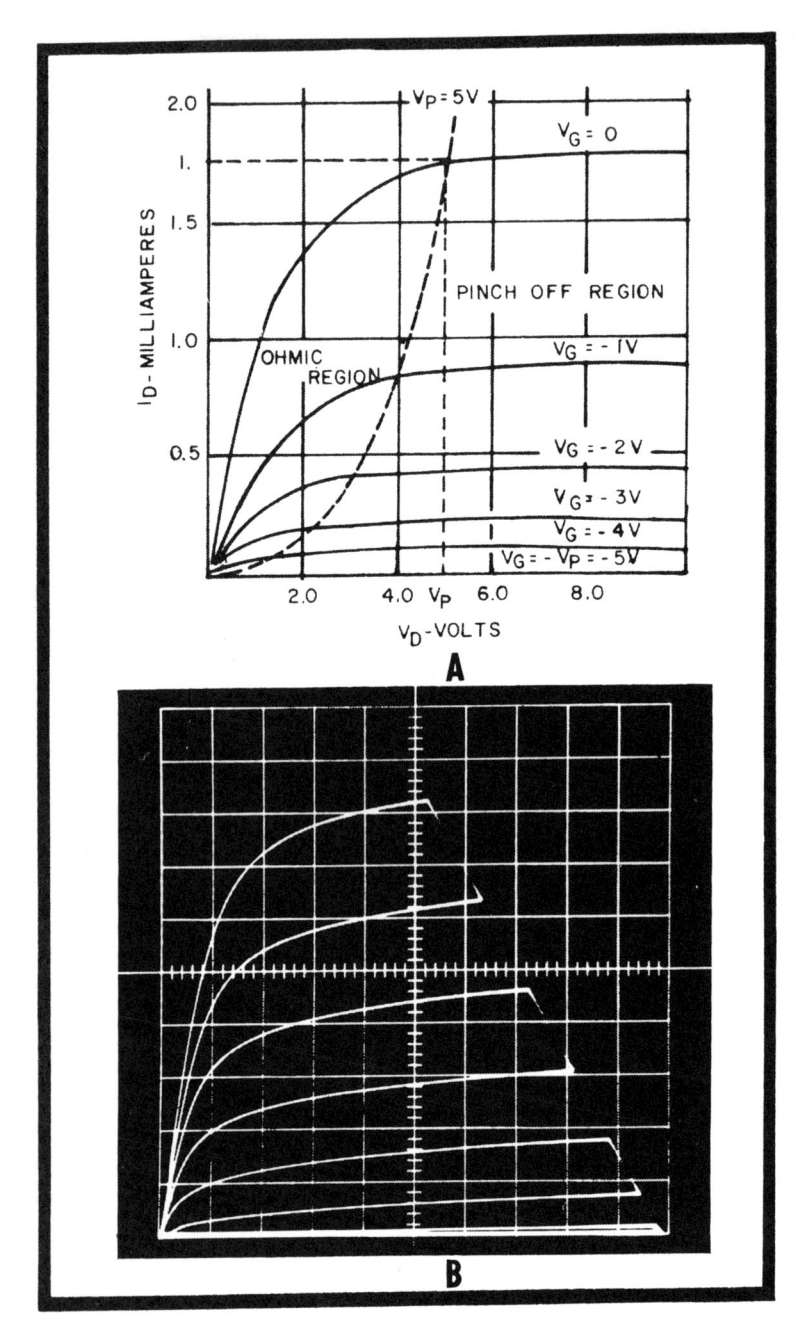

Fig. 5-17. FET characteristic graphs. (Courtesy Tektronix, Inc.)

emitter and the collector will be much more negative to attract the + charges that are coming from the emitter, and our transistor is turned on.

In "B," we have what is called an N-Channel FET. It is made up of the same old PNP sandwich, but it is hooked up differently. The two sections of P material are interconnected and operate as the effective grid of this device. The application of a + voltage to the effective plate of this device results in the right-hand part of this now single-junction device being highly reverse-biased, which cuts down drastically on the capability of the right-hand end of the N-channel to conduct current (by building a depletion region into the channel). The conducting area of the channel that is left becomes saturated and we are left with a pentode-like curve that does not increase much in current as voltage increases. If the effective grid of this device is connected to the effective cathode, we will have the zero-bias curve of a normally ON device. (We had the same thing in a Field Effect Diode, remember?) However, with this device we can disconnect the effective grid from the cathode and begin to bias it more negatively with respect to the cathode this time (reverse-biasing the remainder of the single junction). This will cut down further on the conductivity of the channel (N material with current flowing from left to right), and it will saturate at a much lower level of current. If we continue to drive the effective grid of this N-Channel FET to a more and more negative level of voltage, we will cut off the conductivity of the channel entirely.

It is interesting to note (and this is the place to do it) that the level of grid bias necessary to cut this device off is the same magnitude of voltage as that reached by the effective plate where saturation occurred on the zero bias curve. This is called Pinch-Off Voltage, and is labelled (V_p). Study Fig. 5-17.

Drawings "C" and "D" of Fig. 5-16 compare the NPN structures of the NPN Transistor and the P-Channel FET.

FET manufacturers have settled on a new series of names for the three basic leads of this device; so, once again we encounter a change in terminology. Fig. 5-18 compares the familiar vacuum-tube triode, a conventional PNP transistor and a FET to show this change in basic-lead terminology.

As with conventional transistors, which are represented by two types of devices (NPN and PNP), the FET is also represented by two basic types of devices. These are designated the N-Channel and the P-Channel types and each of these is further subdivided into three types of active devices, depending upon how they are made (this also determines how

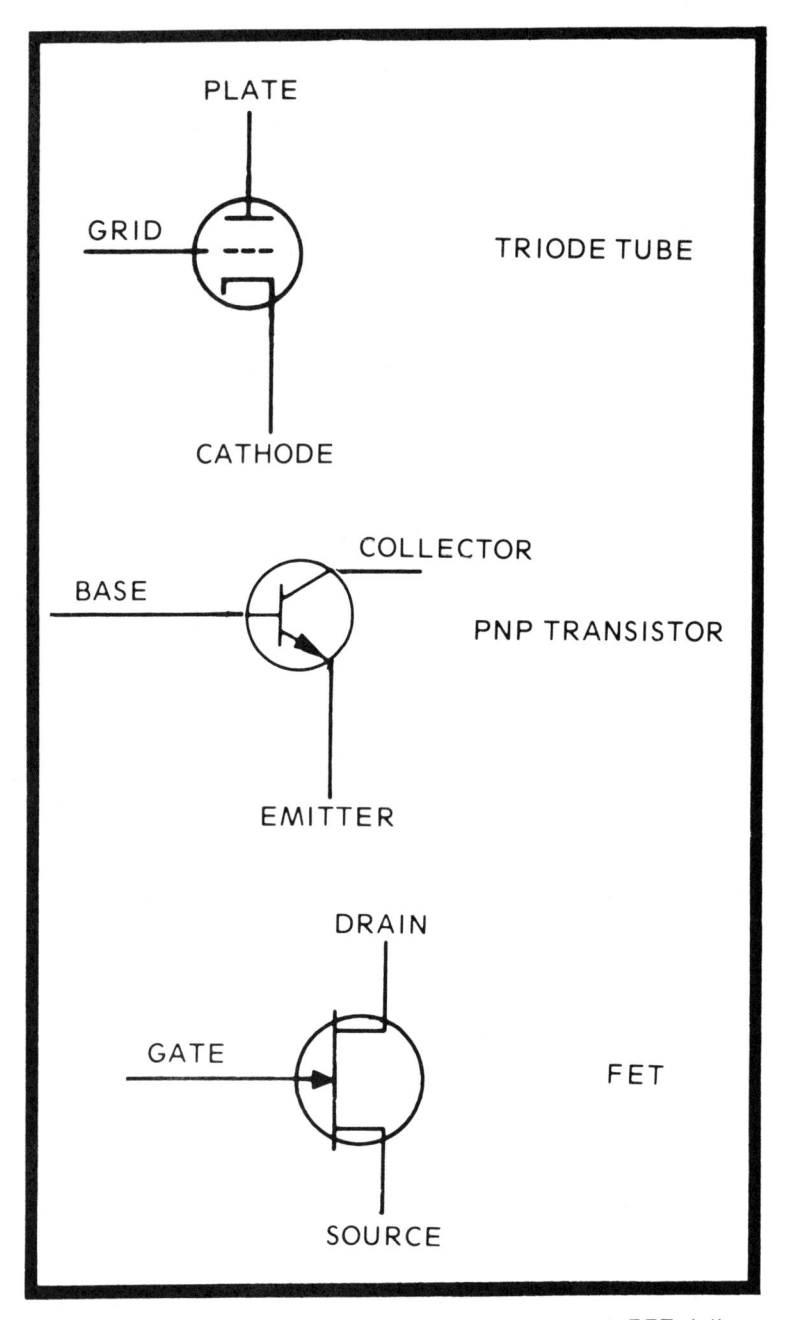

Fig. 5-18. Symbol comparison of tube, transistor, and FET (all as "triodes").

they should be biased). You see, some operate as I have described, with negative bias (reverse bias) only. Others are so constructed as to operate with a limited amount of either reverse or forward bias of the gate with respect to the source, and still others are designed to operate with forward bias of the gate-source junction only. The manufacturers changed their way of telling you this also:

Reverse bias of the G-S junction = Depletion Mode of operation

Reverse or Forward bias = Depletion or Enhanced Mode

Forward bias of the G-S junction = Enhanced Mode only

Fig. 5-19 shows a little more accurate picture of the structure of the FET and attempts to show the regions of physical measurement that are critical to the characteristics of this device.

The characteristics that we are interested in are much better represented in the curves shown as parts of Fig. 5-20 and 5-21 (the slope of the saturated portion of the Drain waveforms and the common slope of the reverse-operated Source Waveforms). The slope of the Drain waveform is usually expressed as an output conductance parameter called G_{oss}, and is a value usually in the neighborhood of 10 micromhos or less. If we take the reciprocal of this rating (1/10 equals 0.1 and the reciprocal of micro is Meg with the reciprocal of Mhos as Ohms), we get 0.1 megohms as its characteristic r_p or—in this case—Drain resistance (r_D). The common slope of the reverse-operated Source waveforms is the characteristic Source resistance (cathode resistance or reciprocal of G_m). This is usually expressed again as a Conductance (like a G_m characteristic), but it is NOT called G_m. It is termed G_{fss}, and is usually expressed in so many thousands of micromhos. Let's say that this one had a G_{fss} equals 5,000 umhos. We could use this in just the same fashion as we do a vacuum tube G_m, or we could take the reciprocal of this number to obtain a value of Source resistance (cathode resistance) for this device of 0.2K (200 ohms).

Of course, the Drain resistance divided by the Source resistance can also be thought of as the u of the device, but the manufacturers seldom point this out.

What lead is which in the actual device you might hold in your hand is just about anybody's guess. There is absolutely no standard by which these things are based and—believe me—they have turned them every way but loose. All that I can say about this is that you have to look it up either in the manufacturer's specification sheets or in a reference manual. They are making them in all kinds of shapes and sizes and they

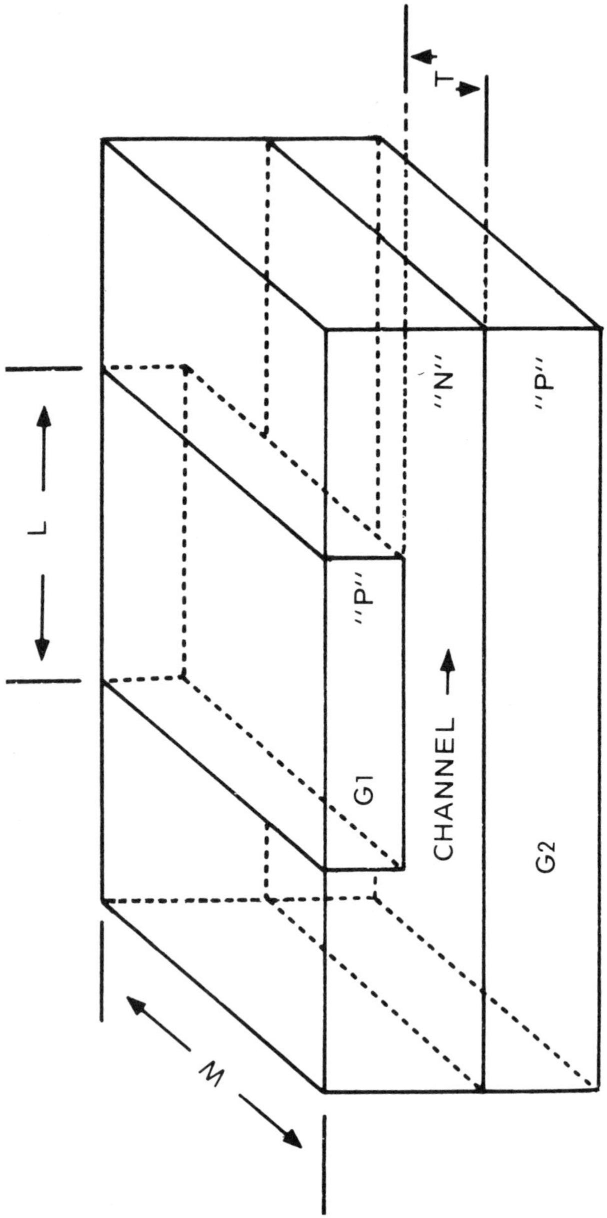

T
L
"N"
"P"
"P"
G1
CHANNEL →
G2
W

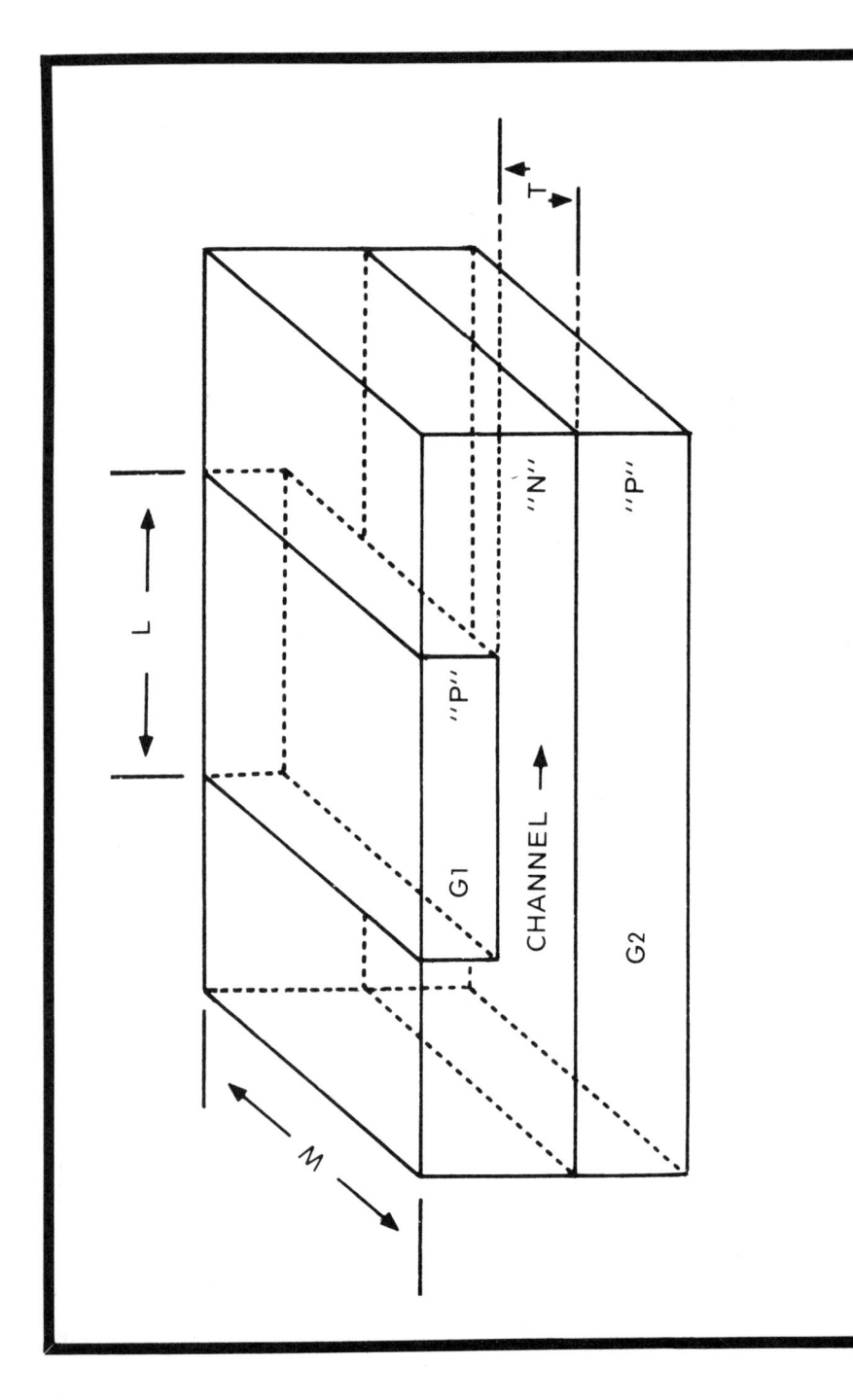

$$Gm :: \frac{k\ I_{dss}}{V_P} :: \frac{T\ W\ n}{L}$$

$$V_P :: T^2$$

$$I_{dss} :: T^3 \text{ \& also } \frac{W}{L}$$

Each of these characteristics is proportional to the "doping level" equals n.

$$I_{gss} :: WL$$

$$BV_{gss} :: \frac{1}{n}$$

$$C_{gs} + gd :: WL\ n^{1/2}$$

$$V^2_{noise} :: \frac{1}{L}$$

$$V_{noise\ hi\ freq} :: \frac{1}{Gm}$$

$$rp :: \frac{L}{T}$$

W	Width of channel
L	Length of channel
T	Thickness of channel
n	Doping level
I_{dss}	∅ bias drain current
I_{gss}	Gate Leakage Current

Fig. 5-19. N-channel FET construction and characteristics.

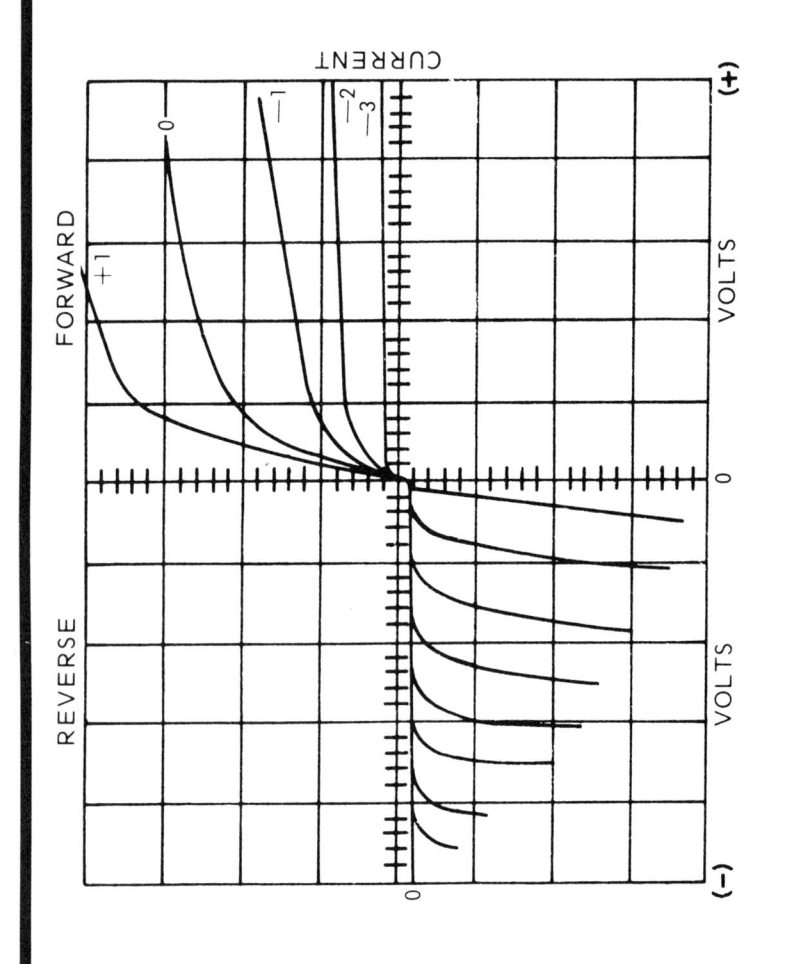

192

FIELD EFFECT TRANSISTOR

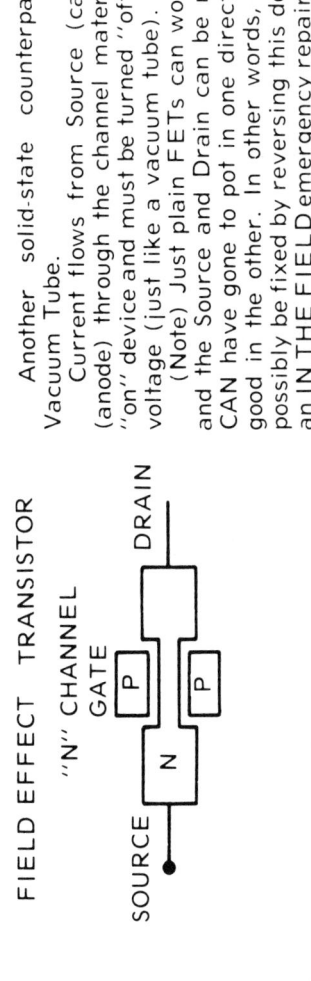

"N" CHANNEL

Another solid-state counterpart to the Pentode Vacuum Tube.

Current flows from Source (cathode) to the Drain (anode) through the channel material. It is a normally "on" device and must be turned "off" by the applied bias voltage (just like a vacuum tube).

(Note) Just plain FETs can work in either direction and the Source and Drain can be reversed. The curves CAN have gone to pot in one direction and be perfectly good in the other. In other words, a sick circuit might possibly be fixed by reversing this device in its socket. As an IN THE FIELD emergency repair procedure this might be OK, but it is NOT recommended as an in-the-factory procedure.

There is a crazy triode-like set of curves that shows up in the reverse voltage direction but these are not used. The Pentode-like family shows up only in the forward voltage direction for the "N" Channel device.

These devices have an exceptionally high r_p, a bit lower than vacuum tube Gm characteristic and the resultant u, which is a bit lower than the vacuum tube pentode has.

Fig. 5-20. N-channel FET as solid-state counterpart of pentode vacuum tube.

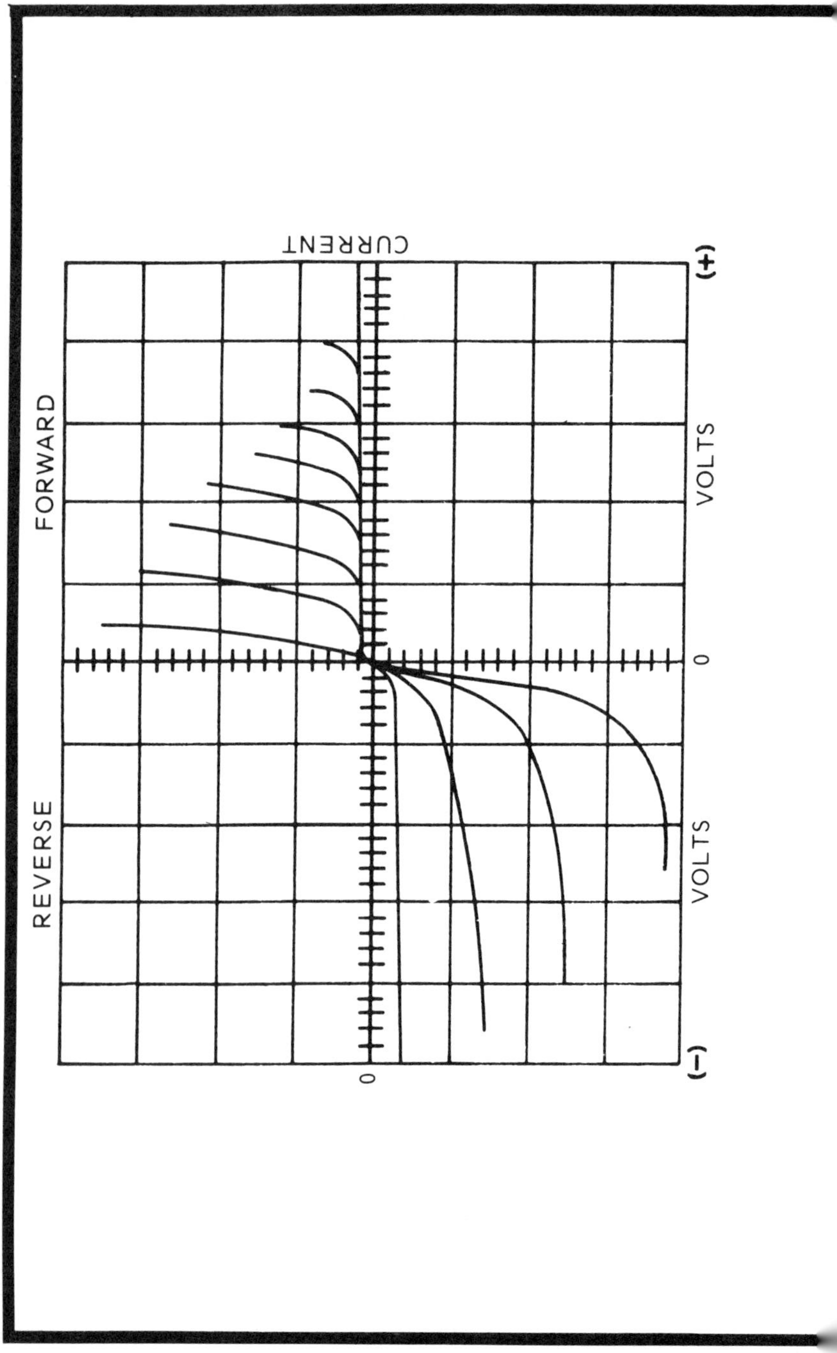
CURRENT
FORWARD
REVERSE
VOLTS
VOLTS
0
0
(+)
(−)

FIELD EFFECT TRANSISTOR

"P" CHANNEL

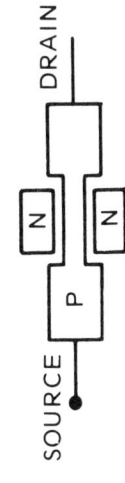

With this, the "P" Channel FET, we actually have the counterpart of the BACKWARD PENTODE since it actually conducts electron current from the equivalent Plate to the equivalent Cathode (Drain to Source).

The gate of this device is positively biased with voltage to turn this "normally on" device to its "off" condition.

(For further details refer to the N channel FET in Fig. 5-20.)

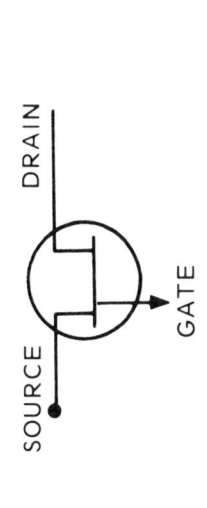

Fig. 5-21. P-channel FET as solid-state counterpart of backward pentode.

have hooked them up in every possible sequence you can dream up.

So—remember—all active devices can be thought of in the same type of terms. Each will have an effective plate resistance, an effective G_m and an effective mu. If we stick to these terms for bench work we'll find that we can make a lot more sense out of circuits that we never even saw before. We'll take a look at what to do with these measurements in the next chapter.

Amplifier Circuits

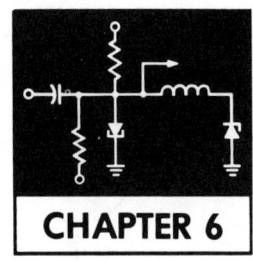

CHAPTER 6

The whole purpose of this chapter will be to get you to learn a simplified system of thinking about an amplifier circuit such that you will be able to figure its input resistance, output resistance, and voltage gain with the least trouble.

We spent a whole chapter showing you that all the different amplifying devices could be thought of as having the same type of characteristics as the old vacuum tube. That is, each has a (u) mu, a G_m, and an r_p type characteristic. We also found that each had a characteristic that could be thought of as its cathode resistance. We found that the mu could be thought of as doing the same thing in all cases as well; it divided R_L when we viewed it from the cathode, and it multipled R_K when we viewed it from the plate. And these ideas become more and more important as we begin to look at these active devices in circuits.

For simplicity's sake, we will start out using a triode vacuum tube in a plate-loaded stage so we can find out where the basic voltage gain formulas come from (see Fig. 6-1).

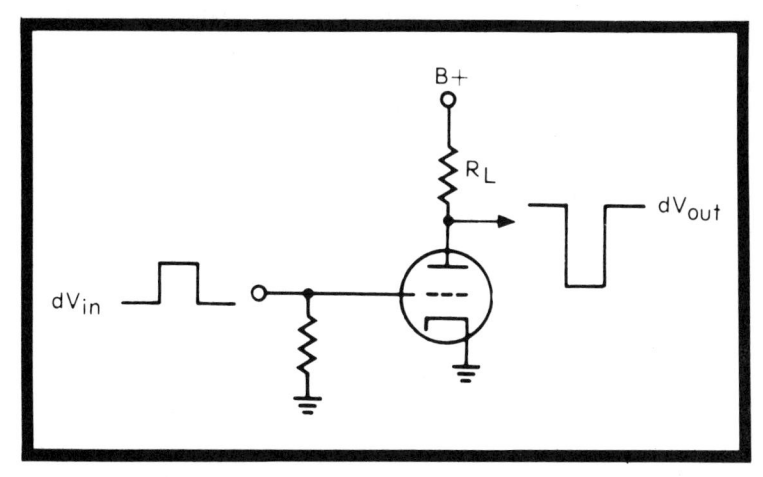

Fig. 6-1. Triode vacuum tube voltage gain measurement.

197

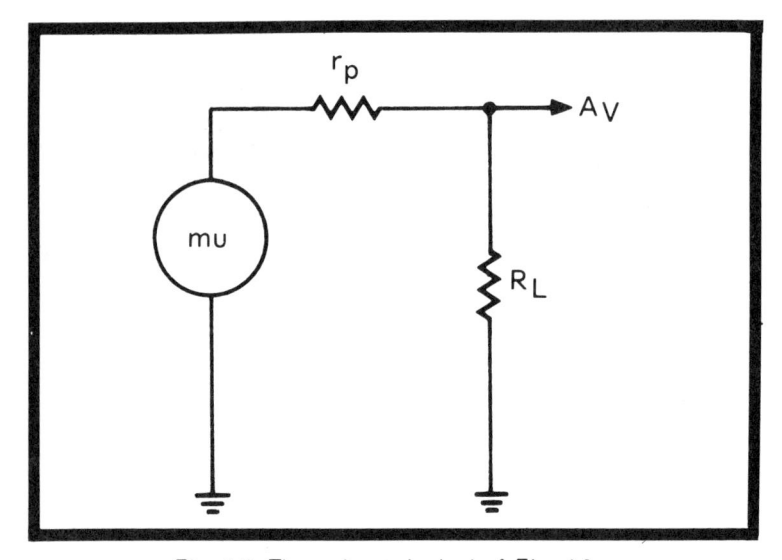

Fig. 6-2. Thevenin equivalent of Fig. 6-1.

Now, quite obviously if I change the voltage on the grid of this amplifier, I will change the amount of current going through it and thus the plate will change voltage too. The question we wish to answer is, "Just how much bigger will the change of voltage on the plate be than that which caused it to begin with?" This is what we mean by Voltage Gain (A_V). If I move the grid 1 volt, then the plate should move a certain number of times more than 1 volt. If I move the grid a half volt, then the plate should move the same number of times greater than the half volt that caused it. This number that stays the same is known as the Voltage Gain of the circuit. It is primarily determined by dividing the change of voltage on the plate by the change of voltage on the grid that caused it. In other words, A_V equals dV_{out}/dV_{in}. This figure will remain essentially constant unless we cut off the tube or drive it to saturation. After all, the plate can't move any further in the negative direction than its own cathode, nor can it move in the positive direction any further than its own B supply voltage. You see, if the A_V of the stage times the input voltage change contemplated indicates that the plate will have to go beyond these limits, it just isn't going to happen. So, what we are talking about is a small signal on the grid (input)—something usually less than a 1-volt change, and for the moment we'll leave it at that.

Do you remember what a Thevenin Equivalent Circuit was? It was a voltage generator (source of voltage) and a series resistance—wasn't it? OK—now—what do you think my (u) is? Isn't it a triggered voltage generator? It always kicks out—within the vacuum tube—a signal so many times greater than the input signal. And if we are going to use this in a Thevenin Equivalent of the vacuum tube, it will be our voltage generator. But the Thevenin Equivalent had a series resistor. Well, what about r_p? This should show up in series with it, don't you think? Then R_L could be drawn from the open end of r_p to the AC ground point represented by B+. (An AC ground point is a given level of voltage—different than ground—that will **not** change voltage.) See Fig. 6-2. Now we have a simple picture of our problem. We will have mu number of volts applied to the top of two resistors in series: r_p on top and R_L on the bottom to ground. And, we want to know how many volts change we will see at the top of R_L. It's as simple as that. Now, in Fig. 6-3, we have a two-resistor voltage divider and R_L divided by R_L plus r_p will multiply the voltage applied (u) to

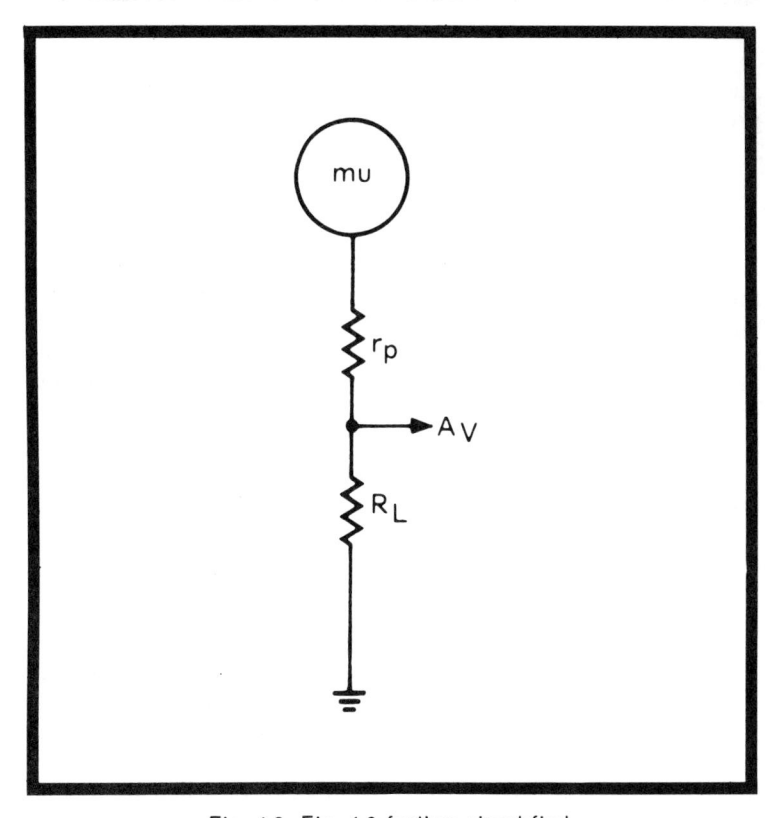

Fig. 6-3. Fig. 6-2 further simplified.

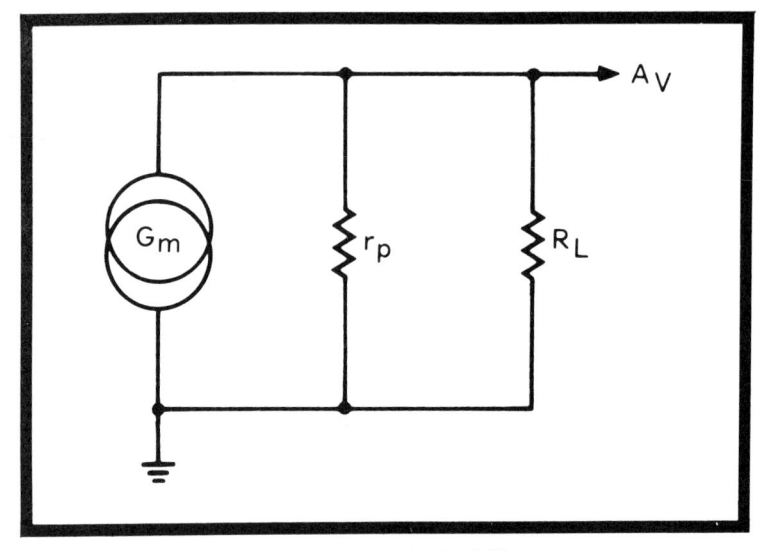

Fig. 6-4. Norton equivalent of Fig. 6-1.

give us the standard formula for the voltage gain of a grounded-cathode plate-loaded triode stage.

$$A_V = \frac{u \times R_L}{r_p + R_L}$$

Let's carry this investigation of this formula a few steps further along the way to see if we can't gain a little bit greater understanding of what we have here. In the preceding chapter we discovered that u was equal to r_p divided by r_K. We also found that r_K was one over G_m and that thus we could also say that u was equal to $G_m \times r_p$.

Let's substitute for mu in our formula for A_V.

i.e., $$A_V = \frac{G_m \times r_p \times R_L}{r_p + R_L}$$ OK? Good.

Now let's think of it like this: $$A_V = \frac{G_m \times r_p \times R_L}{r_p + R_L}$$

In other words, we now have G_m times the product over the sum of r_p and R_L. Well, the product over the sum of two resistors was their parallel combination—wasn't it? And, if we are now looking at an equivalent circuit with r_p and R_L seen in parallel, then it must be a Norton Equivalent Circuit with a current generator represented by G_m!

200

I wonder if this stands up to Ohm's Law. Let's find out. If G_m times dV_{in} is a change of current, then it must be correct. Ohm's Law says that I is equal to V divided by R. Well, our dV_{in} is a voltage for sure. Is G_m a resistance? Well, in a way, it is. G_m is equal to one over r_K, isn't it? And, if we multiply a fraction by a number, it is only the numerator (number on top) that gets multiplied—right? Well then, our current generator must put out a signal current equal to the change of voltage-in divided by r_K. What do you know; it does obey Ohm's Law! G_m times 1 volt change is a current and G_m times one is still G_m, so G_m can be our current generator in the Norton Equivalent of our vacuum tube circuit, Fig. 6-4.

Now then, there is a key phrase in that last paragraph that can make your analysis of an amplifier circuit easier than ever before. It's the one that says, "—our current generator must put out a signal current equal to the change of voltage-in divded by r_K." Within this statement is the idea that r_K is the limiting resistance that determines how much signal current will be generated by the change of voltage-in. It's the resistance that will meter out just so much current for so much voltage input to the circuit. And, on this basis, I am going to suggest that we call this resistance The Metering Resistance, or $R_{metering}$.

In the case of the grounded-cathode configuration (Fig. 6-1) that we are considering, this Metering Resistance can be thought of as simply the reciprocal of G_m (or just r_K) as long as we use R_0 (r_p and R_L in parallel) as the resistor we divide this metering resistance into to find the voltage gain of the stage.

But, you say, didn't you just say in the last chapter that R_L reflects into the cathode? Isn't this part of the Metering Resistance too?

And my answer is: "Yes, you can think of it this way IF you use only R_L in place of R_0 in determining the voltage gain of the stage!"

As I have told many of my students, you have to look through the active device ONCE in setting up the ratio of resistances which will give you the voltage gain of the stage. (Active device equals vacuum tube, transistor, or FET). Do this to determine either the metering resistance or the resistance that V_{out} will be seen across, but NOT both of them. In other words, you must take the characteristics of the active device into consideration but don't do it twice and over-compensate.

Let's change our circuit a little bit and put in some actual values so that I can explain this idea a little more fully. See

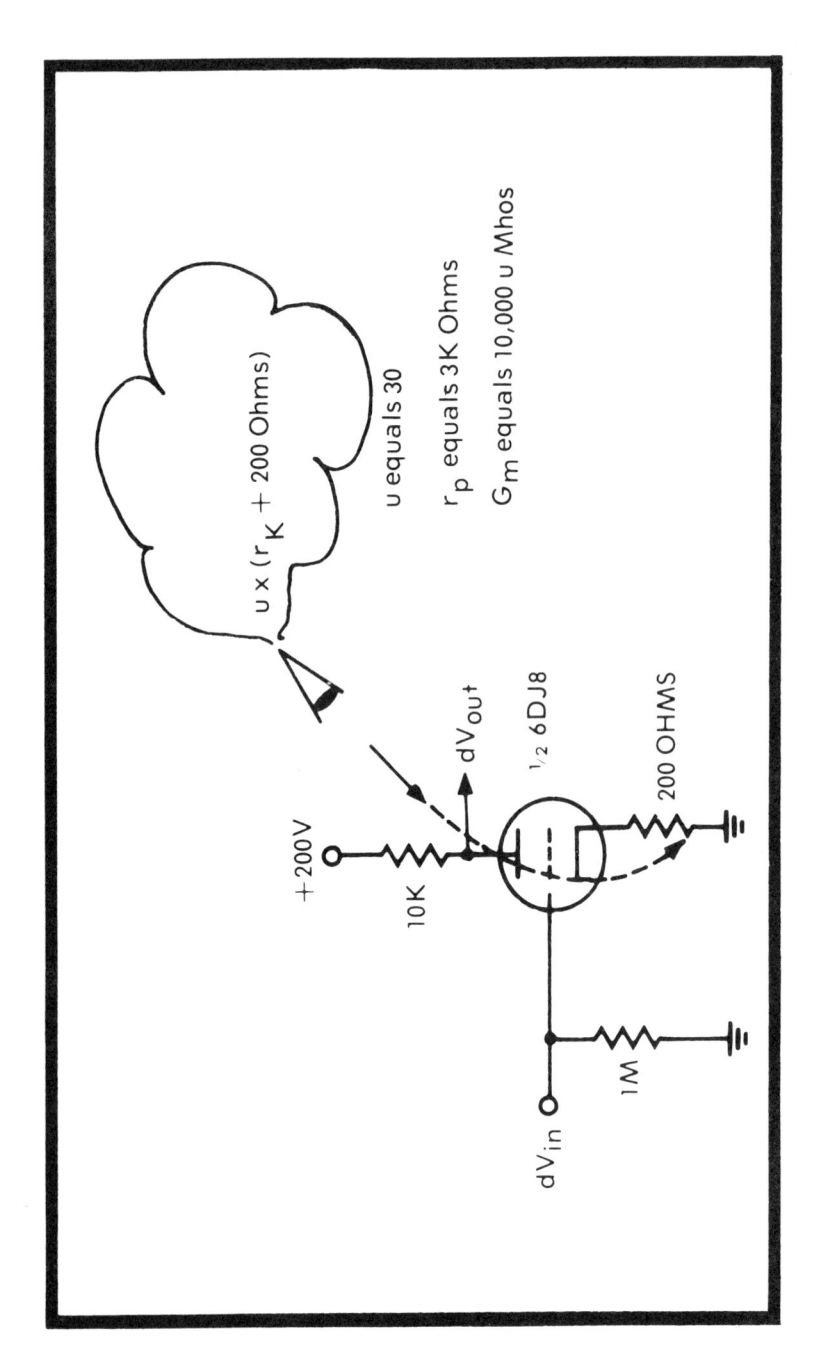

Fig. 6-5. Fig. 6-1 with actual values, figuring A_V.

Fig. 6-5. Note that there has been an external R_K of 200 ohms added to the circuit; the R_L has been given a 10K value and the grid has been referenced to zero volts (ground) through a 1 Meg resistor. Also note that B+ has been given a value of 200 volts. The tube chosen is a 6DJ8 (European ECC88) with the following approximate characteristics: mu is about 30, G_m is near 10,000 u mhos, and r_p is approximately 3K ohms. (I am using approximate values here for number simplicity.)

If you refer to the last chapter where I described how to find the operating point of a vacuum tube, you will find that the grid will be at zero volts (ground), the cathode will be at +2 volts with respect to ground, and the plate (output) will be at +100 volts with respect to ground. The tube will be conducting almost exactly 10 ma.

As for the Voltage Gain of the circuit (the ratio of change of voltage signal out divided by the change of voltage signal in that caused it), I would like to suggest that we use both systems indicated to see if we don't get the same answer each time. See Fig. 6-5.

First, we will use the formula, $$A_V = \frac{R_O}{R_{metering}}$$

In this case, $R_{metering}$ consists of the reciprocal of G_m plus the external 200 ohm cathode resistor since signal current flows from this resistor also because it is not bypassed by a large capacitor. So, one over 10,000 u mhos is 100 ohms. Therefore $R_{metering}$ equals 100 ohms + 200 ohms equals 300 ohms. Using this formula, R_O is made up of the 10K load resistor in parallel with the resistance seen looking down into the plate of the vacuum tube to ground. Consequently, we must have the 100 ohm (r_K) plus the 200 ohm (R_K) multiplied by mu, which, as I said, has a value of 30. Therefore, R_O equals 30 x (100 ohms + 200 ohms) or, 9K in parallel with the 10K load resistor. Their product over their sum is 9 x 10/9+10, or 90/19 which is 4.737K ohms. This is the number that we divide by $R_{metering}$ (300 ohms) to get A_V.

$$A_V = \frac{4.737K}{.300K} = \frac{47.37}{3} = 15.79.$$

The output signal voltage amplitude will be 15.79 times greater than the input signal voltage to this circuit then. Now look at Fig. 6-6, and let's see if we can get the same answer using the other system—that is, by reflecting the plate load (the 10K ohms resistance) into the cathode and using it as part of $R_{metering}$. We will have to divide 10K by 30 (mu) to deter-

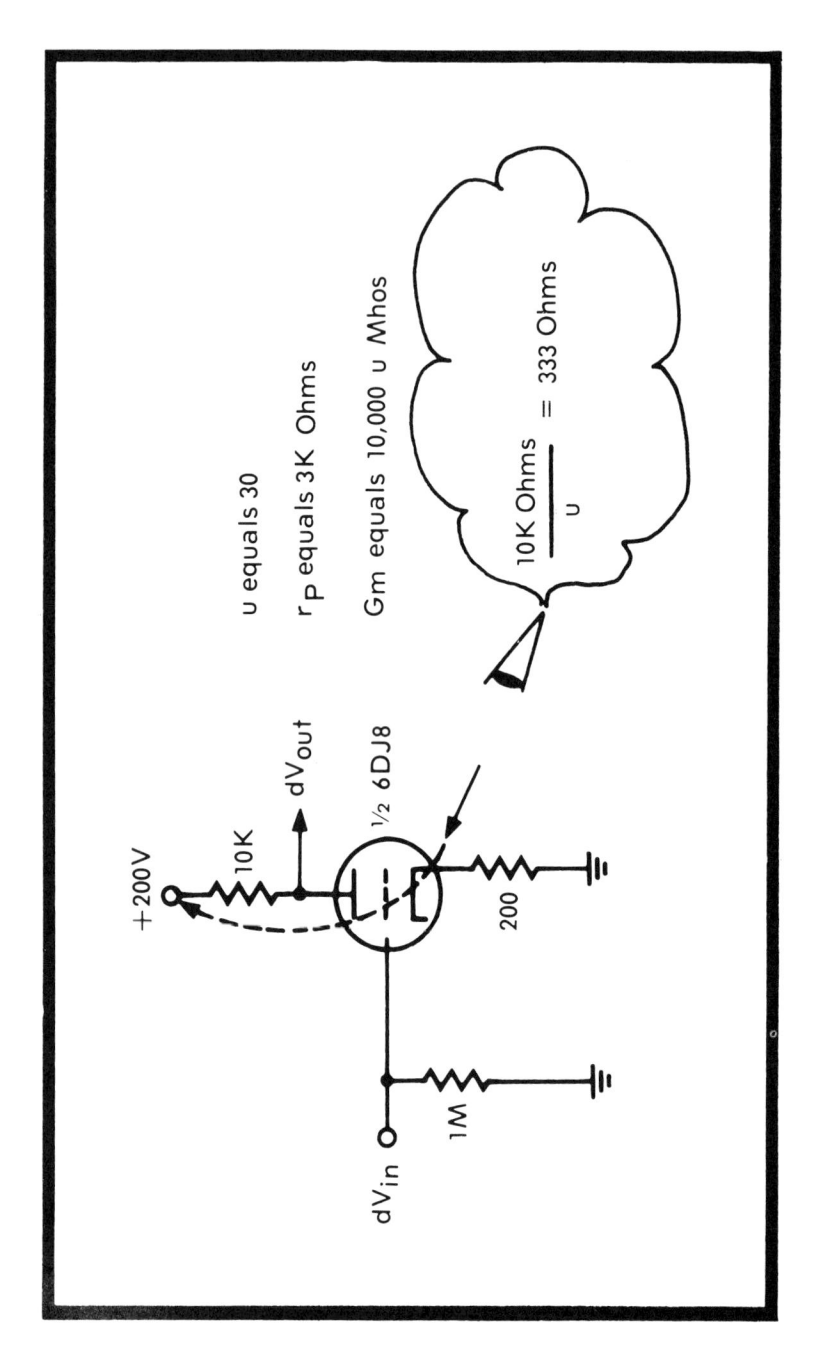

Fig. 6-6. Another viewpoint of Fig. 6-5.

mine that we will have to add an extra 333 ohms to the 300 ohms we already figured we had to get this form of $R_{metering}$. $R_{metering}$ (looking through the tube) is 633 ohms then. This is the form of $R_{metering}$ we use in the formula:

$$A_V = \frac{R_L}{R_{metering}}$$

that you wanted to find out about. Well, let's try it:

$$A_V \text{ should equal} \frac{10K \text{ ohms}}{.633K \text{ ohms}} = \frac{100}{6.33} = 15.79;$$

the same answer we got by the previous method. How about that!

On this basis then, I think it's fair to say that you can use either approach. Just don't mix them up: Look through the active device only once. This is true, no matter what the configuration or active device.

Looking at the other two basic circuits (the cathode follower and grounded-grid configurations) with our concept of $R_{metering}$ in mind, we find that they are equally simplified.

The cathode follower circuit (Fig. 6-7) will have a voltage gain relative to R_K divided by $(R_K + r_K)$. Quite obviously this will be less than 1. However, there is power gain since there is a much greater current-driving capability due to the low output resistance of this circuit consisting of r_K in parallel with R_K. Also, the larger that R_K is in comparison with r_K, the closer the voltage gain of this circuit approaches unity. However, we will not get any closer than mu divided by (mu + 1). Power, after all, is voltage times current and if the signal voltage stays just about the same while signal current increases due to the low impedance source it is coming from, then power is increased almost by the magnitude of the increase of signal current. I say "almost" because this circuit does lose a little bit of signal voltage. If we didn't lose any signal voltage amplitude then the power gain of this type of circuit would be the current gain of the stage.

The grounded-grid configuration also lends itself to this type of thinking just as well as the others (Fig. 6-8). We find it to our advantage here to again reflect the plate load resistance into the cathode. $R_{metering}$ here is determined by

$$\frac{R_L}{mu} + r_K$$

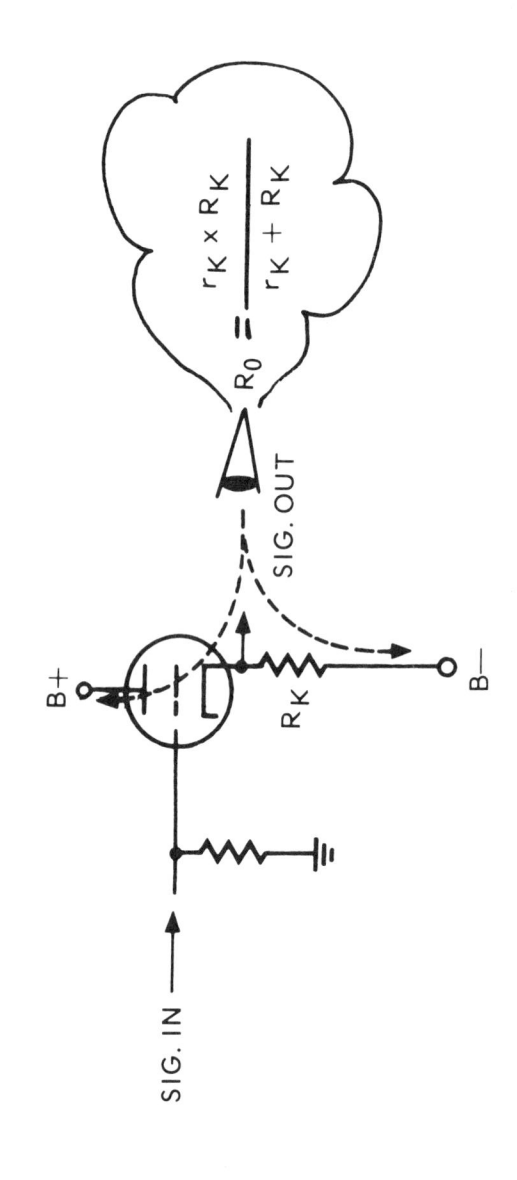

The Cathode Follower Circuit showing R_0.

Equivalent circuit for voltage gain purposes.

Remember—voltage gain is less than one.

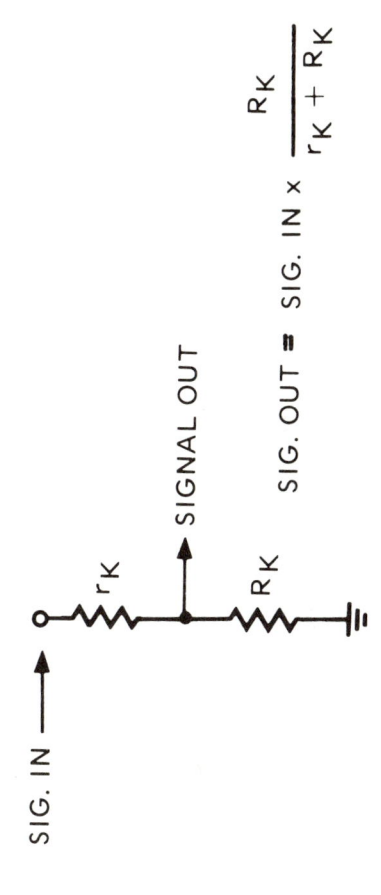

$$\text{SIG. OUT} = \text{SIG. IN} \times \frac{R_K}{r_K + R_K}$$

Fig. 6-7. Cathode follower analysis.

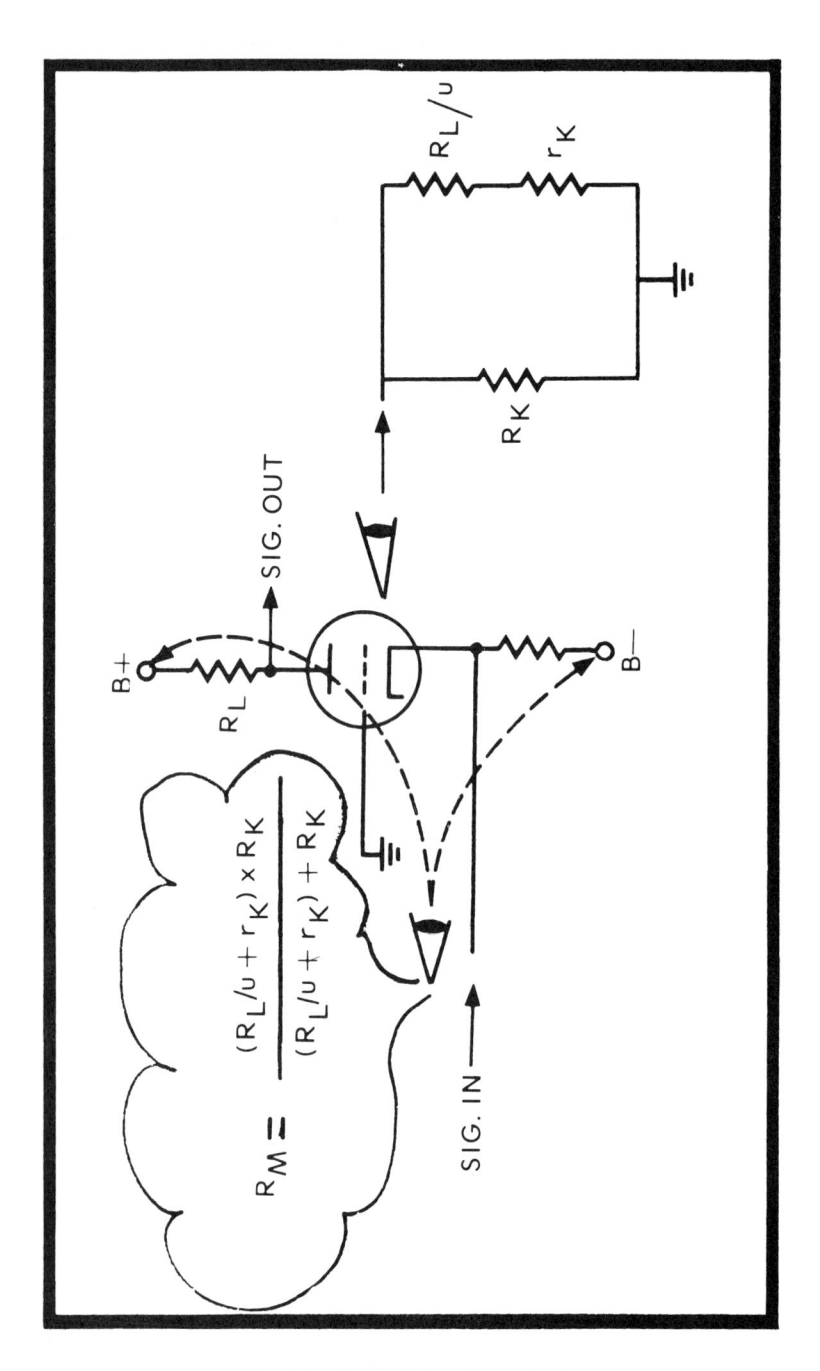

Fig. 6-8. Grounded grid analysis.

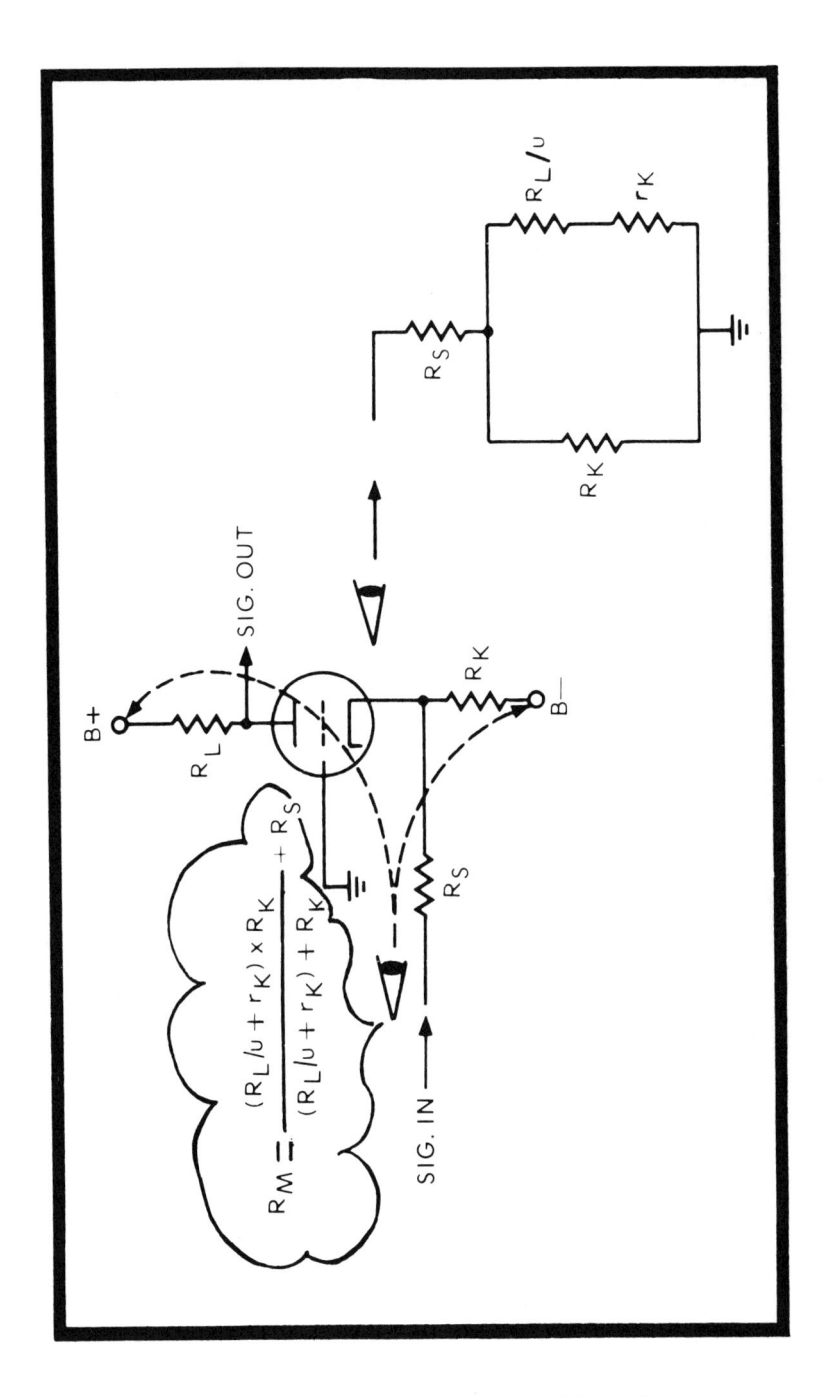

Fig. 6-9. Grounded grid with series grid resistor.

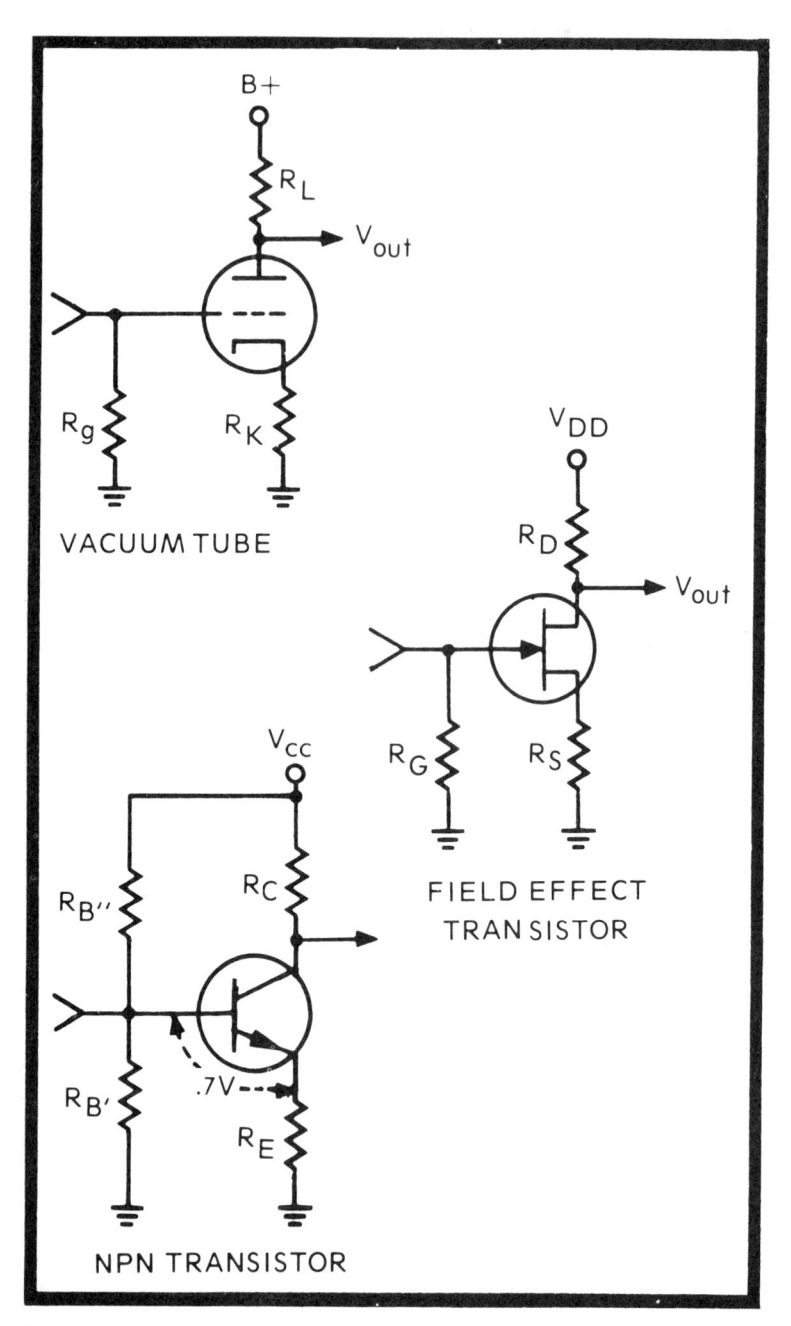

Fig. 6-10. Plate-loaded, drain-loaded, and collector-loaded amplifier circuits (comparing triode, FET, and NPN transistor).

figured in parallel with R_K. If R_K is more than ten times the size of "R_L over mu, plus r_K," most times I don't even bother figuring it in unless I'm working for extreme accuracy.

Many times (see Fig. 6-9) a grounded-grid stage is driven through a series resistor, and—of course—this resistor must be considered as part of the $R_{metering}$ of this stage.

A few pages back, I said that it didn't matter what the configuration nor what type of active device was used in the circuit. And that our $R_{metering}$ way of thinking would simplify the circuit analysis.

Well, we've explored the three most basic circuits. So now, let's take a look-see at what changing the active device in each type of circuit allows us to do.

Turn to Fig. 6-10. Here we have the plate-loaded vacuum tube type amplifier with its two counterparts using the Field Effect Transistor and a regular NPN transistor.

As you learned when we studied the parameters of the FET, we will probably need the G_{fss} (Y_{fss}) counterpart to the vacuum tube G_m and G_{oss} (Y_{oss} — as the case may be), the reciprocal of its plate resistance. G_{fss} divided by G_{oss} is mu for the device, and the reciprocal of G_{fss} is the source resistance (counterpart to cathode resistance). So, R_D divided by mu added to the reciprocal of G_{fss} plus R_S becomes the Metering Resistance which we divide into R_D to get the voltage gain (A_V) for this stage.

The transistor becomes a little more complicated, but we don't need any published specifications for it. We can solve for the A_V of its stage with a simple, straightforward approach. The first thing to do is find out where the base exists in voltage between ground and V_{cc}. We simply multiply V_{cc} by R_B, divided by the sum of $R_{B'}$ and $R_{B''}$. Second, subtract 0.7 volt from the voltage on the base. The remaining voltage is that which exists across R_E to give us our DC emitter current. This answer will usually be in milliamps. Divide this into 26 to get the emitter resistance (cathode resistance) (don't forget to make this a few ohms bigger than 26 over I_e in mills). This value of Transresistance we now add to R_E to get the metering resistance of this circuit and just divide it into R_C for the voltage gain of the stage. In this case, as you have probably noticed, we do not have to reflect any of R_C into the metering resistance. You see, the mu of a transistor is so large that if we divided R_D by its value, we'd probably be adding something between a fraction of an ohm to 2 ohms to an already estimated value of Transresistance. There is really no need to go to all this trouble. Your accuracy won't be that much greater. Likewise, there is little or no need to bother with estimating base current. You can if you wish. You could solve for the Thevenin Equivalent of the base divider circuit made up of R_B' and R_B'', put this resistance in series between the V_{oc} (open circuit voltage of the divider) and the base of the transistor which you can show as a simple diode (base-emitter junction) and then multiply R_E by 50 (an estimated value of Beta) between the emitter and ground. This circuit can be solved by a simple Ohm's Law approach for a slightly more accurate emitter-to-ground voltage and a slightly more accurate I_e, but it's hardly worthwhile. The larger R_E is with respect to Transresistance the less worthwhile this exact procedure becomes.

What about the other characteristics of these three circuits like Input Resistance or Output Resistance, you ask?

Well, that's easy to answer. Let's take Input Resistance first. With the vacuum tube, it'll be R_g since the grid isn't drawing any cathode current (of course, if the grid went positive with respect to the cathode, this input resistance would go down—but vacuum tubes don't usually operate that way).

As for the FET, it's Gate to Source junction is a reverse-biased diode which probably represents something in the neighborhood of 100K ohms. And with a very highly multiplied equivalent R_S in series with it, we can probably (99 times out of a 100) ignore its effect on the value of R_G. The input resistance is R_G then.

In the transistor circuit, of course, R_B' and R_B'' are seen in parallel. If this value is fairly large, it might pay to multiply Transresistance plus R_E by Beta (actually Beta + 1, but this is too picky again) and figure this in parallel with the two biasing resistors. However, this usually isn't necessary either.

As for Output Resistance, well, we go right back to the same old line of thought again.

In the vacuum tube stage of the plate-loaded amplifier, you will look back into V_{out} noting that you have two paths to ground or equivalent ground (B+; remember a Voltage Supply has extremely low back impedance, and its other end is at ground). This says that we will have two resistances in parallel. The first resistor is R_L to B+. The second resistor is made up of r_p (which is mu times r_K) in series with mu times R_K to ground. The product over the sum of these two resistances is R_O for this stage.

R_O for the Field Effect Transistor drain-loaded stage is figured in exactly the same fashion as the vacuum tube stage. You will have R_D in parallel with the reciprocal of G_{oss} added

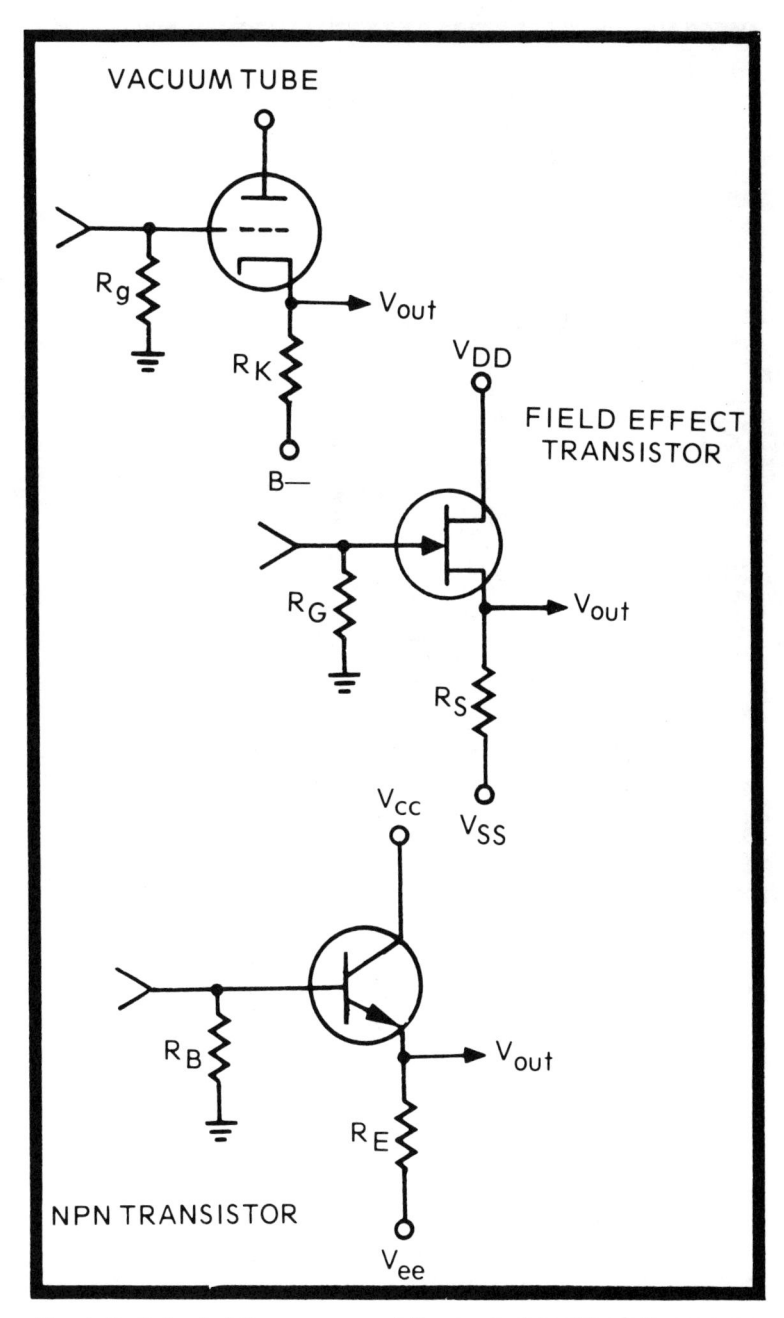

Fig. 6-11. Cathode-follower, source-follower, and emitter-follower amplifier circuits (comparing triode, FET, and NPN transistor).

213

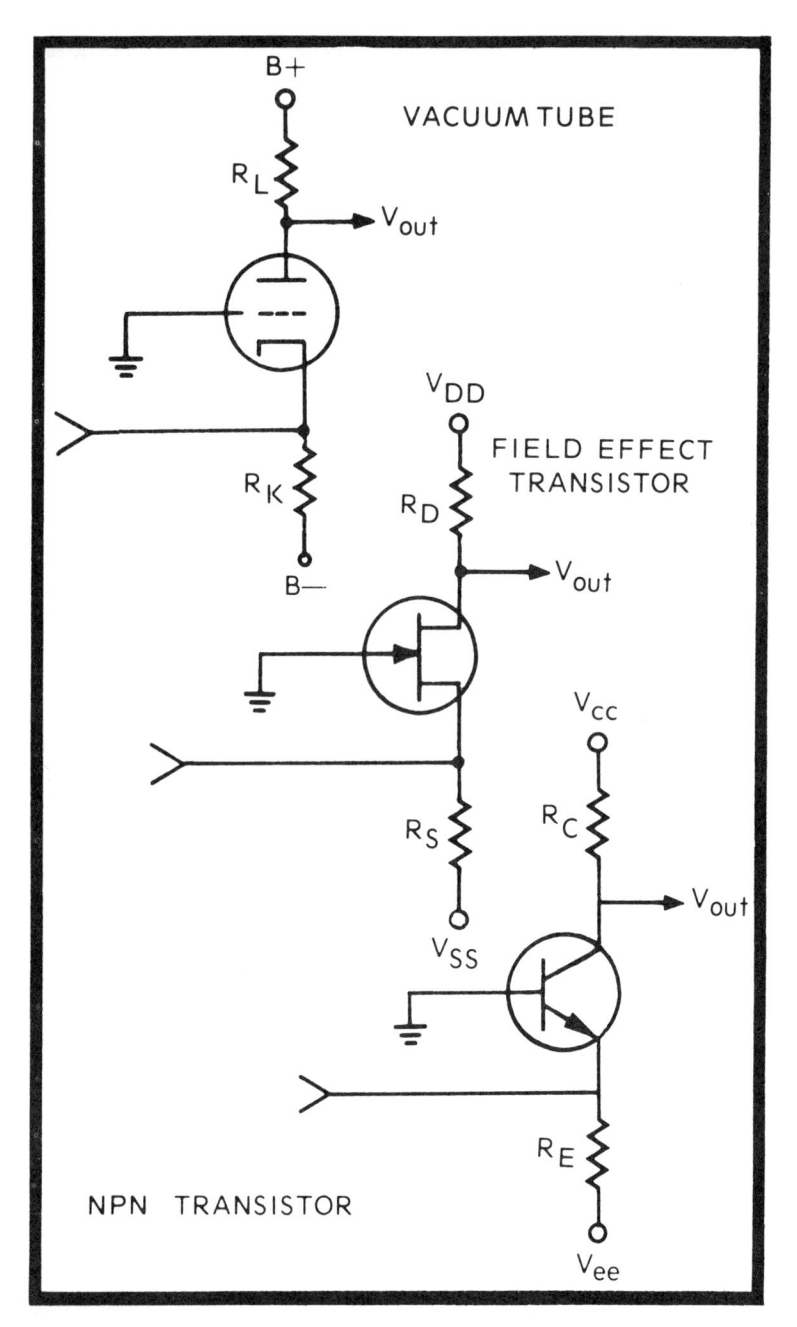

Fig. 6-12. Grounded-grid, grounded-gate, and common-base amplifier circuits.

to mu (which is G_{fss}/G_{oss}) times R_S. Again, it's the product over the sum of these two resistances.

In the case of the transistor collector-loaded stage, R_O is for most practical purposes considered the same as R_C. You see, the mu of the transistor, (1/Hoe) Transresistance, is usually so large (in the neighborhood of a couple of thousand) that by the time you have multiplied R_E by this factor and added it to the effective r_P (1/Hoe) you have a resistance many times larger than ten times R_C. This means that the Output resistance will be less than 10 percent away from the actual value of R_C, and R_C is probably a 10 percent resistor anyhow. So it's usually a lot of work for nothing.

Again with the cathode-follower stage (Fig. 6-11) when we substitute a FET for the vacuum tube to get a source-follower, or substitute a transistor for the vacuum tube to get an emitter-follower, we think of it in exactly the same kind of terms we did when the vacuum tube was there. The names for these terms change (Cathode resistance—r_K—becomes Source resistance—r_S—or Transresistance—r_{tr}—in the case of the transistor and R_K becomes R_S becomes R_E) but the system of their use remains the same. As long as we remember: The Effective-Cathode FOLLOWS the Effective-Grid, we have one of the main keys to the so called Long Tailed Circuit. The cathode follower is our first real introduction to this type of circuit. It's the first one where the effective R_K is a large value of resistance. The larger the value of this R_K resistor, the less the magnitude of current going through the active device will change for the signal applied, and the more constant its G_m type of characteristic will remain. There are other circuits that do this more effectively, but we'll talk about them when we come to them.

In the grounded-grid, grounded-gate, and common-base amplifier circuits (see Fig. 6-12), we find again that changing the active device doesn't change the way we think about it. It does change the language we think about it in, but not the method of thought.

Note that in each case, a small positive-going signal on the input reduces the amount of current going through the active device. Consequently, the current through the effective R_L will be reduced and the voltage across it will get smaller. Since the Effective B supply cannot change voltage, the output of the active device must move toward the supply (in these cases, positive).

In the case of the vacuum tube, the cathode is already positive with respect to the grid by an amount determined by the amount of current being allowed to go through the device

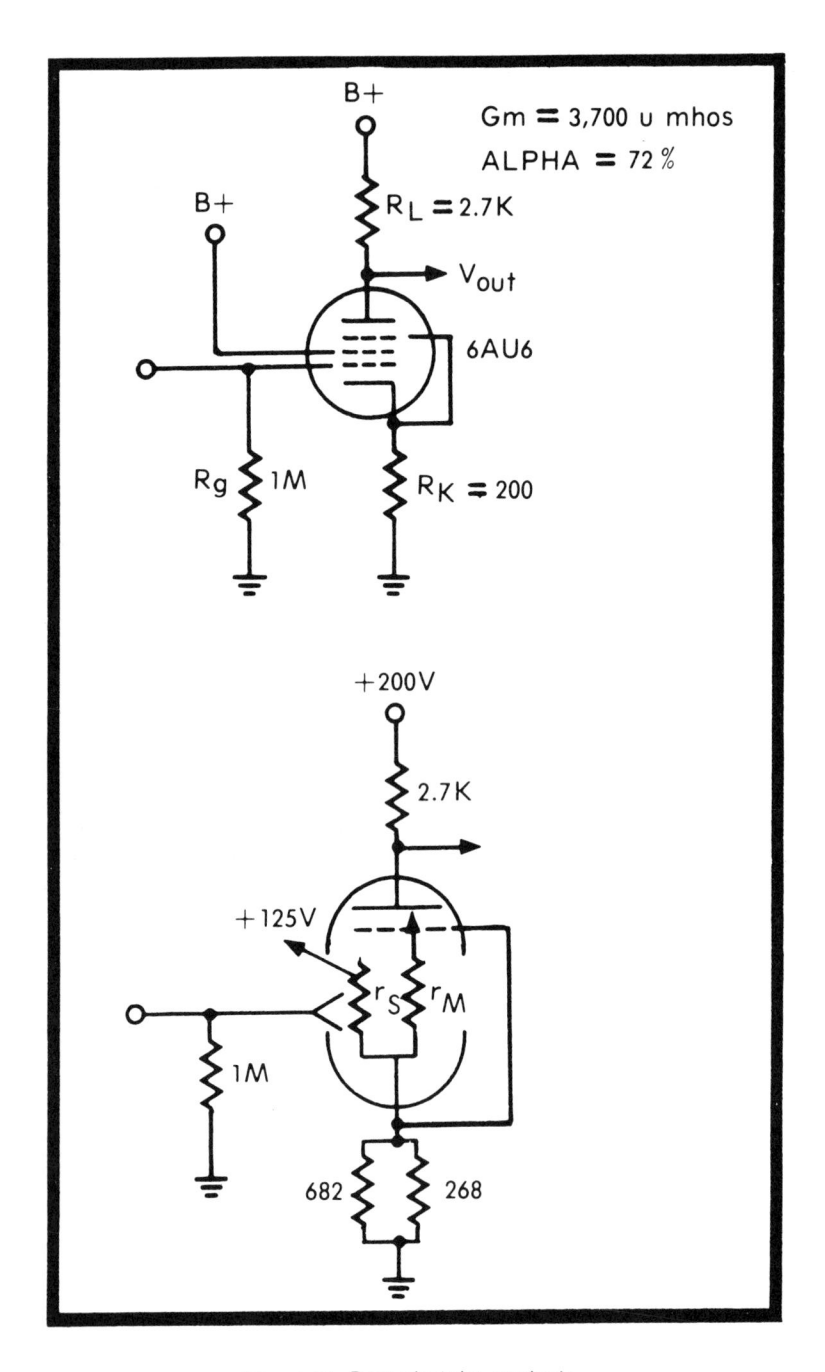

Fig. 6-13. Pentode tube analysis.

and the resultant plate to cathode voltage. Remember, any two of these three things (Plate **Current**, Plate to Cathode **Voltage** and Grid to Cathode **Bias**) determine the third. We talked about this in the discussion on Tube Curves. Consequently, a positive-going cathode will mean a higher bias and less current.

The same thing is true of the Field Effect Transistor. A positive-going Source means a greater bias and less current.

As for the transistor, in spite of the fact that we have a normally "off" device that must be turned on by its bias, we find the same thing happening. A positive-going emitter means that the "on" bias between emitter and base (the base is now positive with respect to the emitter) will be reduced and the device will carry less current, with the same results as before.

In figuring the voltage gain of these stages, we do it in the same fashion as we did with the vacuum tube. The effective plate load resistor is reflected into the effective cathode (divided by mu and added to r_K), figured in parallel with the external effective cathode resistor (the product over the sum), and this answer is divided into the load resistor. With the transistor, of course, we can simplify this process since its mu is so large, and since r_{tr} is so small in comparison with the effective cathode resistance of either of the other two devices. Only on rare occasions do you find a "long-tailed" transistor circuit in this grounded-grid configuration where R_E is so small that it is less than ten times the size of r_{tr} and has to be figured in parallel with it. Most times all that is necessary is to divide R_C by r_{tr} and you have an adequate estimate of the voltage gain of this stage.

It will pay us to note here that the grounded-cathode stage, where the input signal is applied to the effective grid and the output is taken from the effective plate, is the only configuration that turns the input signal upside down (a 180 degree phase shift). In both the cathode-follower and grounded-grid type of stages, we have the same phase signal coming out as is going in for normal frequencies of operation. When it comes to the really high frequencies we do encounter other forms of phase shift that manifest themselves as loss of voltage gain or oscillation but we needn't concern ourselves with that at this time. This is covered in Chapter 9.

Pentode vacuum tubes present a very interesting case when it comes to this general line of reasoning that we have been using. See Fig. 6-13. They actually look a bit like a transistor with a very, very low value of Beta. If you had a transistor whose collector got only 72 percent of the emitter current, its Beta would be just a bit less than 3. Now, this wouldn't be a very good transistor, but for a pentode tube, it's pretty good. And, if you will check back into the last chapter where we were discussing pentodes and look at the curves for the 6AU6, you'll find that this is fairly accurate. For a screen voltage of +125 volts and a bias of 1.5 volts, the curves will tell you that the plate current is 5.6 ma and the screen current is 2.2 ma. Now with the use of a little arithmetic it's fairly easy to see that 5.6 ma is about 72 percent of 7.8 ma which is the sum of 2.2 ma and 5.6 ma (the total current that must flow in the cathode).

It's interesting to note that this characteristic of Alpha holds true for signal currents generated in the cathode by an applied signal voltage on the grid. In the case we are considering, the plate will only receive 72 percent of the total signal current generated by a change of voltage on the grid when the total cathode resistance is considered. But most times we don't consider the total cathode resistance. Instead, we just take the reciprocal of G_m (which in this case is the same as r_m) and say that since the r_p of the pentode is so large with respect to R_L we needn't figure the two in parallel, just divide R_L by the reciprocal of G_m. In the case of the older formula, we multiplied R_L by G_m to get the gain instead of going to the trouble of figuring r_M first. AND all of this was great just as long as we had a **grounded cathode** or if there was a big bypass capacitor around any R_K that might be in the circuit. However, in the case we are looking at in Fig. 6-13, the system doesn't quite work out. The second picture underneath shows why. It shows clearly that the reciprocal of G_m is not the cathode resistance. To figure the cathode resistance you can get a reasonable value by multiplying $1/G_m$ by .72 to get about 195 ohms. This, you could add to the 200 ohm (R_K) to make up R metering, which you would divide into the 2.7K (R_L) to get an answer that you would have to modify again with the old 72 percent figure for the final answer.

However, if we use our heads, we can simplify this procedure a bit more. Look at the bottom drawing of our circuit again. To find that portion of R_K that carries only plate signal current, we simply divide its 200 ohm value by .72 to get about 268 ohms. (If we divide 200 by .28, we get 682 ohms—the product over the sum of 268 and 682 gives an approximate 200 ohm value.) This 268 ohms is that parallel part of R_K that we can now add to the reciprocal of G_m to determine the true value of R metering now carrying only plate signal current. This value of R metering (268 ohms + 270 ohms equals 538 ohms) we divide into 2.7K to get the voltage gain of this stage:

2,700/538 equals 5.02; that is, for all practical purposes, a voltage gain of five.

This—to me—is the easiest way to do it. Modify the value of the external R_K to fit that portion of the actual cathode resistance which is already modified to fit our Alpha situation of 72 percent. Since G_m is a measurement of the change in plate current caused by a given change in grid voltage, one over G_m is already that portion of cathode resistance which carries plate current. Thus—in the case of the pentode—$1/G_m$ is NOT r_K. Cathode resistance is the parallel combination of r_S and r_M. However, if we modify the external R_K to go with r_M we don't have to worry about it any further.

I can hear somebody in the background saying, "Well, all this is great, but just how do you figure them out when there is more than one active device in each circuit?"

It's not as hard as you think. You already have all the tools you need in order to do it. Now, by "TOOLS, I mean all the little rules of thumb I've mentioned all along the way through this text, Ohm's Law (including the Thevenin, Norton, and Millman uses of Ohm's Law) as well as what impedances we see in the active devices and what other impedances look like when viewed through (or reflected back into) an active device.

As a good example, see Fig. 6-14, where we have two active devices in series. It doesn't matter what type of active device we have where. Active Device No. 1 and Active Device No. 2 can be Transistors (NPN type of course with the B+ voltage on top), or Vacuum Tubes (you worked with a circuit almost exactly like this in proving that external cathode resistance was multiplied by the mu when viewed from the plate), or FETs (this circuit is a natural for FETs). You may also see a mixture of active devices used in this circuit. This is why I have drawn this circuit without identifying the type or types of active devices used. They could be anything that has a quality similar to r_p, one similar to r_K and, of course, one similar to mu (if it has an r_p and an r_K—it has to have a mu since mu equals r_p divided by r_K—remember?). So let's look at this circuit and see what we can figure out about it.

B+, B—, and Ground are known voltages. We don't have to figure anything in order to find out what they are, but we should take careful note of their actual values. These will be our starting points for everything else we do figure.

R_{B1} and R_{B2} form a divider between B— and ground for the effective grid voltage of active device No. 1. Its cathode (or similar element) will follow its grid. This—you will notice—sets the voltage across R_{K1}, which in turn determines the amount of current each active device will be carrying.

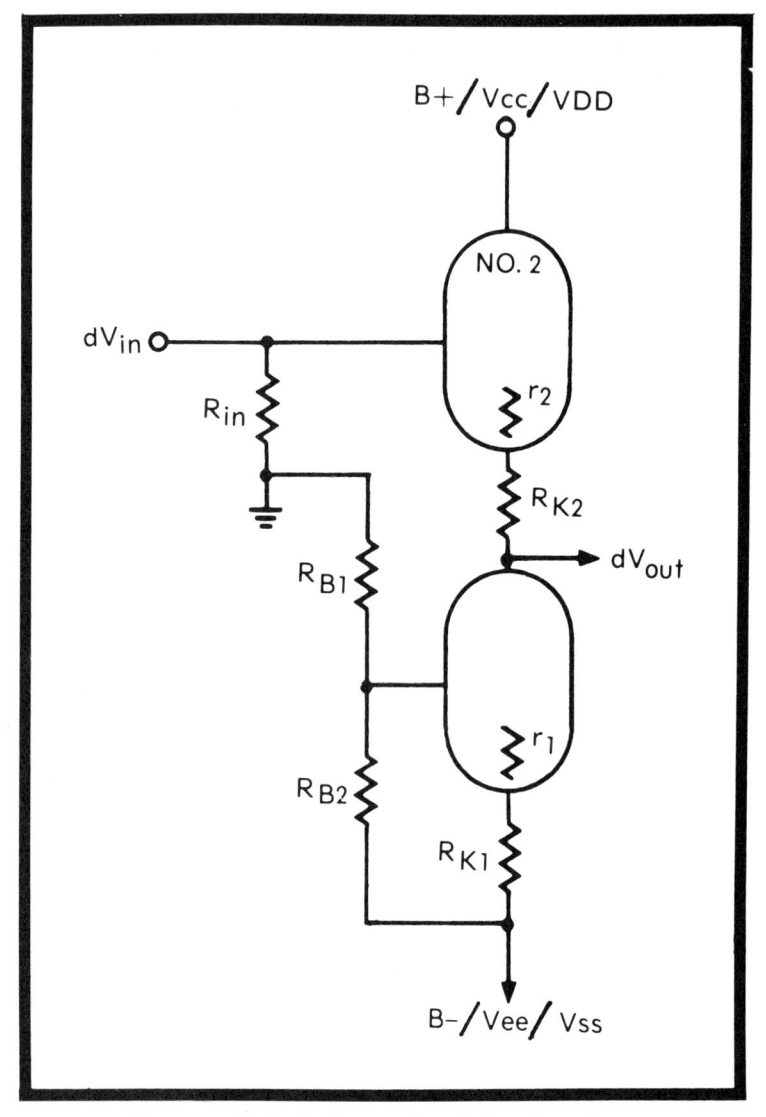

Fig. 6-14. Analysis of two active devices in series.

The dV_{in} (change of voltage, signal, in) grid of active device No. 2 is referenced at ground and may be thought of as being at zero volts with no DC voltage drop across R_{in}. This point has to be movable (voltage-wise) so we can't just tie it to ground. We have to have a resistor between it and ground.

Device No. 2 cathode follows its grid and may be thought of as being at zero volts also (no matter what type the active

220

device is, we can't be wrong by more than a volt or so—and we can correct for this later).

We know the amount of current coming from the cathode of No. 1 and thus we know the amount of current going through R_{K2} and consequently the voltage drop across it, so we must be able to figure the DC voltage level at dV_{out}.

The input resistance of the circuit will be made up of R_{in} seen in parallel with a resistance made up of $r_2 + R_{K2}$ added to the No. 1's mu of times $(r_1 + R_{K1})$; all of this multiplied by Beta of No. 2 if it has one. Otherwise, it has to be bigger yet. I don't think it's going to change the value of R_{in} at all. So, R_{in} is R_{in} for the circuit.

R_{out} is a little different story, but not much. If you look back into dV_{out}, you will see $(R_{K2} + r_2)$ in parallel with No. 1 mu times $(r_1 + R_{K1})$. This impedance seen looking down into the plate of No. 1 is usually so great in comparison with $(R_{K2} + r_2)$ that you hardly ever have to figure it out. R_{out} then is usually considered to be $(R_{K2} + r_2)$.

We obviously have a cathode-follower (emitter-follower or source-follower; the same thing as a common-collector or common-drain) circuit. If we swing the voltage at dV_{in} with a signal, No. 2 cathode will follow. Now, since No. 2 cathode can't change the current through No. 1, the voltage drop across R_{K2} must remain the same and the swing in voltage at dV_{out} will be the same as at dV_{in}. Thus we have practically no loss of signal voltage at all and still maintain the current gain and thus the power gain of the circuit. This type of circuit presents unusually high impedance to its source of signal and thus does not load it down, and it switches the signal to its own output which is a low impedance and thus can drive just about whatever load it must.

This stacking of active devices—one on top of the other—can also be made into a voltage amplifier with quite different characteristics. See Fig. 6-15, where we drive the grid of No. 1, tie the grid of No. 2 to some appropriate level of voltage where it cannot move (making a grounded-grid amplifier out of No. 2) and take the amplified signal voltage from an R_L plate load given to No. 2. This is called a cascode circuit.

Note in particular that the input resistance of the cascode circuit is predominantly dictated by R_{in} and that the output resistance of the circuit is for all practical purposes the same as R_L. The voltage gain of the stage is, of course, R_L divided by $R_{metering}$.

As I mentioned earlier, the stage voltage gain is easiest to obtain by simply dividing R_L by the sum of $r_1 + R_1$. Not that you can't get there by figuring the active devices individually.

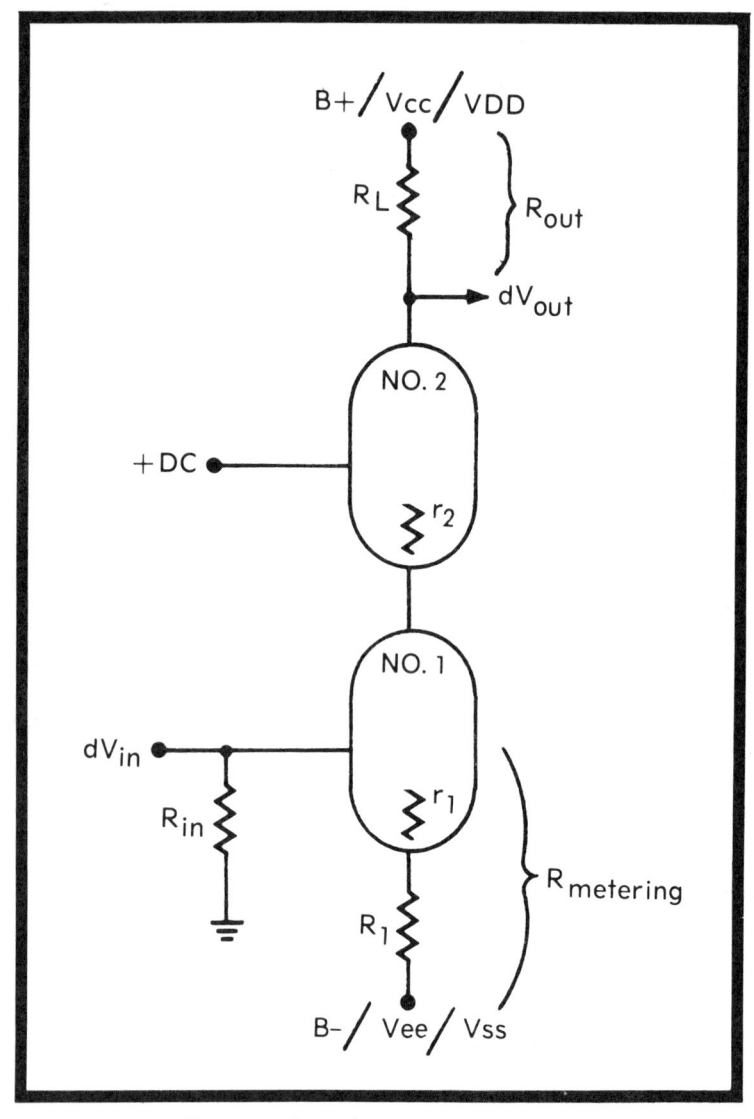

Fig. 6-15. Cascode amplifier analysis.

You can. The amplitude of signal seen at the plate of No. 1 is $(R_L/mu_2) + 1/G_{m2}$, divided by $R_{metering}$. This is the voltage gain of active device No. 1 and will in all probability be much less than 1 (probably near 0.5 or less). This gain times the dV_{in} signal will be the amplitude of signal driving active device No. 2. This voltage signal occurs across R_L/mu_2 added to $1/G_{m2}$

which resistive sum now becomes the $R_{metering}$ for No. 2 and should be divided into R_L to tell us the A_V from cathode to plate of this grounded-grid amplifier. The A_V of stage No. 2 times the A_V for stage No. 1 should be the total gain of the cascode amplifier and should be the same as we get before when we divided R_L by $(r_1 + R_1)$.

After all, knowing what amplitude of signal to expect where in a circuit like this is half the battle of troubleshooting this circuit. For instance, suppose you saw (with the use of an oscilloscope) a voltage signal on the plate of No. 1 that was five times larger than the signal at dV_{in}. Unless r_2 is five times the size of $(r_1 + R_1)$, this is something that should not be if active device No. 2 is any good at all. It would be telling you that something was wrong with No. 2.

On the other hand, suppose you saw a voltage signal on the plate of No. 1 that was much smaller than it ought to be. Would this tell you anything? It should. It could be telling you that R_1 had started to open up, that the cathode emission of No. 1 (its G_m, in other words) had gone sour. Either one of these things could be a cause of this. So, knowing what you should be able to expect from a given circuit is of importance to you.

The cascode circuit is extremely nice with respect to its frequency response. The A_V of No. 1 being less than 1 gives it an excellent bandwidth capability while the grounded-grid configuration of No. 2 gives it the best (lowest input capacitance) frequency response capability for the voltage gain needed that you can rightfully expect.

Another multiple active device single-ended stage is the distributed amplifier stage, Fig. 6-17. (Single-ended means having one signal input and one signal output, see Fig. 6-16.) In Fig. 6-17 all the active devices are in parallel and their grids (or what passes for a grid) are driven from as many different points down a transmission line as there are active devices. In this manner, for all practical purposes each grid sees the same signal at slightly different times. The voltage gain for each active device should probably be between 1 and 2, giving good bandwidth characteristics (since each device will have the same gain x bandwidth product).

If the speed of propagation down the plate transmission line is the same as that for the grid transmission line, the voltage gains of each device will be additive and the voltage gain for the stage will be greater than the gain for any individual device. For that matter, the voltage gain of the stage should be equal to the voltage gain of the first active device times the number of active devices in the stage. I have seen as many as twelve tubes in a row in this kind of a circuit. The

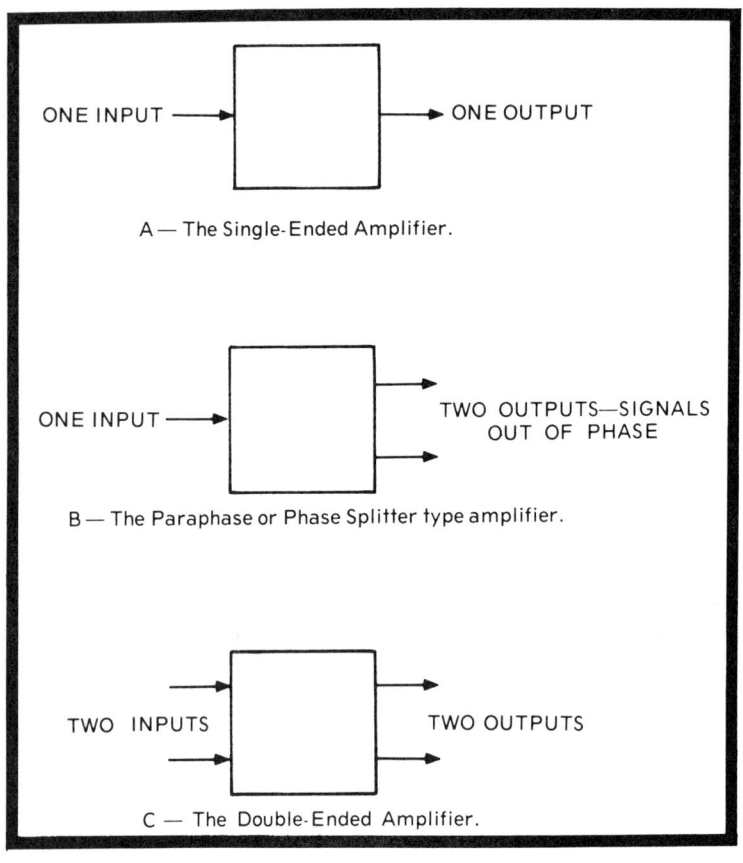

Fig. 6-16. Block diagram of single-ended, paraphase, and double-ended amplifiers.

actual voltage gain of the stage wasn't quite equal to 12 x 1.5, but it was plenty close enough to use this as a general rule of thumb to figure the gain.

Of course, the transmission lines must be terminated in their own impedances (R_L must equal the plate transmission line impedance, and R_T must equal the grid transmission line impedance). This prevents any reflections from lousing up the signals.

The voltage gain for each active device in this stage can be figured using one over G_m as the metering resistance (as long as the cathode bypass capacitors are there and large enough— 30 uf or a bit more) divided into R_L.

NOTE: There are conditions which can change this. If the plate transmission line is terminated on BOTH ends in its own impedance, then the circuit will act as though the actual load is the two R_Ls in parallel and the gain will only be half what we figured in the first place.

Don't expect to see the true signal gain on the plate transmission line with an oscilloscope even with all but one active device removed. The actual plate signal can be seen on the transmission line side of R_L and with all but one active device removed should illustrate the gain of the single device remaining. If the right-hand end of the plate line is NOT terminated in its own impedance even this might be distorted by the reflected wave coming back down the line from right to left. This then says that the best place to view the plate waveform is at the right-hand end of the plate transmission line. This, after all, is the signal that drives whatever it is that comes next.

Any of the stages we have talked about can be doubled, cathode coupled, driven 180 degrees out of phase with each other and thought of as a single push-pull stage (see Fig. 6-16).

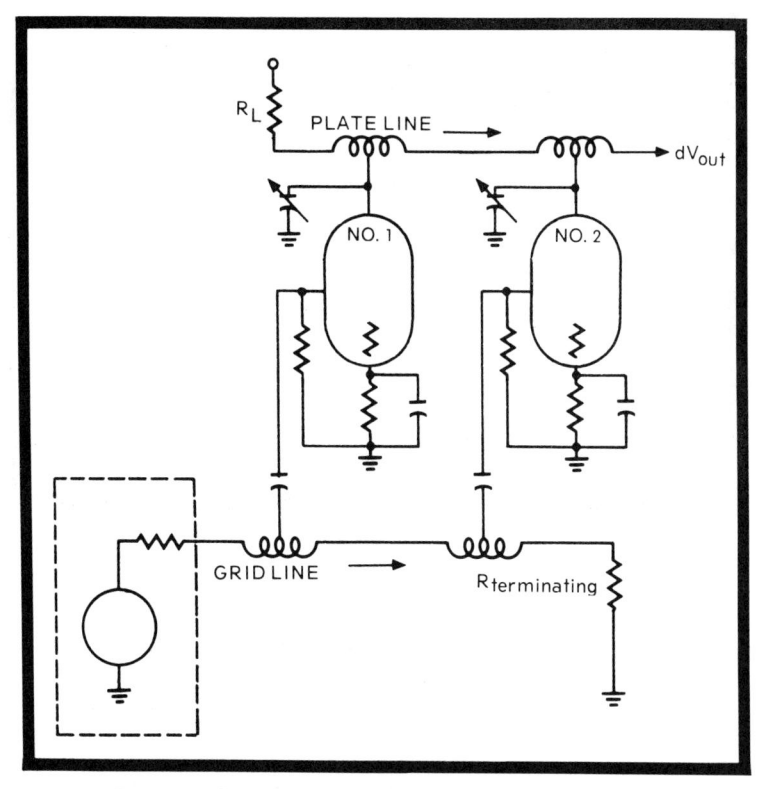

Fig. 6-17. Single-ended distributed amplifier analysis.

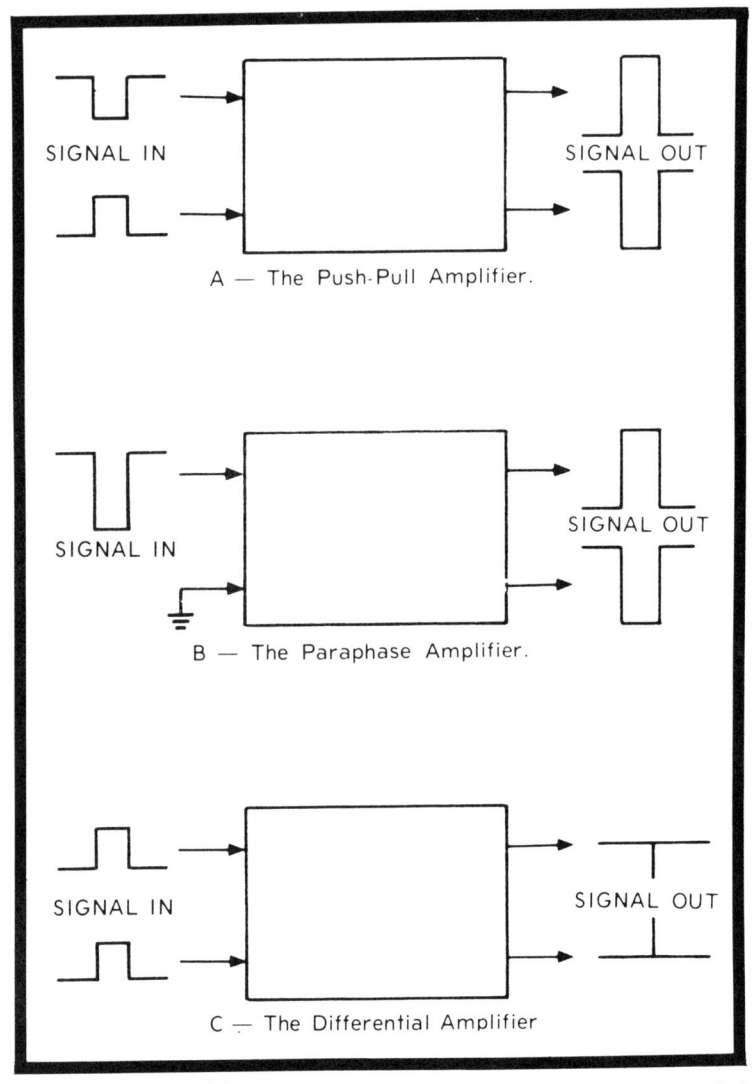

Fig. 6-18. Block diagrams of push-pull, paraphase, and differential amplifiers.

However, this brings up the cathode coupling of two active devices which is a whole family of amplifiers unto itself. These circuits are characterized by having two points where the input signals can be introduced and two outputs where the resultant signals are seen as being out of phase with each other. If one goes positive, the other goes negative; that is, they are 180 degrees out of phase with each other. See Fig. 6-18.

It is interesting to note that the names of these circuits are largely determined by the relationship of the input signals, one to the other.

If the input signals are the same, but 180 degrees out of phase with each other, then the circuit is called a **push-pull amplifier**. The total input signal amplitude is thought of as the sum of the two amplitudes. Thus if each signal is 1 volt high, the total input voltage is 2 volts even though they are out of phase.

If the 2 volts of drive is applied to one input as a single signal 2 volts high with the other input effectively held stable in voltage, the circuit now becomes a **Paraphase Amplifier**. This also requires a fairly stable source of current for the two coupled cathodes.

If the two signals are in phase but just slightly different and the two coupled cathodes have a well regulated source of current, then this circuit is called a **Differential Amplifier**. It may also be called a Difference Amplifier because it is only the difference between the two input signals that will be amplified and show up as two out-of-phase signals at the outputs.

To understand how all this comes about, let's study the circuit in Fig. 6-19 under each set of conditions.

First, let's examine the Push-Pull Circuit (where R_{L1} equals R_{L2}) and (R_{K1} equals R_{K2}):

The changes of voltage that occur at input No. 1 and input No. 2 are equal and opposite. As one active device gets turned off, the other gets turned on by the same amount. NOTE: this means that the total circuit must still be drawing the same amount of electrons up through the cathode circuitry. R_C must be going slightly negative on one end as it goes positive on the other; just like a voltage teeter-totter might be expected to act, with the middle of R_C being thought of as staying at the same voltage all the time. In other words, R_C looks like it has a ground point in the middle of it as far as each active device is concerned. Thus the voltage gain for the No. 1 side will be R_{L1} divided by an $R_{metering}$ made up of R_{L1}/mu added to its $1/G_m$ characteristic added to one half of R_C in parallel with R_{K1}.

The voltage gain for side No. 2 can be figured in the same fashion, r_2 being made up of $R_{L2}/mu + 1/G_m$; this is in series with the other part of $R_{metering}$ made up of half of R_C in parallel with R_{K2}. This Metering Resistance is then divided into R_{L2} for the gain of the right-hand side.

Pay attention to the idea that says that the voltage gain of either side can be thought of as being the voltage gain of the whole stage. Since we figured the gain of one side by using half

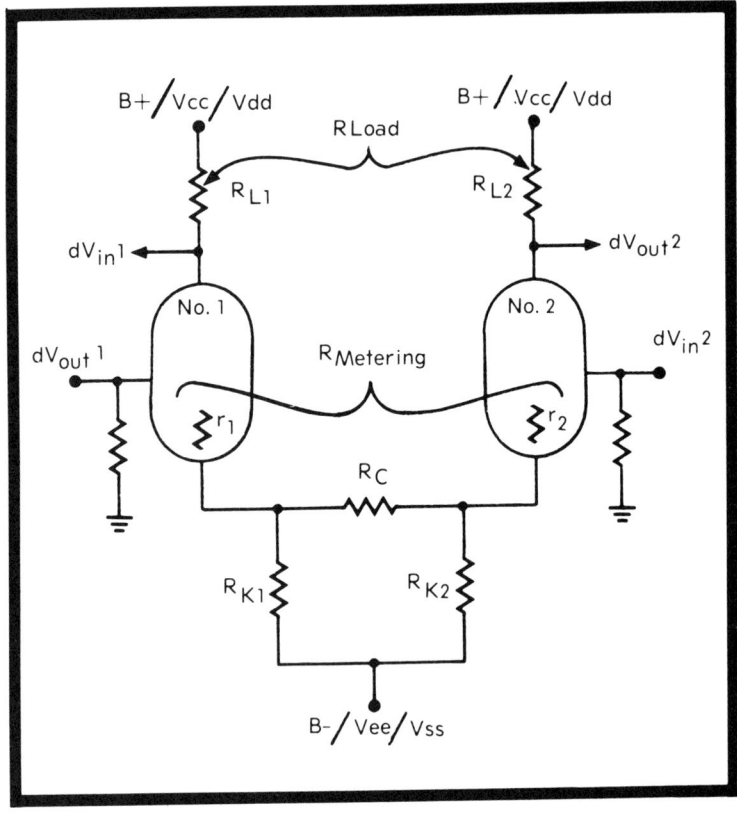

Fig. 6-19. Analysis of simplified circuit exemplifying configurations of Fig. 6-18.

the input signal in relation to half the output signal, then it stands to reason that when we use the total output signal divided by the total input signal we'll have the same answer. OK? OK!

If R_{K1} and R_{K2} are 10 to 100 times the value of R_C, their effect on the circuit may be ignored and the voltage gain of the whole stage can be figured by dividing the sum of R_{L1} and R_{L2} by the sum of r_1, R_C and r_2. Remember though, r_1 and r_2 represent not only the cathode resistance but also that part of the load resistance reflected into the cathode.

Keep in mind that two things about this circuit have to be true: corresponding parts on opposite sides of this circuit must be equal in value and the input signals must be equal in amplitude but opposite in phase. If these two conditions are not met, then this circuit comes out under a different name. Keep reading and I'll show you what I mean.

Now let's look at the Paraphase Amplifier (where the total input voltage is applied to one input only).

The change of voltage applied to input No. 1 will now be equal to the sum of the two signal amplitudes applied to the Push-Pull circuit.

The input to No. 2 will be anchored solidly to the same DC level of voltage as we have at input No. 1 (it might even be made adjustable so that we can match the operating point of the two sides exactly).

Assuming the input signal to be negative-going as shown in Fig. 6-18B, active device No. 1 will start to turn off and its output will move positive. But, the interesting thing is what happens to No. 2. Active device No. 2 effective cathode will move in the negative direction. This is going to either increase the ON bias of a transistor, or decrease the OFF bias of a vacuum tube or FET. In either case, No. 2 is going to conduct more current and its output will go negative.

The output change of voltage seen at No. 2 will be equal in amplitude and of opposite phase to that seen at No. 1 as long as r_2 equals r_1 and each active device changes its conduction by the same amount of current.

Note that the voltage gain of this amplifier is figured the same as for the Push-Pull Amplifier. The sum of R_{L1} and R_{L2} is divided by a Metering Resistance made up of the sum of r_1, R_C and r_2 (where r_1 and r_2 are made up of the $1/G_m$ characteristic of the active device and R_L/mu in each case).

This then is the basic Paraphase Amplifier circuit. I must admit that once in a while we find this circuit such that r_1 and r_2 are not matched and the two active devices do not see the same change of voltage. In this case, usually, the load resistors are mismatched as well but in the opposite direction such that the two output signals are of equal amplitude anyhow.

The third use of this same basic circuit, dependent upon how it is driven, is as a Differential Amplifier (a circuit that amplifies only the difference between two similar signals). In this case, we usually find that R_{K1} and R_{K2} are quite large with respect to R_{L1} and R_{L2}. This is done to eliminate, as much as possible, that portion of the input signals that is the same.

Let's see how this one works. In Fig. 6-18C, the two inputs move positive (in this example) and the two effective cathodes follow, increasing the voltage across R_{K1} and R_{K2} (as seen in Fig. 6-19) by just about the magnitude of the input signal (cathode follower action). If R_{L1} and R_{L2} are smaller than R_{K1} and R_{K2}, the output change of voltage will be smaller by almost the same factor. Thus the voltage gain for the common parts of the two input signals is kept quite a bit less than 1. In the meantime, that part of the two incoming signals representing their difference may be thought of as driving this circuit either as a Push-Pull Amplifier or as a Paraphase Amplifier. In either case, the voltage gain to this part of the signal is figured in the same way:

$$\text{Difference Signal } A_V = \frac{R_{L1} + R_{L2}}{r_1 + R_C + r_2} = \frac{R_{\text{Load Total}}}{R_{\text{metering}}}$$

And, if the circuit does what it is supposed to do, this will be a number of times greater than 1.

The ratio of these two gains is usually taken as a measure of the effectiveness of this type of circuit and is called The Rejection Ratio. Many times it is not so much something to be figured as it is something to be measured. We drive the circuit at one input only (like a Paraphase Amplifier) and measure the output change that occurs. Then we drive both inputs with an increased amplitude of the same signal, increasing the input amplitude until we obtain the same output change as we measured under Paraphase drive. The ratio of the Common Mode Drive to the Paraphase Drive is now the Rejection Ratio and is stated like maybe 10,000 to 1 or something of this nature. (There's a real good possibility that it might not be this good though.)

Now, please don't misunderstand me. I don't mean to imply that these are all the different types of amplifiers around. There are loads of other different types of amplifiers. However, this has been a fairly good series of samples of different types of amplifiers, each figured out by the use of the same old reasoning. If you will apply it, I think you will find that it works for you just as it does for me.

Look for a Metering Resistance that dV_{in} appears across to generate a signal current which the active device passes along to a Load Resistance to give a dV_{out}. This is the major secret of Amplifier Analysis.

Complete Analysis of an Amplifier

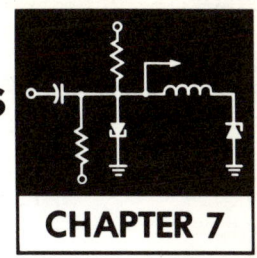

CHAPTER 7

This chapter is one of the main focal points of this book since it requires a basic understanding of everything that I have tried to present so far. The idea is to have you analyze, in as complete a fashion as this system allows, a commercially manufactured amplifier circuit. The circuit I have chosen is the Horizontal Amplifier from the 3B3 Plug-in for a Tektronix Oscilloscope. It uses vacuum tubes, NPN transistors and a PNP transistor. The vacuum tube is the 6DJ8 for which we already have all the data (in Chapter 5 on Active Devices and also Fig. 7-7) and the transistors are all silicon devices.

You will find that you have a choice as to how you wish to go through this circuit.

(1) You may try it completely on your own. Start with the assumption that the outputs are at equal voltages (at terminals 17 and 21 on the right side of Fig. 7-1). In other words, the spot on the screen of the cathode ray tube which this circuit drives is being held somewhere on the vertical axis going through the center of the screen. (Note: if the spot on the screen is at exact electrical center, then the vertical deflection plates are at equal voltages also.) To place the spot at the left hand edge of the screen (preparatory to presenting a display) terminal 17 would have to be 50 volts more positive than it is and terminal 21 would have to be 50 volts negative from where it is now. One half of the total peak-to-peak change of voltage at the output (17 and 21) will be required from each half of this amplifier and each output will proceed to the voltage the other started from. If this change of voltage takes the shape of a ramp waveform, then the progress of the spot from one side of the display to the other will be constant all the way across the screen. A positive-going ramp will be seen at terminal 21 with an oscilloscope, while a negative-going ramp will be seen at terminal 17. The sum of the two changes of voltage will be the peak-to-peak output voltage.

INITIAL METHODOLOGY

My advice is to figure the DC voltage levels present when the spot is centered. First, note all points of known voltage.

CONDITIONS

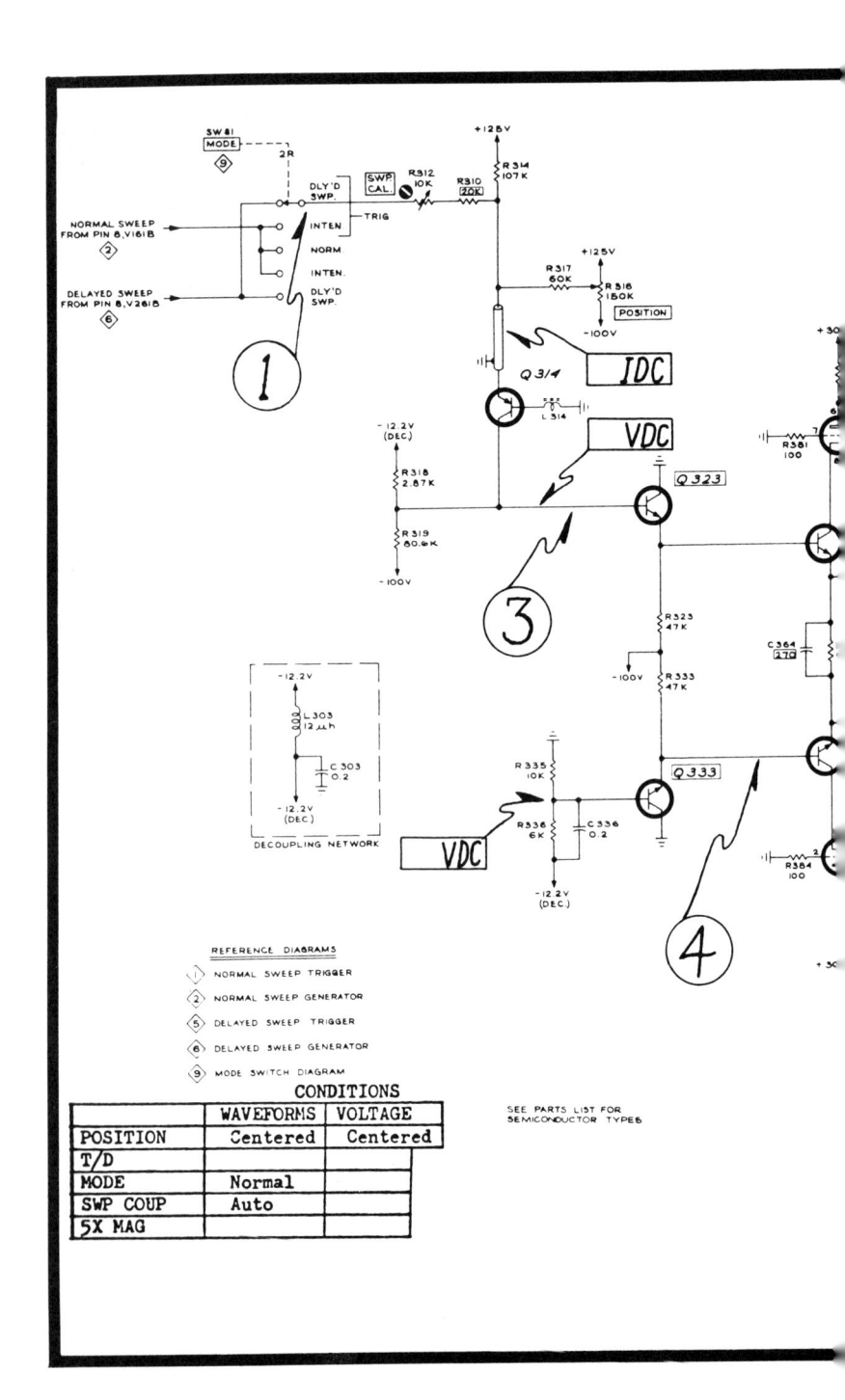

	WAVEFORMS	VOLTAGE
POSITION	Centered	Centered
T/D		
MODE	Normal	
SWP COUP	Auto	
5X MAG		

REFERENCE DIAGRAMS
1. NORMAL SWEEP TRIGGER
2. NORMAL SWEEP GENERATOR
5. DELAYED SWEEP TRIGGER
6. DELAYED SWEEP GENERATOR
9. MODE SWITCH DIAGRAM

SEE PARTS LIST FOR SEMICONDUCTOR TYPES

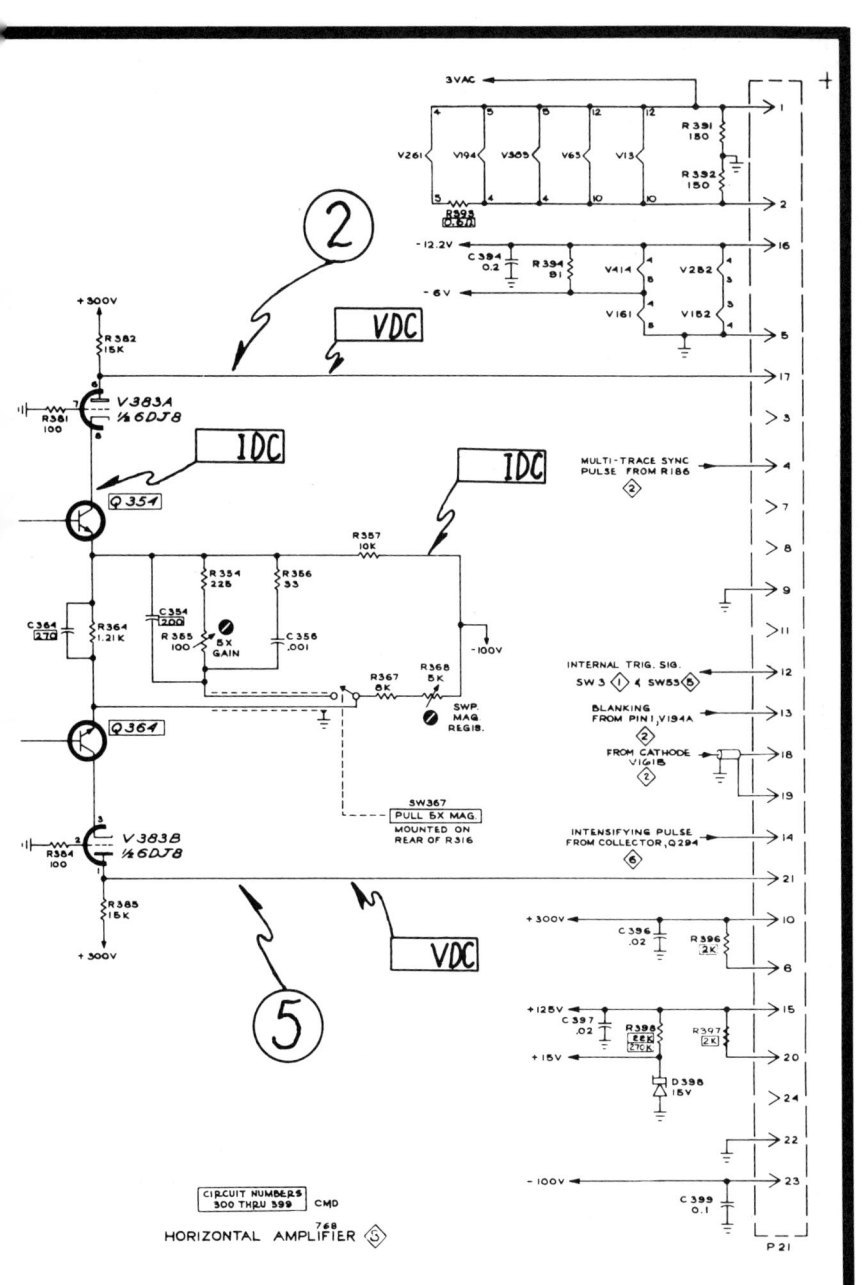

Fig. 7-1. Schematic of 3B3 plug-in horizontal amplifier of Tektronix oscilloscope (worksheet). (Courtesy Tektronix, Inc.)

Then locate all points of figurable voltage and figure them. Figure all the currents that you can by making use of the voltages you have found. Now figure the voltages you couldn't figure before because you didn't know the currents. By this time you should have enough information to find the voltage gains of the respective amplifiers in the circuit. Combine these different gains and figure how large a signal must come into this amplifier via the Mode Switch to give you a 200-volt peak-to-peak signal at the outputs.

(2) You may analyze the circuit by answering the questions in sequence. Be sure to answer each question as you go because some of the answers later on depend upon some of the answers you get in the beginning.

(3) Or you may allow me to guide your way through this circuit, problem by problem, until we obtain a complete analysis of this circuit.

Actually, I urge you to use all three approaches to the analysis of this circuit:

First: Find out what you can about this circuit by yourself. Try to fill in all the squares on Fig. 7-1 with the proper voltage to be found at the points indicated. Determine, for yourself, the signal path and types of amplifier circuits involved and predict the voltage gain for each. (Note: you can expect some surprises here.)

I will tell you this much: Q333 is not exactly an emitter follower. It is used to establish the DC level of voltage on the base of Q364 and filter out any signal from reaching this point.

Second: Try the sequence of questions I have prepared to either verify what you have already found out about the circuit or to take you a bit further through it. (Half the battle is knowing what questions to ask.)

Third: Read the detailed analysis to further your understanding of the circuit and maybe learn better how to use the tools of circuit analysis to which I may have introduced you.

PRELIMINARIES AND SELF-TEST

(INSTRUCTIONS) Answer these questions in sequence, treating this 3B3 Horizontal Amplifier as though it were a balanced amplifier. In other words, figure it as though the voltage on the Plate of V383A is the same as the voltage on the Plate of V383B.

(1) Note: First, we pick out the points of KNOWN VOLTAGE: so, what is the voltage with respect to ground on:

a. The Grids of V383A & B?
  b. The Collectors of Q323 & Q333?
  c. The Emitter of Q314?
  d. The junction of R317, R314 & R310?

(2) Assuming zero base current for Q333, what voltage with respect to ground does the divider made up of R335 and R336 put on the base of this transistor? Ans: ....

(3) Solve for the THEVENIN EQUIVALENT of R318, R319 & their respective voltage sources. V_{oc} .... R_{th} ....

(4) If the output voltages of this amplifier are equal, then the base of Q323 and Q333 are the same. What is the voltage on the base of Q323? Ans: ....

(5) If we know the voltage on both ends of a resistor and the value of resistance, then we should be able to determine the current through it.
  a. How much current is the collector of Q314 getting? Ans: ....
  b. How much current is going through R314? Ans: ....
  c. If the voltage on the right side of R317 is the same as that on the left side, then how much current does it carry? Ans: ....

(6) If all the current that flows into the collector of Q314 flows out the emitter, how much current is left to flow out R310 & R312? Ans: ....

(7) With R312 set at 5K ohms, what level of voltage must exist at the input labelled "DELAYED SWEEP FROM PIN 8, V261B?" Ans: ....

(8) Since we know the voltage on the bases of Q323 & Q333, what is the voltage on their emitters? Ans: ....

(9) How much current enters the emitters of Q323 & Q333? Ans: ....

(10) Since we know the voltage on the emitters of Q323 & Q333, what is the voltage on the bases of Q354 & Q364? Ans: ....

(11) What is the voltage on the emitters of Q354 & Q364? Ans: ....

(12) Again assuming zero base current, this time for Q354 & Q364, how much emitter current does each of these transistors carry? Ans: ....

(13) What is the voltage with respect to ground on the plates of V383A and V383B? Ans: ....

(14) Since we now know the approximate plate voltage and plate current for both V383A & V383B, how much bias will each tube have? (Note: Refer to the plate curves for the 6DJ8) Ans: ....

(15) By this time you should have enough information to quote V_{ce} and I_e for each transistor in this circuit. Fill in the following table:

	V_{ce}	I_e
Q314		
Q323		
Q333		
Q354		
Q364		

(16) What kind of an amplifier circuit is Q314 in? (check correct answer)
    Common Emitter
    Common Base
    Common Collector

(17) What kind of an amplifier circuit is Q323 in? Ans: ....

(18) Since an AC signal cannot get past either a DC or AC ground point, does Q333 have any drive? Ans: ....
    If your answer is "yes," where does it come from? Ans: ....

(19) How much impedance does the base of Q364 see looking back into Q333 and R333? Ans: ....

(20) What would you call the amplifier circuit made up of Q354, Q364, V383A & V383B? Ans: ....

(21) What approximate impedance do you see:
    a. Looking back into pin 17 of the output board? Ans: ....
    b. Looking back into pin 21 of the output board? Ans: ....

(22) Assuming the 5X MAG switch to be open, what is the approximate value of $R_{metering}$ for the amplifier made up of Q354, Q364, V383A & V383B? Ans: ....

(23) If I see a 1 volt signal on the base of Q354, what will be the amplitude of signal seen on the plate of V383B? Ans: ....

(24) What is the peak-to-peak voltage gain of the stage made up of Q354, Q364, V383A & V383B? Ans: ....

(25) What is the voltage gain from the base of Q323 to the emitter of Q323? Ans: ....

(26) What is the voltage gain from the DELAYED SWEEP terminal to the base of Q323? Ans: ....

(27) What is the total voltage gain of this circuit (from DELAYED SWEEP input to Peak-to-Peak output)? Ans: ....

(28) Assuming that the output signal is 200 volts peak to peak, how big a signal will we see at the DELAYED SWEEP input? Ans: ....

(29) Assuming that the output—peak to peak—of this amplifier is 200 volts, and that the signal on the plate of V383A is negative-going, what is its amplitude? Ans: ....

(30) What will be the amplitude and polarity of the sweep sawtooth waveform seen on the base of Q354?
    Amplitude: ....
    Polarity: ....

(31) Describe the signal seen on the base of Q333.

(32) What will be the polarity and amplitude of the DELAYED SWEEP signal that drives this board from pin 8, V261B? Ans: .... & ....

(33) At quiescence, the spot on the CRT is at the left edge of the faceplate and this amplifier is in an unbalanced state. What DC levels of voltage with respect to ground will you read:
    a. on the plate of V383A? Ans: ....
    b. on the plate of V383B? Ans: ....
    c. on the base of Q354? Ans: ....
    d. on the base of Q323? Ans: ....
    e. at the DELAYED SWEEP input to this board? Ans: ....

If you have tried (1) and (2) and are still doubtful—please understand this. I do not expect you to be able to go independently through this circuit finding out everything that I shall point out to you or ask you for at this stage of the game.

You are capable now of determining for yourself more about this circuit than you could before with much greater ease. You are also capable of understanding a fairly complete analysis of this circuit, so let me guide the way.

Check your answers to the preceding list of questions against my analysis and conclusions that follows. Fig. 7-6 is again the schematic, with my answers on it.

AUTHOR'S ANALYSIS IN DETAIL

Now that you have tried analyzing this circuit for yourself and then gotten an idea of the questions I feel are important with regard to the circuit as well as try to answer them, let's go through the circuit together.

You will find that I will make use of a couple of Rules of Thumb that both time and usage have proved to be of great assistance.

RULE OF THUMB NO. 1 is for Transistors in the "ON" condition. With a germanium transistor the base will be 0.2 volt forward-biased from the emitter. With a silicon transistor the base will be 0.6 volt forward-biased from the emitter.

RULE OF THUMB NO. 2 is always go from the known to the unknown. Thus, we should pick out the points of known voltages as the first step in our analysis.

Regardless of what our amplifier is doing or where it is in the process of amplifying its input waveform, there are five points that will not change voltage which we can locate now (there are more, but we'll get to those later). These five are:

(1) The base of Q314 at 0.0 volt.
(2) The collector of Q323 at 0.0 volt.
(3) The collector of Q333 at 0.0 volt.
(4) The grid of V383A at 0.0 volt.
(5) The grid of V383B at 0.0 volt.

Since all these transistors are made of silicon, Rule of Thumb No. 1 comes into play to predict the voltage on the emitter of Q314. It will be at +0.6 volt.

The voltages on the collectors of Q323 and Q333 don't tell us much except that their bases and emitters will be at some negative voltage.

The zero volts on the grids of V383A and B allows us to estimate that the collectors of Q354 and Q364 will be slightly positive (bias for the vacuum tubes) and that their remaining elements will also be negative.

We have a start. Now let's see where we can apply Thevenin's concept to simplify our circuit and help us further along the way. There is a DC divider on the base of Q333 and another one on the collector of Q314. Let's take the one on the base of Q333 to see what it can tell us. The Thevenin Equivalent resistance for these two resistances is their product over their sum. So, 60K divided by 16K equals about 3.75K. And, 10K over 16K is ⅝. And ⅝ of —12.2 volts is —7.5 volts. NOW THEN: since it would take something between .2 & .3 ma to drop another volt across this 3.75K resistor, and we couldn't possibly have that much base current, we will assume —7.5 volts on the base of Q333. This allows us to estimate the emitter to be 0.6 volt more negative, at —8.1 volts, which is also the base of Q364 and we carry on from here, saying that the emitter of Q364 is another 0.6 volt more negative, yet at —8.7 volts.

If we can assume that we are at THAT POINT IN TIME when the driving waveform is half way through its change of voltage and the output of our **Paraphase—Cascode—Amplifier** has the spot located at the CRT electrical center horizontally, then we can carry this line of thought even further with useful results. This would demand balanced voltages all the way back to the bases of Q323 and Q333.

I gave you a clue, now I'll state this: Q333 doesn't handle any AC signal at all! It is a voltage regulator and impedance balancing circuit which establishes the one voltage around which everything else operates, and we have already established what voltages should exist around it. This means then that at this point in time, the voltages around Q323 must be the same; with —7.5 volts on the base and —8.1 volts on the emitter establishing the base voltage for Q354. This, in its turn, means that the voltage on the base of Q354 must be —8.1 volts and its emitter must be close to —8.7 volts.

We now have most of our "balanced condition" voltages which can tell us what the "balanced condition" currents should be.

Let's start at the emitter of Q354. With —8.7 volts to ground, we should have 91.3 volts across R357—a 10K resistor—for an emitter current of 9.13 ma. (After all, Ohm's Law does say that I equals V/R.) It should also be safe to assume that R368 is set at such a value as to allow the same amount of current to flow as emitter current in Q364 (9.13 ma). These same two currents will be cathode currents for V383A and B and will thus be the currents that will cause the voltage drop across the two output plate load resistors, R382 and R385 (the 15K resistors to +300 volts). So, 9.13 ma through 15K gives us a voltage drop of 137 volts; thus the output midsignal voltage must be equal to 300 — 137 or +163 volts on both sides.

A quick inspection of the emitter circuits of Q323 and Q333 shows us that we have 91.9 volts across R323 and also across R333 for an emitter current to each of their respective transistors equal to 91.9 divided by 47K or close to 2 ma (1.95 ma to be more exact).

At this stage of the game, we have Q314 left to figure out, and Thevenin comes to the rescue again. Looking at R318 and R319 from a —12.2 volts to —100 volts we find that we have an 87.8-volt drop across the pair if we momentarily eliminate Q314 as a load. This will give us an open-circuit voltage equal to the drop across R318 added to —12.2 volts and a Thevenin Equivalent Resistance equal to their parallel combination. So, —2.85 divided by 82.85 (the quantity) times 87.8 volts equals 2.9 volts. And, —12.2 added to —2.9 gives us a V_{oc} of —15.1 volts.

Let's see now—the Thevenin Equivalent Resistance of 2.85 times 80 (which is 228), divided by 82.85 gives an R_{th} of about 2.75K. This gives us the effective V_{cc} and the load resistance that Q314 is working into, along with the voltage on both ends of R_L. Checking this, we'll find that we have 7.6 volts across the equivalent 2.75K load resistance for a collector current of about 2.76 ma. This will also be almost exactly equal to the emitter current for Q314 and we may use it as such.

It will pay us to note here that if R316 is set close to its electrical center, we will have a V_{oc} of about +12.5 volts and an R_{th} consisting of two 75K resistors in parallel (from the center of 150K to AC ground at the two supplies), or 37.5K which shows up in series with R317 to the effective voltage we've just found. This adds up to almost 100K with +0.6V on one end and +12.5 volts on the other for a current close to 12V divided by 100K or 0.12 ma. Since it's quite possible we might be gaining this amount of current from the grounded base, let's just ignore this current and the base current and carry on as though they didn't exist. They are there, but I think that they are so small we can forget them.

However, R314 to +125 volts is a different story. With +0.6 volt on one end of it and +125V on the other, it will carry a good part of the Q314 emitter current. 124 volts divided by 107K gives us a current of 1.16 ma. When we take this away from the Q314 emitter current of 2.76 ma we have 1.6 ma left to tell us where the driving waveform has to be on the other end of R312 in series with R310.

We will use half of the R312 10K value as a "Design Center" resistance and add it to the 20K found in R310 for a total of 25K through which the 1.6 ma will travel to tell us what the vertical midpoint of the driving waveform is. This voltage will be equal to our 1.6 ma times the 25K, which gives us a positive voltage of 40 volts.

At this point in our analysis we can closely examine Position Control R316 to see if it is capable of centering the unblanked CRT beam when the sweep generator is not running. We have proved that we must demand an extra 1.6 ma from Q314 to achieve equal voltages throughout this amplifier, side to side. If the sweep generator is not putting out a signal, it's output will be down near ground, and the 1.6 ma will have to take the path through R317. The question facing us here is: "Can we put 1.6 ma of current through 60K and not require more than +125 volts which is as high in voltage as this resistor can be set?" Let's see, 1.6 times 60 with millis and kilos cancelling gives +96 volts. Well, what do you know, this adjustment obviously should be capable of actually moving the spot a ways beyond the center of the CRT since we still have 29 volts to go. Note also that R317 can be referenced to a negative voltage which would actually cut Q314 off by supplying more than enough current to R314, which normally carries 1.16 ma of the transistor emitter current. If this much current or a little more can be supplied from a different source, then no current in this network can come from the transistor and it is in the "off" condition. This would move the spot way off to the left of the CRT face, actually positioning the right-hand end of the sweep waveform to the left of center.

Well, our "Balanced Condition" DC analysis of this circuit is complete. We have found just about everything except the bias voltage on the output vacuum tubes and this isn't critical to us. It will be whatever the plate to cathode voltage and plate current demands that it be for the condition the tube is in. In other words, it is a result rather than a cause and as such—for analysis purposes—we can get by with ignoring it. However, a 6DJ8 with a plate voltage of +160 volts and 9 ma of plate current ought to have about —4 volts of bias if you want to include it. (See Fig. 7-7.)

The CRT on which this circuit produces a sweep trace has a 20 volt per division horizontal sensitivity. This means that for 10 divisions of sweep, we must have 200 volts of peak to peak change on the output leads, which tells us that we will have a 100 volt sweep waveform on each output plate, with V383A going negative and V383B going positive. And, in order to find out how large a sweep generator waveform we will need to achieve this, we must proceed with an AC analysis of this amplifier.

Before we forget it though, we'd better find the extreme of voltage found at the output, 50 volts up and 50 volts down from the calculated center voltage of +163 volts. V383A will start from 163 plus 50 or +213 volts and proceed down to 163 minus 50 or +113 volts. V383B down at +113 volts proceeds up to the +213 volt level before falling off. Now it is up to our driving waveform to give us enough signal to achieve this.

Since the base of Q364 is stationary, all of the driving waveform to this output amplifier will be seen on the base of Q354. The resistance over which this e_{in} appears is essentially made up of Transresistance of Q354 (26 divided by 9 ma of emitter current plus a few ohms R_{eb}, or about 8 ohms of impedance), R364 (the 1.21K resistor) and Transresistance for Q364 (about 8 ohms or the same as for Q354); all of this means a total resistance of 1.226K. The signal current so developed

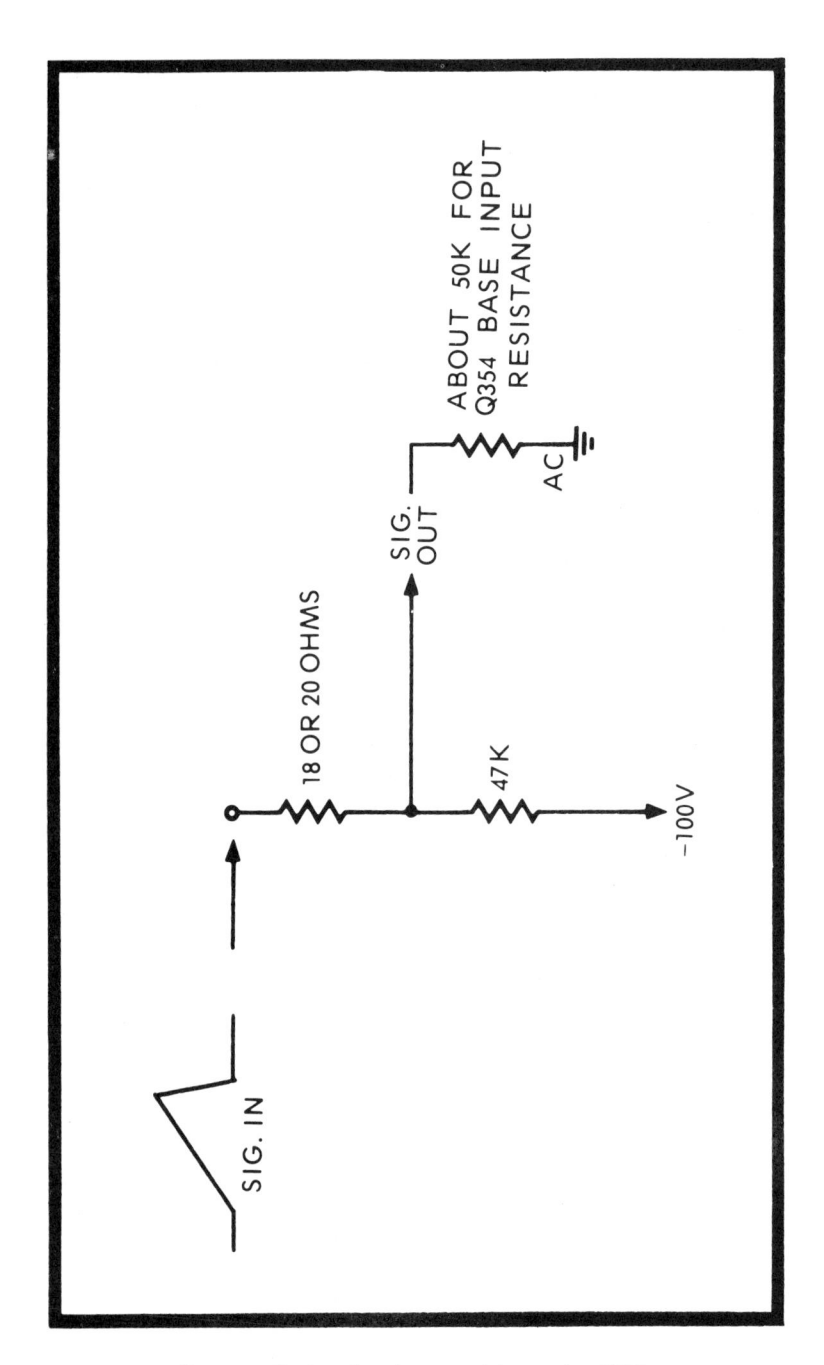

Fig. 7-2. Estimating transresistance for Q323.

will essentially come from the plate load to V383B and will be delivered to the plate load resistor (R382) for V383B to achieve our output signal. Thus the A_V from the base of Q354 to the plate of V383A must be about equal to R382—the 15K resistor divided by the composite resistance that e_{in} appears across, or the 1.226K we figured it to be. 15 divided by 1.226 is about 12.25. So, now I can say that the 100-volt change seen on the output is 12¼ times greater than the signal we should have on the base of Q354. (100 divided by 12.25 is close to 8.16 volts drive at this point.) The emitter of Q323 must succeed in giving us this 8.16 volts of signal over whatever its AC emitter resistor is. And, this emitter resistor is a composite. It will be the parallel combination of R323 and the input resistance seen looking into the base of Q354. The input resistance to this output stage is the 1.226K resistor that e_{in} appeared across, magnified (more correctly, multiplied) by beta for Q354. (It is not out in left field to use a general figure of 50 for the beta of an unknown transistor in this day and age, since this value is usually quite close to average or even worst-case condition for modern transistors. I suppose I should have listed this as RULE OF THUMB No. 3—but I didn't.) So now we can say that Q323 has an AC emitter resistor made up of 47K in parallel with about 50K, or something close to 23K in value. If we take a look at Transresistance for Q323 (26 divided by 2 gives us 13— add a few ohms R_{eb} and we estimate it to be 18 or 20 ohms, worst case), we can see that, in series with this 23K emitter resistor, it won't lose any amount of signal worth worrying about. See Fig. 7-2.

So, we can now figure that the full 8.16 volts of drive will also be found on the base of Q323, and since an emitter follower doesn't give us a phase change, it should look like the same signal found on the base of Q354. Since the signal will go through a 180-degree phase shift from the base of Q354 to the plate of V383A, we can now note that we must have a positive-going signal at the two points in question (the bases of Q354 and Q323). At this point in the analysis we can say that the driving signal seen on the base of Q323 will start at a point 4.08 volts more negative than the design center of —7.5 volts, which means at —11.58 volts, and the driving signal will move to a point 4.08 volts more positive than the —7.5 level or —3.42 volts to achieve the change needed. The base of Q354 will do the same thing around a center of —8.1 volts rather than —7.5V.

Finally there is the grounded-base stage in which we find Q314. See Fig. 7-3. This transistor has a transresistance of about 15 ohms when we figure it by dividing 26 by 2.76 ma and adding something close to 5 ohms for R_{eb} to estimate it. It

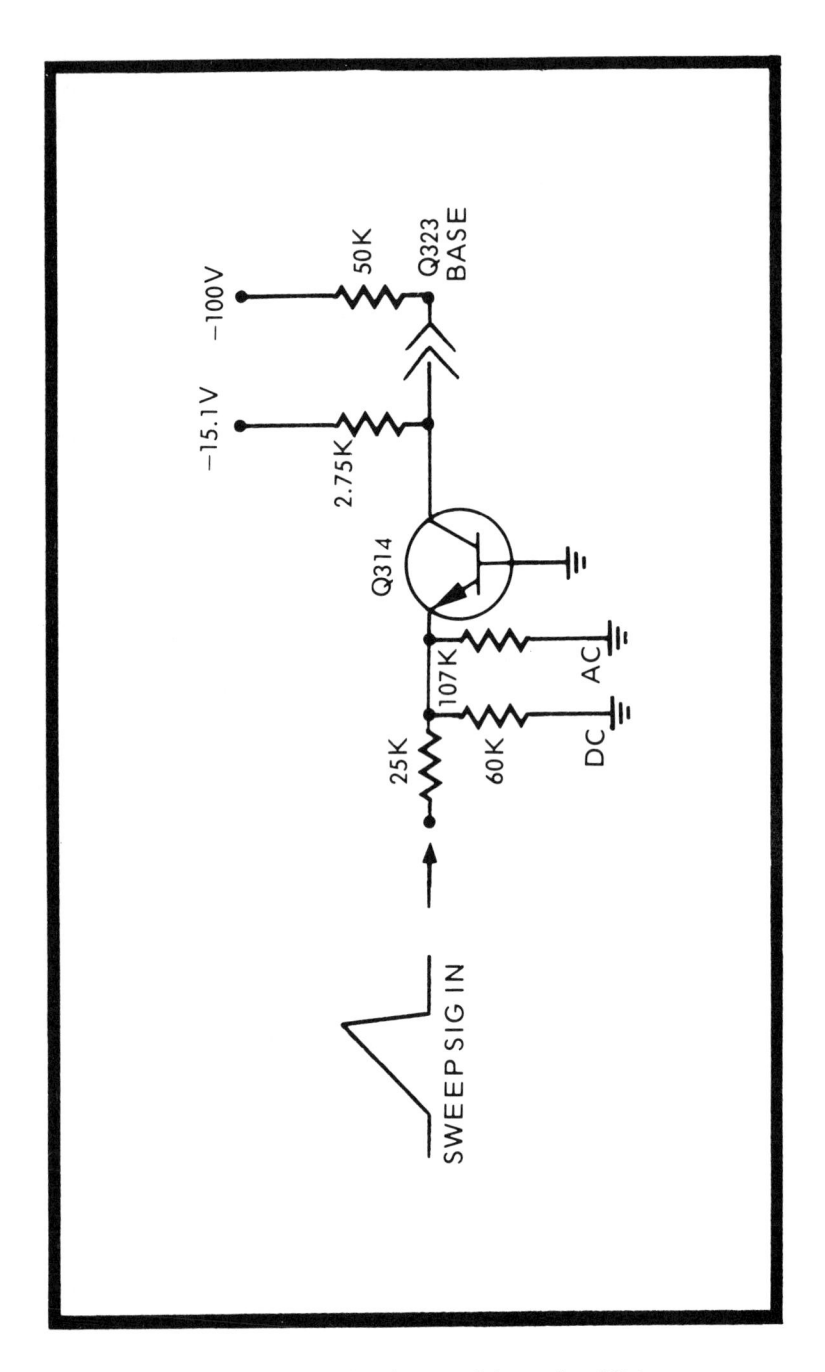

Fig. 7-3. Estimating transresistance for Q314.

really isn't too critical as long as it offers us a comparatively low impedance point to drive from several generators working through resistance that is many times higher (this prevents one generator from distorting the other generator signal and results more in an addition of their signals).

The positioning control will have a minimum of 60K through which it drives this point, and the Delayed Sweep Generator (the one in use according to the switch position) has 25K approximately, through which it drives this same point; and, of course, we can also use the Normal Sweep Generator through the same 25K the Delayed Sweep Generator works through. Quite obviously, none of these generators will load each other down due primarily to the 15 ohms transresistance seen in the emitter of Q314. (You see, Q314 gets all the signal current from any combination of generators driving it due to its low 15 ohm impedance.)

What's that you say? "The plate load reflects into the emitter also!" Yes, it does, but remember it is divided by the mu of the transistor. Now, mu equals G_m times r_p. And transistors in general have 1/Hoe values estimated on the low side near 10K. Along with a G_m equals $1/(z_K)$ measurement, in this case the reciprocal of 15 ohms, the answer is a bit better than 60,000 micromhos. This would give us an estimation of mu equal to 10,000 multiplied by 0.06, for a value of 600. Now then, 2,570 divided by 600 is in the neighborhood of 4 ohms which we might have neglected to add to the r_{tr} estimate of 15 ohms. This might be almost a 20 percent increase in transresistance, but when considered in series with 25,000 ohms, it doesn't make much difference whether we bother to add it or not. So maybe r_{tr} is 20 ohms. This total impedance is less than .1 percent of 25 thousand ohms and by leaving it out we only incorporate an error of ¼ of that 1/10th of one percent. Why bother about it?

Looking at the voltage gain of this circuit, we find that e_{in} must give us a current change that will create an 8.16 swing in voltage across the Q314 R_L equivalent value of 2.75K. The actual sweep generator will have to do this across an impedance of close to 25K, so the A_V of this stage will approximately equal 2.75K divided by 25K, for something close to 0.11. This says that the sweep generator must give us a swing in voltage equal to our 8.16V divided by 0.11 for a signal very close to 74 volts in amplitude, half of it above the signal center of +40 volts and half of it below that same center. 74 over 2 is 37, and 40 minus 37 is +3 volts (the quiescent starting level of voltage for the sweep generator). And 40 plus 37 is 77 volts (the peak of the input signal).

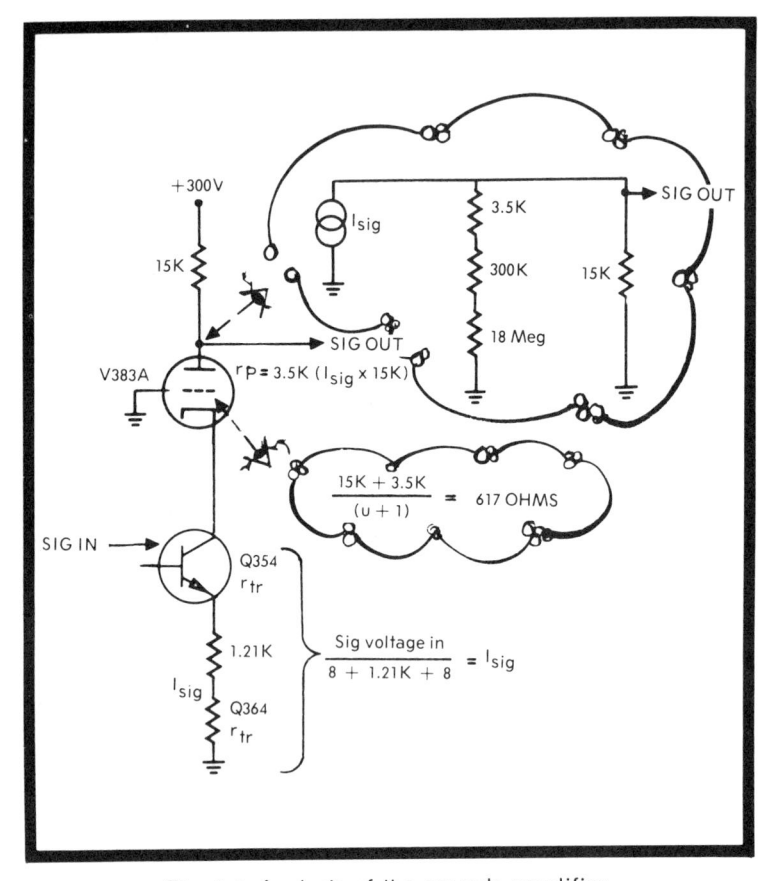

Fig. 7-4. Analysis of the cascode amplifier.

There now! I think our analysis is fairly complete. If you happen to be a user of a type 3B3 TIME BASE plug-in for a Tektronix Oscilloscope, I think you'll find my conclusions quite close to what is actually happening in the instrument (see Fig. 7-6).

This leaves us with just one major point that deserves clearing up for those readers who are not familiar with cascode amplifiers, or are in a situation where they haven't thought about them for a long time. I can hear you saying, "How do you get by ignoring r_p of the triode showing up in parallel with the 15K Load?" (See Fig. 7-4.)

Well, let's take a look at it. The r_p of the 6DJ8 with V_p equals +150V and I_p equals 10 ma is quite close to 3.5K and it will be found in an equivalent circuit path that is parallel to the 15K load resistor. However, there are a number of other

246

composite resistances in series with it. We will have the 1/Hoe value of Q354 (let's use the 10K value for a transistor's r_p even though it may be a bit low). This, we must remember, is also multiplied by the mu of the 6DJ8 (u equals 30) and adds something close to 30 x 10K equals 300K ohms to the r_p of the tube when viewed from the output plate. This would be sufficient in itself to cause us to ignore it, but there is more yet. We will also have to consider the composite resistance made up of the Q354 1.21K (emitter resistor) and the Q364 r_{tr} which was something like 1.21K. This impedance will be multiplied by the Q354 mu and the mu of the vacuum tube (that's about 600 x 30 for a magnification of about 18,000) which adds up to another series resistor close to 18 Meg in value. This 18 Meg resistance will add to our 300K resistor which will add to the 3.5K, the r_p resistance to form the impedance seen in parallel with the 15K resistor R382. Being at least 1,000 times greater in value, the impedance won't shift the output resistance of the

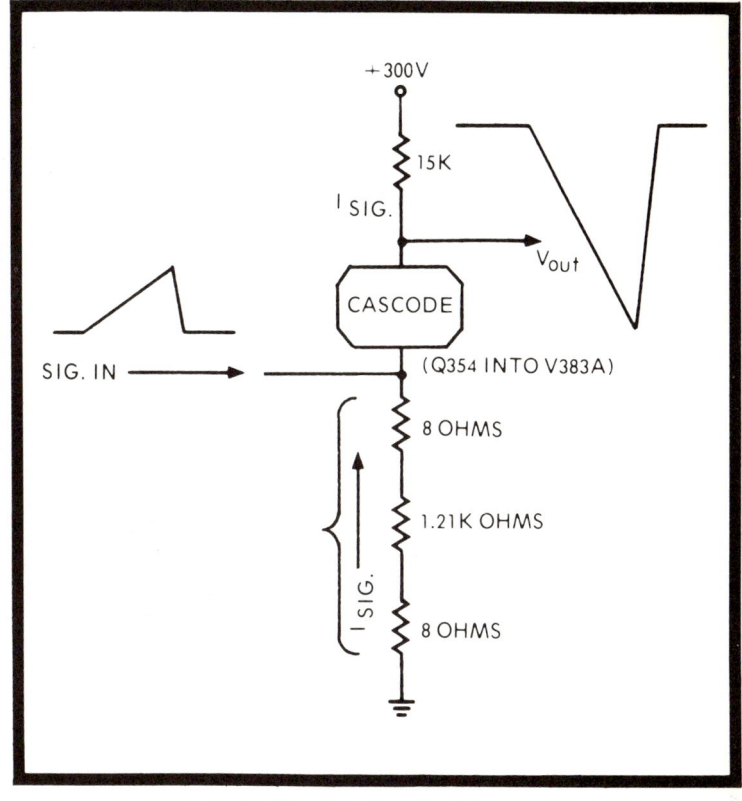

Fig. 7-5. Simplification of Fig. 7-4.

TYPE 3B3 PLUG-IN

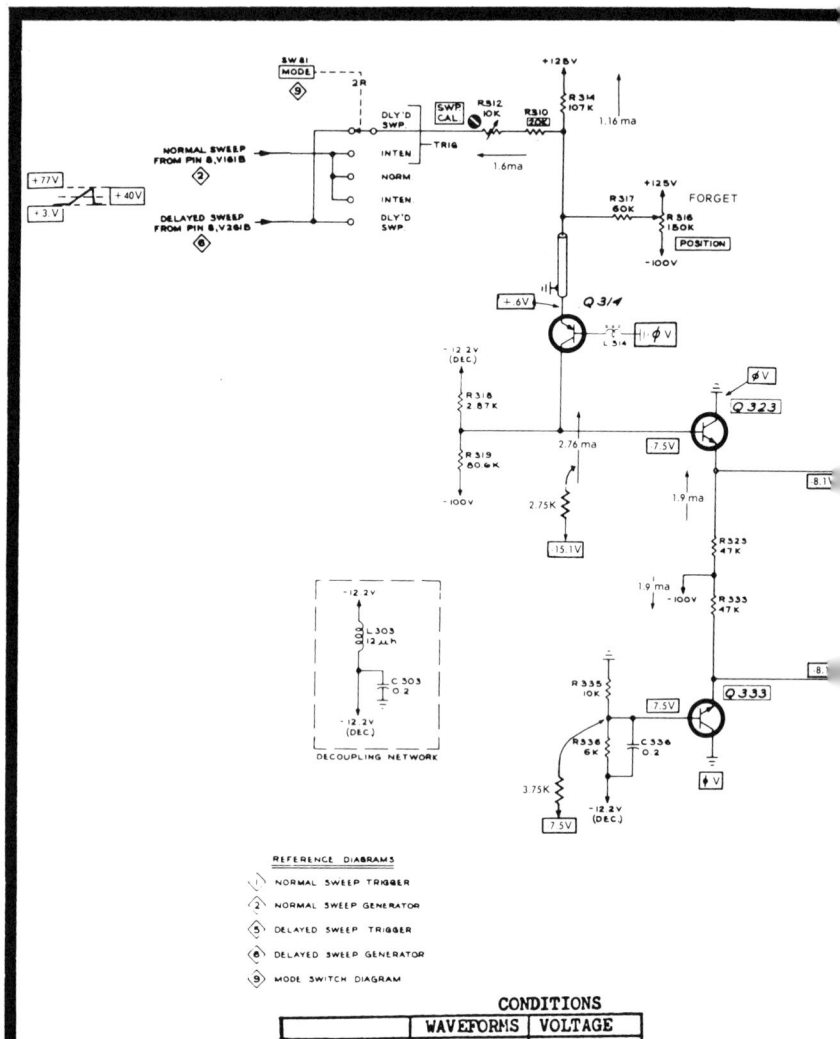

REFERENCE DIAGRAMS
1. NORMAL SWEEP TRIGGER
2. NORMAL SWEEP GENERATOR
5. DELAYED SWEEP TRIGGER
6. DELAYED SWEEP GENERATOR
9. MODE SWITCH DIAGRAM

CONDITIONS

	WAVEFORMS	VOLTAGE
POSITION	Centered	Centered
T/D		
MODE	Normal	
SWP COUP	Auto	
5X MAG		

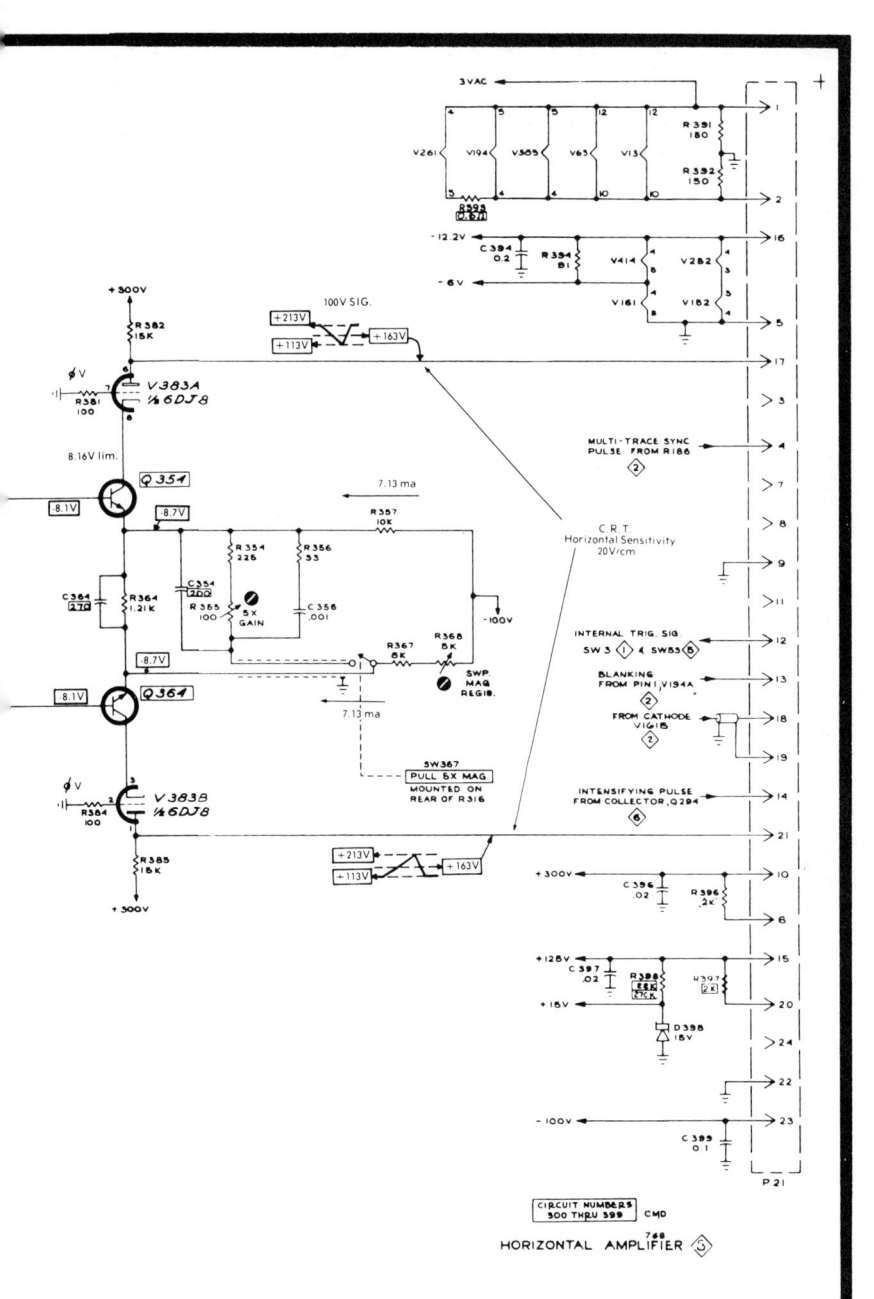

Fig. 7-6. Fig. 7-1 with values as concluded by author. (Courtesy Tektronix, Inc.)

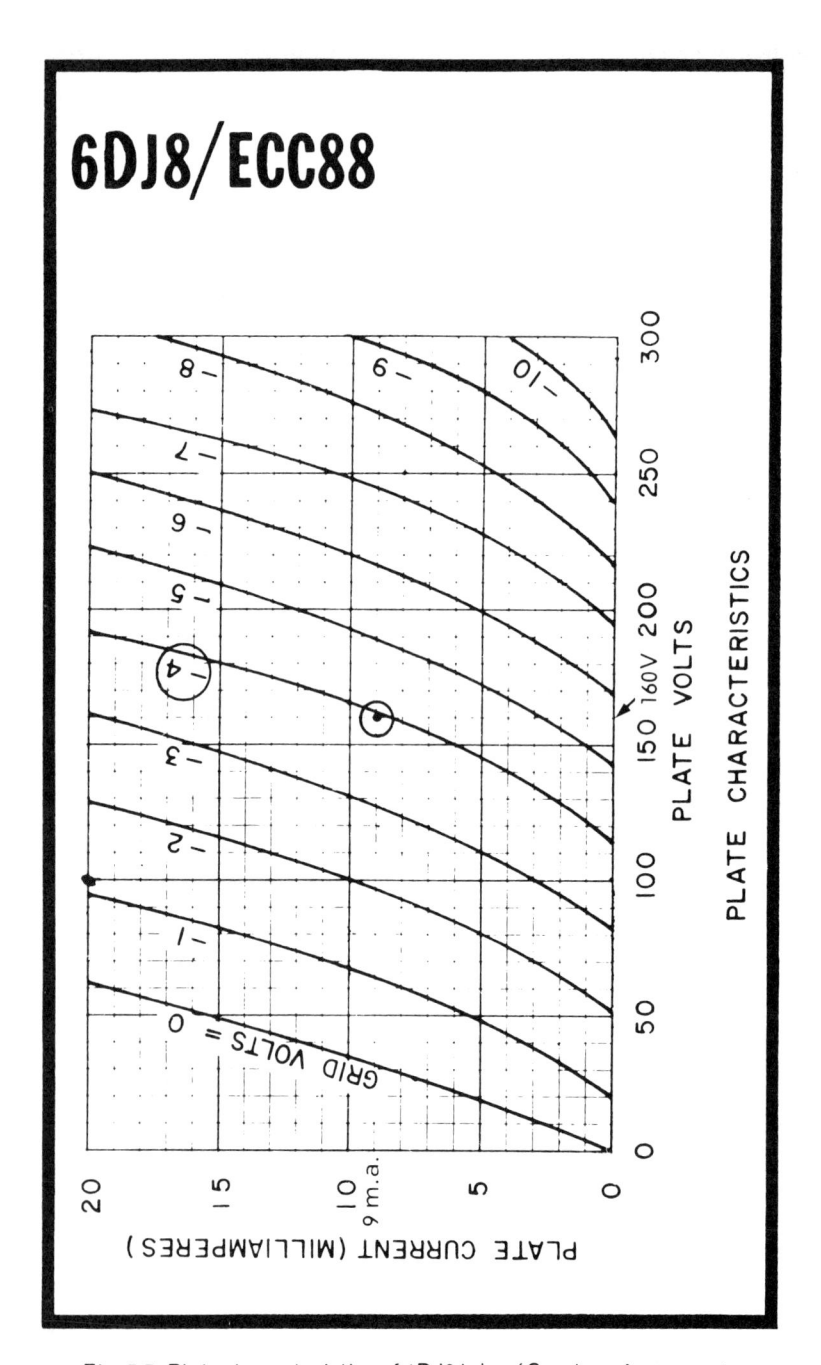

Fig. 7-7. Plate characteristics of 6DJ8 tube. (Courtesy Amperex.)

circuit away from the 15K value seen to the AC ground point in the +300 volt supply by any appreciable amount. It is for this reason that we can ignore it, and it is one of the glories of the cascode type of circuit. Also see Fig. 7-5.

The transistor (Q354) will react as though it were working into a load impedance, not of 15K, but the cathode impedance of the 6DJ8 plus R382 over mu (Z_K equals 117 + 15K/30 equals 617 ohms). This 617-ohm load for the transistor will give us a voltage gain from base to collector of far less than 1 and a frequency response capability beyond the limitation of f_t of the transistor. The grounded-grid amplifier stage represented by the 6DJ8 has a natural high frequency response too, since its C_{gk} and C_{gp} are in series, representing a minimum value even though in parallel with C_{kp}. This total capacitance changes its charge relative to in-phase signals from cathode to plate of the tube, which also helps minimize internal capacitance for good frequency response.

There we have it. I hope that in some way I've been able to point out how some of these ideas can be of assistance to you.

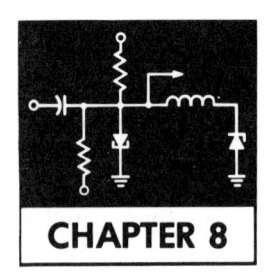

CHAPTER 8
Frequency Response, and Operational Amplifiers

As you no doubt learned a long time ago, no matter what active device is used in any given amplifier, it has input capacity. Looking into the grid (or whatever takes its place) we have the following;
(1) For vacuum tubes, there is grid-to-cathode capacitance and grid-to-plate capacitance,
(2) For transistors, there is the reverse-biased collector-base junction—which is a capacitor, and
(3) For FETs, there is the reversed-bias junction made up of the gate and both ends of the channel—the source and the drain.

This input capacitance is usually small enough so that its RC time constant is quite small in relation to the times given it to change voltage at mid-frequency ranges. However, as we both know, there is bound to be an upper point in frequency where this capacitance is not given adequate time to change voltage before the input signal is already changing back the other way. This in its own way says that the output swing in voltage will be something less than it was at mid-frequency speeds of change even though the input amplitude is exactly the same. This is loss of gain due to frequency response limitations.

Please understand also that this input capacitance for the active device is not the only capacitance that is responsible for this loss of gain under high frequency operation. There is capacitance in the wire that hooks up the circuit. There is capacitance in the type of socket used for the active device, and several other sources of capacitance as well. All these capacitances act together and many of them we cannot predict. BUT, it does pay us to know the limitations placed on a circuit by those we can predict, and that's what this discussion is all about.

Fig. 8-1 is an idealized frequency response curve. The horizontal axis of this graph is in terms of frequency while the vertical axis is in terms of gain (either voltage or current).

252

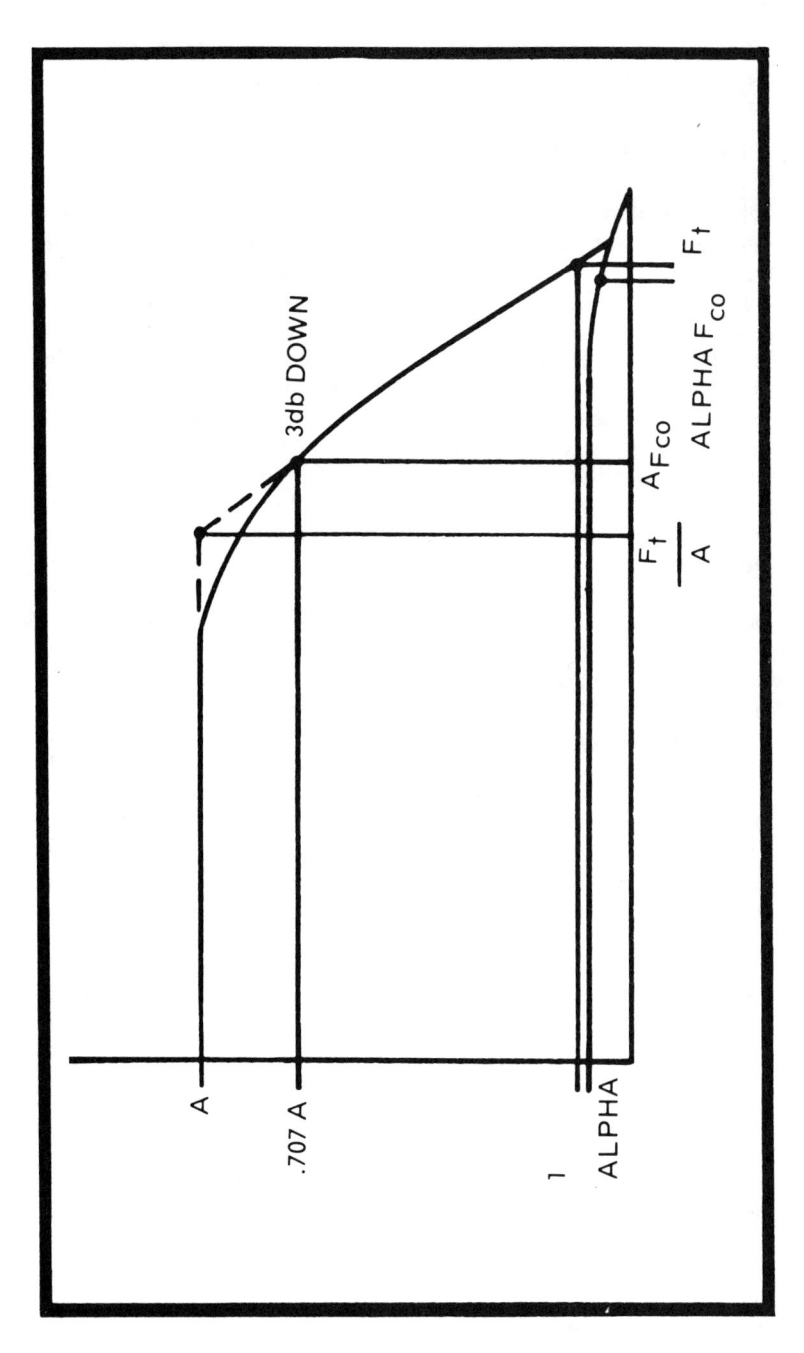

Fig. 8-1. Voltage (current) gain vs frequency, graph.

Sometimes the vertical axis is in terms of power, but very seldom. Our graph, though, is in terms of either voltage gain or current gain.

FREQUENCY RESPONSE

If we follow this response curve out to its right-hand end where we have it showing a gain of one (A equals 1) we have a terminal point in frequency that is of interest to us. With vacuum tubes (vertical axis in terms of mu) this point is called the "Gain Bandwidth Product." With transistors (vertical axis in terms of Beta) this point is called F_t. It is of interest to us in a couple of ways. This frequency can be thought of as being the very highest frequency this device can be thought of as amplifying. Or, we can divide it by the mid-frequency gain value (the height of the middle part of our curve) to predict roughly the slope of frequency fall off. The frequency response may not actually follow this slope exactly, but it is close enough to give us a general idea of what to expect.

The point on this slope where the power output of the device is one half of its mid-frequency level is also of interest to us. Since a 6 dB change of power is a 100 percent change, it stands to reason that a loss of 3 dB is a 50 percent loss, and this half-power point is called the "3 dB down" point.

But, you say, "That point isn't half way down on that curve."

I agree with you. It isn't half way down on the curve AND that curve isn't drawn in terms of POWER either. Electrical POWER is the product of Voltage and Current. Our graph—remember—is in terms of voltage OR current. This does make a difference, so let's see what that difference is.

$$\text{POWER (Watts)} = V \times I$$

Now, since this is the half-power point, let's divide each side of this equation by the factor 2.

$$\frac{W}{2} = \frac{V \times I}{2} = \frac{V}{\sqrt{2}} \times \frac{I}{\sqrt{2}}$$

Remember, when we multiply fractions, the numbers on top multiply each other and the numbers underneath multiply each other. For example, $(2^{-3}) \times (2^{-3})$ equals (4^{-9}) and in the case we have here, the square root of 2 times the square root of 2 is 2. It's the only way we can break down the half-power point into terms of just voltage or current alone rather than their product. Now, we can divide the mid-frequency value of gain by the square root of 2 or multiply it by the reciprocal of the square root of 2. This will give us the value of gain on the slope of our curve at the half-power point or "3 dB down" point.

What's that? How do we find the square root of 2? Oh that's easy on your slide rule. Set the cross-hair over 2 on the left hand side of the A scale (its an odd number of digits in front of the decimal). Read the square root of 2 under the cross-hair on the D scale. To find its reciprocal, move the sliding scales (B, C, and, usually, C1) to a position where the B scale matches the A scale and the C scale matches the D scale. Now the reciprocal of the square root of 2 is 1.41 and its reciprocal is 0.707. This gives rise to calling the half-power point on a curve the "point seven oh seven" point as well as the "3 dB down" point.

GAIN TIMES BANDWIDTH

For purposes of standard bandwidth measurement, the 3 dB down point is usually used as the terminal point. Thus for a DC-coupled device, the 3 dB down point is also the bandwidth of the device. However, for an AC-coupled circuit (like a given amplifier) there is a low-frequency 3 dB down point also and the bandwidth is the frequency range between the two.

All of this gives us a couple of new ideas that can be written down as equations that are related to each other.

(No. 1) is:

$$A \times BW = F_t$$ (Gain times Bandwidth

equals the unity gain frequency). This is true only so long as the gain is equal to or less than the maximum characteristic of the device in question.

(No. 2) is: $A_{3dB} = A_{mid\ freq} \times 0.707$

and (No. 3) is: $$Freq_{3dB} = \frac{F_t}{A_{3dB}}$$

Now, all of this is great, except: How do we find the parts of that first equation? We know that A for vacuum tubes is the mu of the tube and is listed by the manufacturer. For FETs it's G_{fss} divided by G_{oss}. For transistors, A is Beta, of course. But how do we find this "F_t" bit? Well, sometimes for transistors it is listed as a characteristic of the device and once in a while we find it listed for vacuum tubes under the heading of "Gain Bandwidth Product." However, we can estimate it for ourselves if we have an idea of the capacitance involved, and these are most times more available than the F_t characteristic.

All of this hinges on the idea that the Gain Bandwidth Product is the same at the 3 dB point as it is at the F_t point. For that matter, this is true all the way down the slope in our response curve and determines the slope.

We have to go back to the universal gain formula

$$(A = G_m \times R_o),$$

and note that from this equation we can get

$$R_o = \frac{A}{G_m} \quad \text{(We'll use this in a minute.)}$$

At the 3dB down point, the capacitance involved in our situation must be using up half the power available to R_o and thus must be equal to R_o. Thus we can say:

$$R_o = X_C = \frac{1}{2\pi fC}$$

With this approach, I must think of "f" as being the 3 dB down frequency and for a DC-coupled amplifier this is also the bandwidth. So, when I transpose the quantities "f" and "R_o," I get:

$$F = \frac{1}{2\pi R_o C} = BW \quad \text{or bandwidth.}$$

Now remember R_o equals A over G_m, and I can substitute equals for equals. This gives:

$$BW = \frac{1}{2\pi \frac{A}{G_m} C}$$

And, when I divided with a fraction, I invert it and multiply. So, I can say:

$$BW = \frac{1}{2\pi C} \times \frac{G_m}{A} \quad \text{which simplifies to:}$$

$$BW = \frac{G_m}{2\pi CA}$$

Now, if we transpose the factor for Gain to the left side of the equation, we have the Gain Bandwidth Product in terms of the active device in question:

$$A \times BW = \frac{G_m}{2\pi C}$$

The only thing mysterious about it is the value of C. It is that part of the capacitance that changes voltage.

DETERMINING STRAY CAPACITIES AND THEIR EFFECTS

Let's take a cathode follower circuit for instance, and assume that the cathode follows the grid by 95 percent of the input signal. The only part of the grid-to-cathode capacity that changes voltage is 5 percent of it and thus it looks like only 5 percent of its actual size. This value (.05 x C_{gk}) has to be added to the grid-to-plate capacity which (since the plate voltage does NOT move) only changes voltage equal to one times its value. So, in this case C equals (.05 x C_{gk}) + (1 x C_{gp}). See Fig. 8-2A.

However, if we have a plate-loaded amplifier stage where the cathode is grounded, Fig. 8-2B, C takes on a completely different value. In this case, we will have all of C_{gk} since the cathode remains stable and it has to change voltage equal to the input signal amplitude. This adds to the grid-to-plate capacitance which is magnified many times since it has to change voltage by an amount equal to the gain of the stage plus one times the input signal amplitude. (This is called Miller Effect.)

You will note from the Gain BW formula,

$$A \times BW = \frac{G_m}{2\pi C}$$

that the bigger the value of C, the lower the frequency response. Comparing the two dynamic values of C in the percentage formula explains why the frequency response of the cathode follower is so much better than the frequency response of the plate-loaded amplifier.

These formulas may not be perfect, but they can tell us if we are expecting more out of a circuit than it can supply. Also, if we are not getting anything near what a circuit ought to be capable of, the formulas can tell us something is wrong.

Since the proof of the pudding is in the eating rather than in the anticipation, perhaps we'd best measure our circuit. And the most obvious way to do this would be to put in a constant amplitude signal (note the gain of the signal at the output) boost the frequency a bit and note the gain again, etc. However, a constant-amplitude wide-frequency-range generator is sometimes hard to come by, and such a process takes time and patience.

RELATING BANDWIDTH, RISE TIME, AND RC

Perhaps it has occurred to you that this business of Frequency Response can be related to Rise Time (T_r) somehow. And, it can. So let's see how.

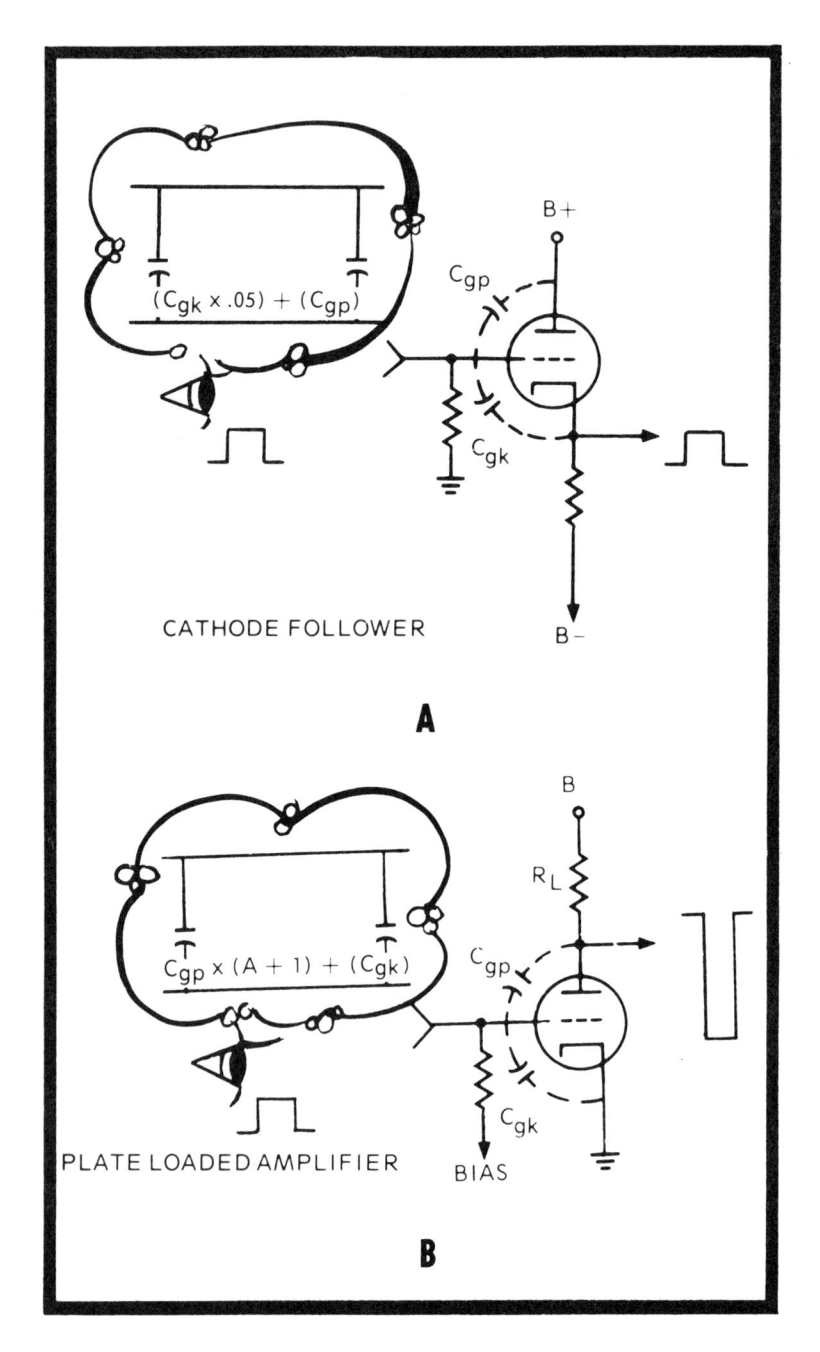

Fig. 8-2. Analyzing stray capacitances.

Part way through the last derivation a couple of pages back we noted that at the 3dB point (the half-power point) we had a condition where:

$$R_o = X_C = \frac{1}{2\pi fC}$$

so let's start here. Perhaps we can build an equation that has the factor 2.2 RC. This, after all, was the definition of Rise Time when we were studying the RC curve in Chapter 2. Remember?

Well,

$$R_o = \frac{1}{2\pi fC}$$

and if we transpose Ro and f, we get;

$$f = \frac{1}{2\pi R_o C}$$

And since f equals BW

$$BW = \frac{1}{2\pi R_o C}$$

Now if I multiply this by one, I won't change its value any, and 2.2/2.2 equals 1, so I should be able to say:

$$BW = \frac{1}{2\pi R_o C} \times \frac{2.2}{2.2}$$

This gives us:

$$BW = \frac{2.2}{2\pi \; 2.2 \; R_o C} \text{ or } BW = \frac{2.2}{2\pi T_r}$$

Now we can transpose the Rise Time factor in our equation to the left side to get something of interest to us:

$$BW \times T_r = \frac{2.2}{2\pi}$$

The right side of our equation now is a numerical constant. It won't change with the type of the device nor the circuit it is in. The value for "pi" (π) is 3.1416 and two times this is 6.2832. 2.2/6.2832 is 0.35 and we can write: BW x Tr = .35

And, of course,

$$BW = \frac{.35}{T_r}$$

A

This Transistor Op Amp can be thought of like this:

B

C

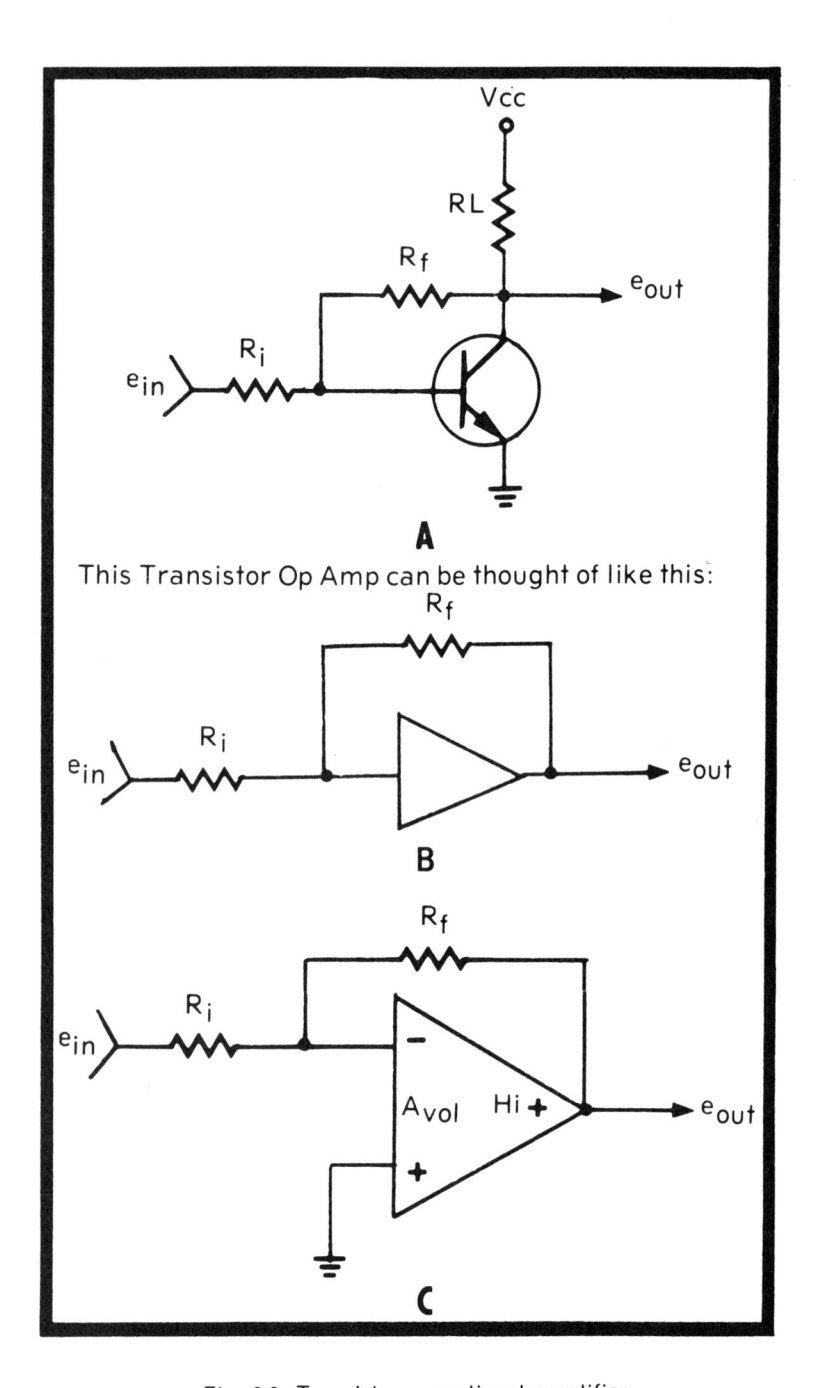

Fig. 8-3. Transistor operational amplifier.

This says that if I can put a STEP WAVEFORM into my amplifier circuit and measure its rise time at the output, I can divide its rise time into point three five to find the 3 dB down point in frequency. How about that? Believe me—it's useful.

OPERATIONAL AMPLIFIERS

The next step in circuitry that seems to need explanation is the variation where the output signal of the amplifier is fed back in such a manner as to obtain a reduction in effective driving signal and thus limit the amplifier's output signal amplitude. I will call this type of circuit an Operational Amplifier for the remainder of this text.

Incidentally, if you want to start a good argument among engineers, just ask them to define an operational amplifier. You are liable to get as many different definitions as you have engineers. They will agree on the qualities an op amp has to have, but each will have his own definition. Don't let this bug you.

This line of thinking (the operational amplifier concept) has even given rise to its own equivalent circuit system of drawing schematics so as to show only the critical parts of the circuit rather than the whole thing. A simple triangle drawn on a horizontal axis becomes the amplifier between the point chosen for signal pick-off and the point to which the signal is fed back. In other words, it is that active device or group of active devices that exist within the feedback loop. Rarely will you find more than three active devices within the feedback loop, but I guarantee nothing.

TRANSISTOR OP AMP

For purposes of basic understanding, let's take the simplest form of this circuit first. We will have an amplifier stage made up of a single NPN transistor in a grounded-emitter configuration with a feedback resistor from the collector (R_f) to the base and another resistor (R_i) in series with input change of voltage (e_{in}) applied to the stage. The output signal is taken from the collector (e_{out}). See Fig. 8-3.

At quiescence, waiting for a signal to come in, the emitter will be at ground with the base at +0.6 volt above ground and the collector more positive yet, somewhere between the base voltage and V_{cc}. We will assume for the moment that no current is flowing in R_i and e_{in} is at +0.6 volt also. The transistor is turned on and waiting. Now then, let us assume a small positive-going step voltage is driven into R_i. At first this signal demands current from the transistor base-emitter junction and starts to increase the collector current by the factor Beta. This causes the collector voltage to go negative, reducing the voltage drop across R_f. When this reduced voltage drop across R_f is large enough to reduce the current through it to the magnitude demanded by the input signal across R_i, the transistor sees no more drive and the voltage change at the output stops. Actually, our transistor is still amplifying the amount of signal it really sees on its base by the factor of transresistance divided into the load, but this drive is reduced (more than you realize) by the feedback. Of course, the load is made up of the parallel combination of R_f, R_L, and the input resistance of whatever this circuit is driving. This in turn determines the gain actually seen across the transistor, which is now called the open loop gain (A_{Vol}). However, as I pointed out, the actual amount the base voltage will move is much smaller than would be the case if R_f were not connected to the collector.

In this circuit, the input side of R_f moves one direction in voltage while the collector end moves the other way relative to the open-loop gain of the transistor. This means that there is a point just a short distance into R_f that carries signal current, yet does not change voltage (an equivalent ground point). If R_f were divided into ($A_{Vol} + 1$) parts, one of these parts would appear in parallel with the input of the transistor base-emitter junction. The remaining portions of R_f would appear in parallel with R_L and the load to determine the actual open-loop gain. For all practical purposes, we can ignore the difference in open-loop gain that this reduction of R_f would give us since it is so small. However, bear in mind the fact that it will only be small if the open-loop gain is large with respect to the total gain of the stage.

Since the signal current either supplied or demanded by the input signal is driven through R_i and then shunted around the active device (effectively) by the output change of voltage across the great majority of R_f, the output change of voltage is limited to the ratio of two resistors involved. A_V for the stage is equal to R_f divided by R_i. And this is close enough for most practical purposes. Actually, that small part of R_f shunting the input of the transistor to the equivalent ground point in this resistor should be added to R_i and taken away from R_f to make our calculation more exact. However, in most general cases this is not necessary.

With the base of the transistor so close to an equivalent ground point, R_i is usually quite large with respect to that small part of R_f that shows up here, and so we can usually consider R_i to be the input resistance of this stage.

FREQUENCY-COMPENSATED OP AMP

Output resistance, though, is a different story. Let's look at it for a minute. This R_O will be an AC impedance relative to the change in voltage and the change in current that caused it. This output change in voltage appears across R_f, causing signal current to travel through this resistor only relative to a current change in the output that is Beta times greater than at the input. This consequently makes R_f appear to be divided by Beta when seen looking back into the output of this circuit, which in turn gives us an AC R_O much smaller than that which dictates the open-loop gain. The high-frequency response on an RC time constant basis is much better because of this. For that matter, if R_f and R_I are compensated with capacitors like a resistive divider such that C_i x R_i equals C_f x R_f, the gain x bandwidth of the stage will approach F_t of the active device. So, if we know F_t for the transistor, we can divide it by the R_f over R_i gain of the stage to determine the frequency response of the stage (or—if you will—its 3dB down point). Now this won't be absolutely exact, but you can use it as a rule of thumb by which you can judge the response of the stage. See Fig. 8-4.

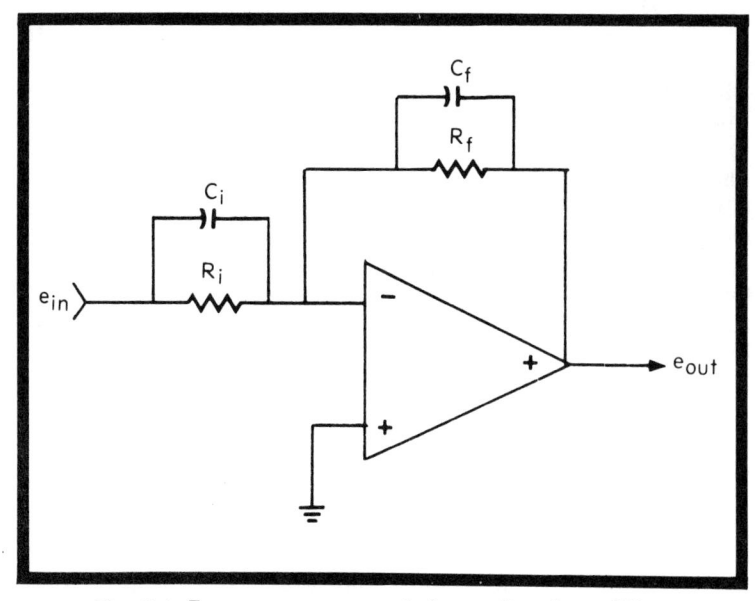

Fig. 8-4. Frequency-compensated operational amplifier.

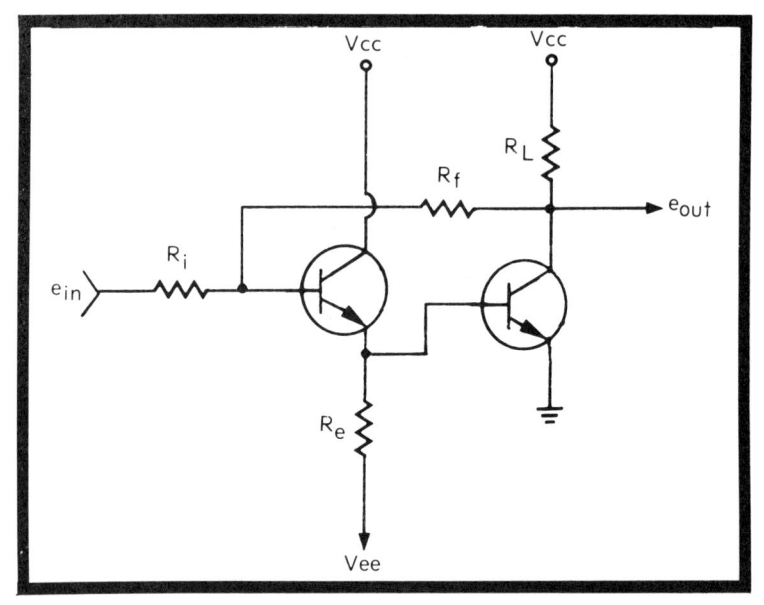

Fig. 8-5. Emitter follower driving common-emitter amplifier in an Op Amp configuration.

EMITTER FOLLOWER—COMMON EMITTER OP AMP

This low output resistance effect isn't too noticeable with just one active device within the operational amplifier, but when we have two transistors within the feedback loop, it can make for an extremely low value of AC R_O. Take an emitter follower driving a common-emitter amplifier and put them inside the feedback loop. (See Fig. 8-5.) Now we have a current-change capability in the output of the circuit more closely related to the product of the two Betas. Even with a relatively average pair of Betas near 50, this would divide R_f by a factor of 50 x 50 or 2,500. In the case of R_f being a 22K resistor, this would mean that the output resistance would be less than 10 ohms, and this can be a big help in maintaining frequency response. The bandwidth-risetime product is still 0.35, and with such low output resistance, the $(2.2 \times R_f/A_i)$ C factor can be quite small, thus making the bandwidth factor quite large and very useful.

PUSH-PULL OP AMPS

More and more, we are finding paraphase amplifiers used to drive two stages of voltage gain using feedback from

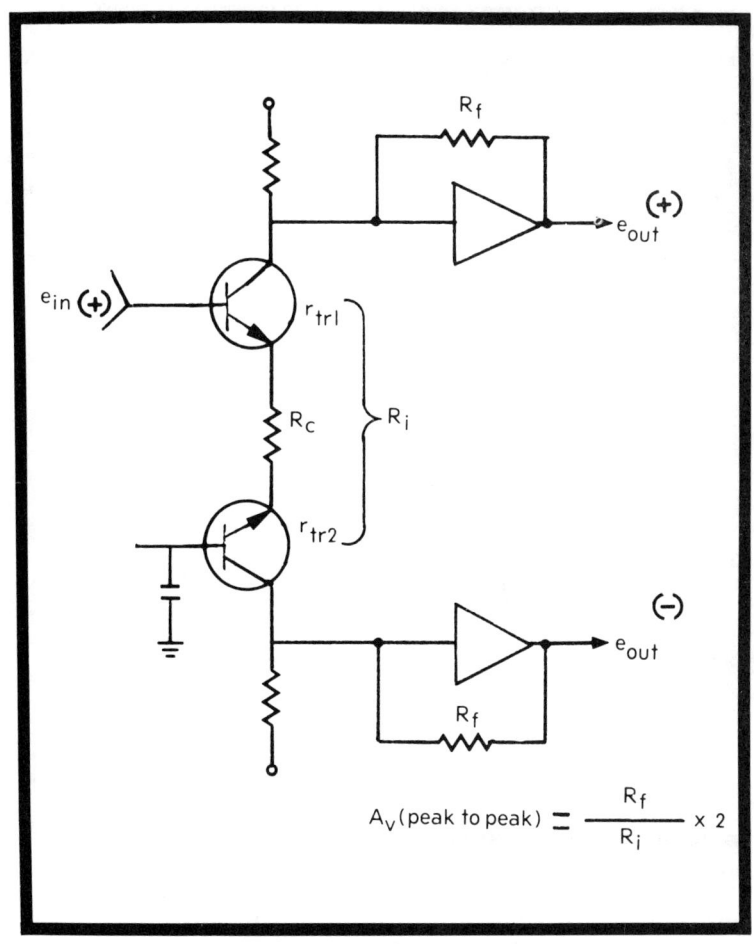

Fig. 8-6. Paraphase amplifiers connected as push-pull Op Amps.

$$A_V(\text{peak to peak}) \cong \frac{R_f}{R_i} \times 2$$

collector to base such that we have an R_f on each side of a push-pull stage. What we effectively have are push-pull op amps that are sharing a common R_i, as long as R_i is defined as the resistance over which the input signal is developed, thus determining the signal current that goes through or from R_f in our operational amp. Note what we have in Fig. 8-6. The resistance over which e_{in} appears is $(r_{tr1} + R_c + r_{tr2})$ with the active devices of the paraphase stage demanding that this current come from one R_f of the push-pull stage and supplying this signal current to the other R_f on the opposite side.

There is one other configuration of the push-pull op amp driven with a paraphase stage which conforms to operational

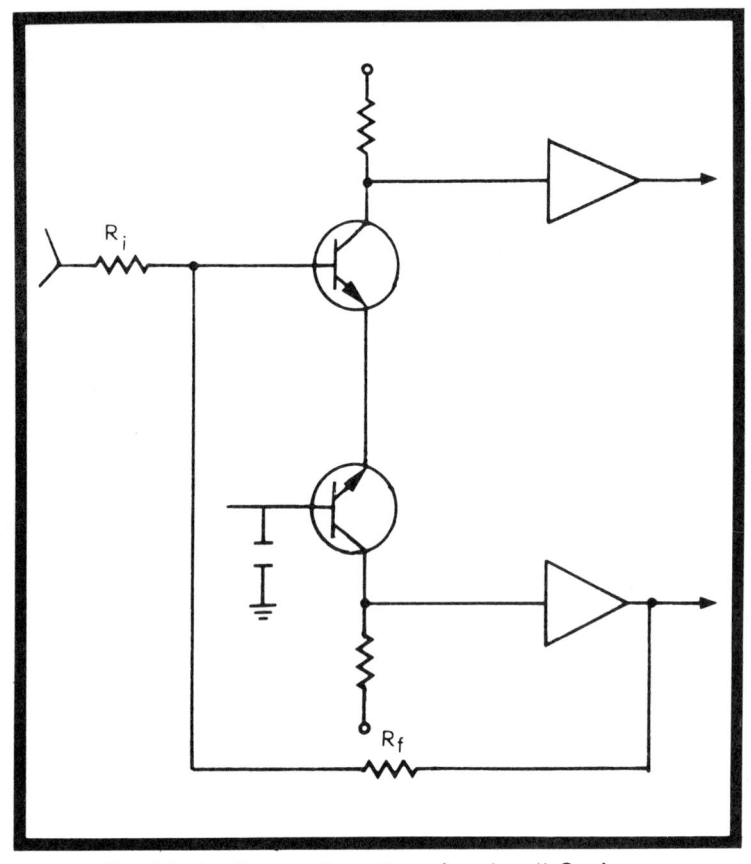

Fig. 8-7. Another configuration of push-pull Op Amps.

amp analysis. See Fig. 8-7. R_i in this case must be whatever resistance is in series with the input signal that is driving the paraphase amplifier, not the sum of the transresistance, etc. There is one difficulty with this last circuit, though; the output resistance will not be exactly the same on each side. With extremely high values of R_f, though, they can be made to approximate each other.

IN-PHASE FEEDBACK

In all of the operational amplifiers we have discussed so far we have used out-of-phase signal feedback to the input base of the first active device, and the A_V of the stage has been R_f divided by R_i. In each case, the output change of voltage robbed the amplifier of driving current. This same idea can be achieved with in-phase feedback as well. See Fig. 8-8. Here we have an extremely useful circuit that takes the control of the actual gain of the stage away from the transistors and puts it into the hands of external resistors again, but in a slightly different way.

R_{e1} will be large with respect to R_2 and will supply all the quiescent emitter current for Q1. The emitter of Q1 should be at zero volts with respect to ground and R_{c1} is usually made adjustable so as to achieve this situation. Note that the voltage across R_{c1} cannot be changed to anything different than the sum of the Zener voltage plus the "on" diode voltage of the Q2

In-phase feedback where $A_v = \dfrac{R1 + R2}{R2}$

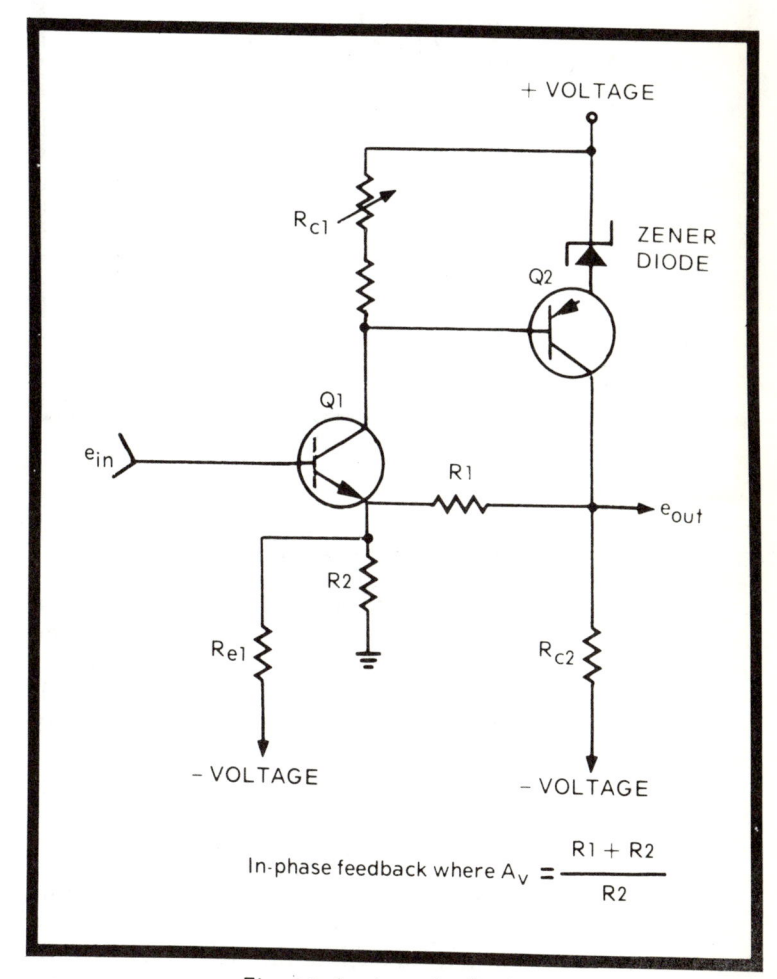

Fig. 8-8. In-phase feedback.

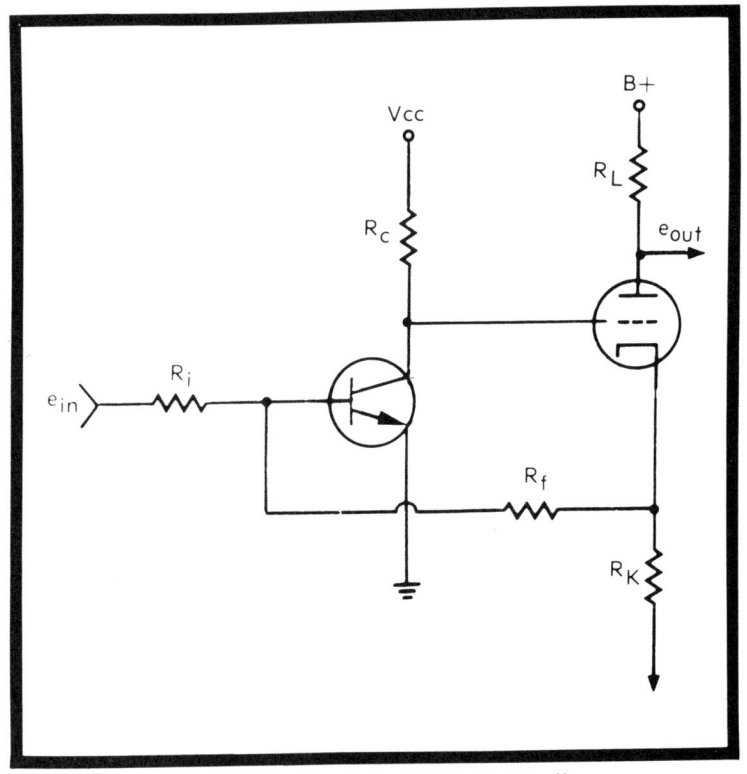

Fig. 8-9. Feedback in a Hybrid circuit.

base-emitter junction. Consequently, all the change of current that Q1 gets will be drive to Q2. The amount of resistance we have in R_{c1} thus determines the amount of current that will have to come from R_{e1} and we adjust its value such that there is no current flowing in R_2. (This is especially true if we are going to be switching different values of R_2 in and out for different gains.)

R_{c2} will be large with respect to R_1 and will carry the full quiescent current for Q2 such that e_{out} will be at zero volts with respect to ground also.

Now we are ready to start. Assume a small positive change of voltage at e_{in}. The voltage across R_2 will just about follow the full input signal. This extra current will first become drive to Q2, causing its collector to move in the positive direction. This will demand current (electrons) left to right through R_1, robbing Q1 of its extra signal current. The signal is inverted twice and fed back to the right end of R_1 by the collector of Q2. Q2 cannot pull this output point any further positive than the point where all signal current generated in R_2 is now called for through R_1. We have a voltage divider in reverse here and the voltage gain of the stage is again determined by a resistance ratio. A_V for the stage is equal to $(R_1 + R_2)$ divided by R_2.

This is an extremely useful circuit where several different gains are needed. Since only signal current is seen in R_1 and R_2, we can have several different values of R_2 set up in a rotary switch and not affect the quiescent operating condition of the transistors at all.

R_1/Beta for Q2 will approximate the value of R_O and Q1 Beta times R_2 will approximate the input resistance of this circuit.

There are times when the feedback loop is a little less obvious than in those op amps that we have looked at so far. See Fig. 8-9.

With the feedback resistor following the collector signal of the transistor, we will find that we have an operational amplifier up to the cathode of the vacuum tube, and the voltage gain to this point will equal R_f divided by R_i. This will be the change in voltage that will appear across the total cathode resistor of the vacuum tube. It will be this change of voltage that will determine the signal current the tube will get. The grid-collector circuit cannot make the tube get any more current than this, and it cannot make it accept any less than this. Therefore, the voltage at the output must be relative to R_L divided by the total cathode resistance. Here's where we must be careful in our thinking about this circuit. The total cathode resistance is not just R_K alone. It is R_K in parallel with that portion of R_f that is on the tube side of the equivalent ground point of the operational amplifier. Signal current in R_f is part of the cathode signal current, not just a trigger for this current. The total voltage gain then equals the quantity R_f over R_i times the quantity R_L divided by the parallel combination of R_K and R_f.

I suppose that you might say that this last circuit has cathode follower action in the feedback loop. It's not quite the same thing as having a cathode follower (emitter follower) in the loop strictly for this purpose though. See Fig. 8-10.

Now here's a circuit that does have an emitter follower actually in the feedback loop. It is used to decouple the feedback divider made up of R_{d1} and R_{d2} from the actual feedback resistor such that only a part of the output signal gets fed back.

The gain to the emitter of the feedback emitter follower is as usual, R_f divided by R_i. If you are willing to go along with

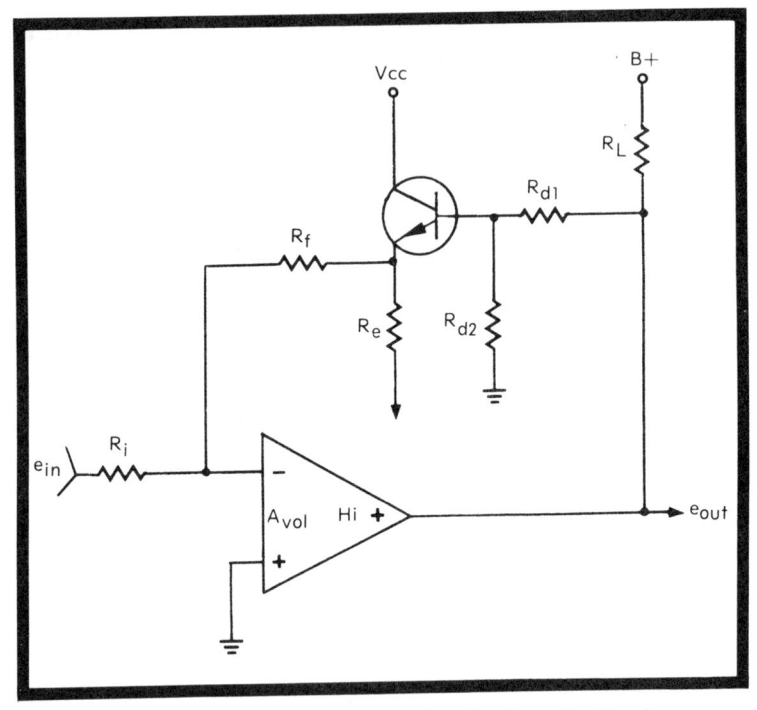

Fig. 8-10. Emitter follower used as in-phase feedback.

the idea that the voltage difference between the feedback emitter follower emitter and base is a constant, then this is also true to the junction point of the feedback divider resistors (R_{d1} and R_{d2}). From here to the output of this circuit then may be figured just as we did the in-phase feedback op amp and the total voltage gain of this stage is:

$$A_V = \frac{R_f}{R_i} \times \frac{R_{d1} + R_{d2}}{R_{d2}}$$

The output resistance of this circuit will not be quite as low as usual since it will be more closely approximated by the parallel combination of R_L and $(R_{d1} + R_{d2})$. However, the equivalent ground on the left end of R_f will be extremely close to the negative input to the amplifier. Since this amp had a much higher than normal A_{Vol} (open-loop voltage gain), all of it was not needed to maintain the closeness to ground at this point and the extra gain was needed at the output.

Occasionally we find the situation where we need to maintain linearity over a wide range of voltage swing. This immediately says that the output device will have to be a vacuum tube since BV_{ce} for transistors is still a limiting factor of these devices. However, this doesn't solve all our problems. With vacuum tubes, G_m can change with operating point, so here's a circuit that overcomes that difficulty. See Fig. 8-11.

Some people like to think of the transistor in this circuit as being an in-phase op amp used to couple the cathode follower tube to the output amplifier tube of this stage. The feedback resistor becomes extremely hard to find since it is the Z_k of the pentode (R_C) and R_2 of this op amp can be thought of as being R_2 in the circuit. Note that if G_m changes, so does R_C. If G_m goes down, the effective R_C gets bigger and the tube is forced to accept the same amount of signal current it did before. If G_m goes up and the effective R_C gets smaller, the tube doesn't call for any more signal current than that which is available with the same old signal across R_2.

R_C will be fairly large to limit the current change for the swing in voltage that will occur across it, thus stabilizing V_{be}

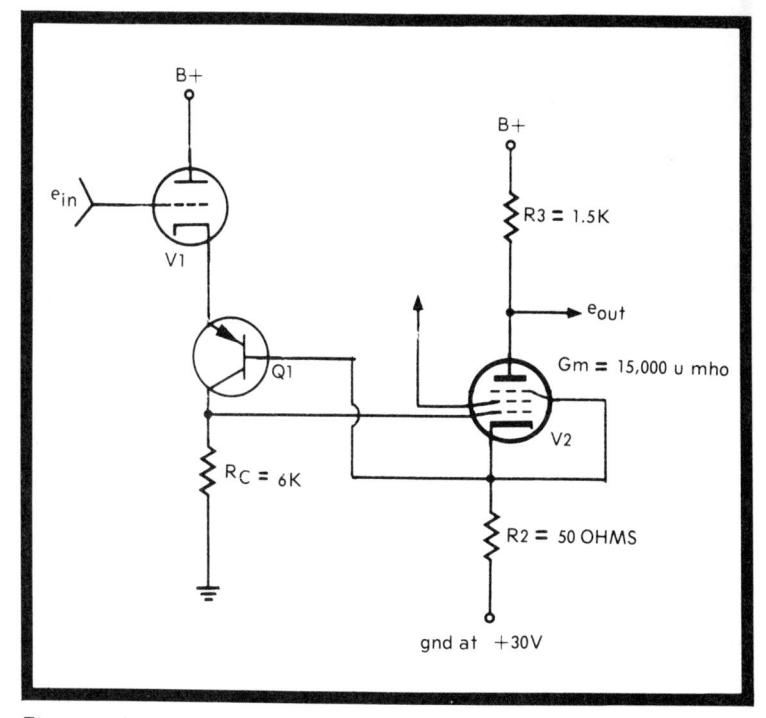

Fig. 8-11. In-phase PNP Op Amp coupling tube cathode follower to pentode output amplifier.

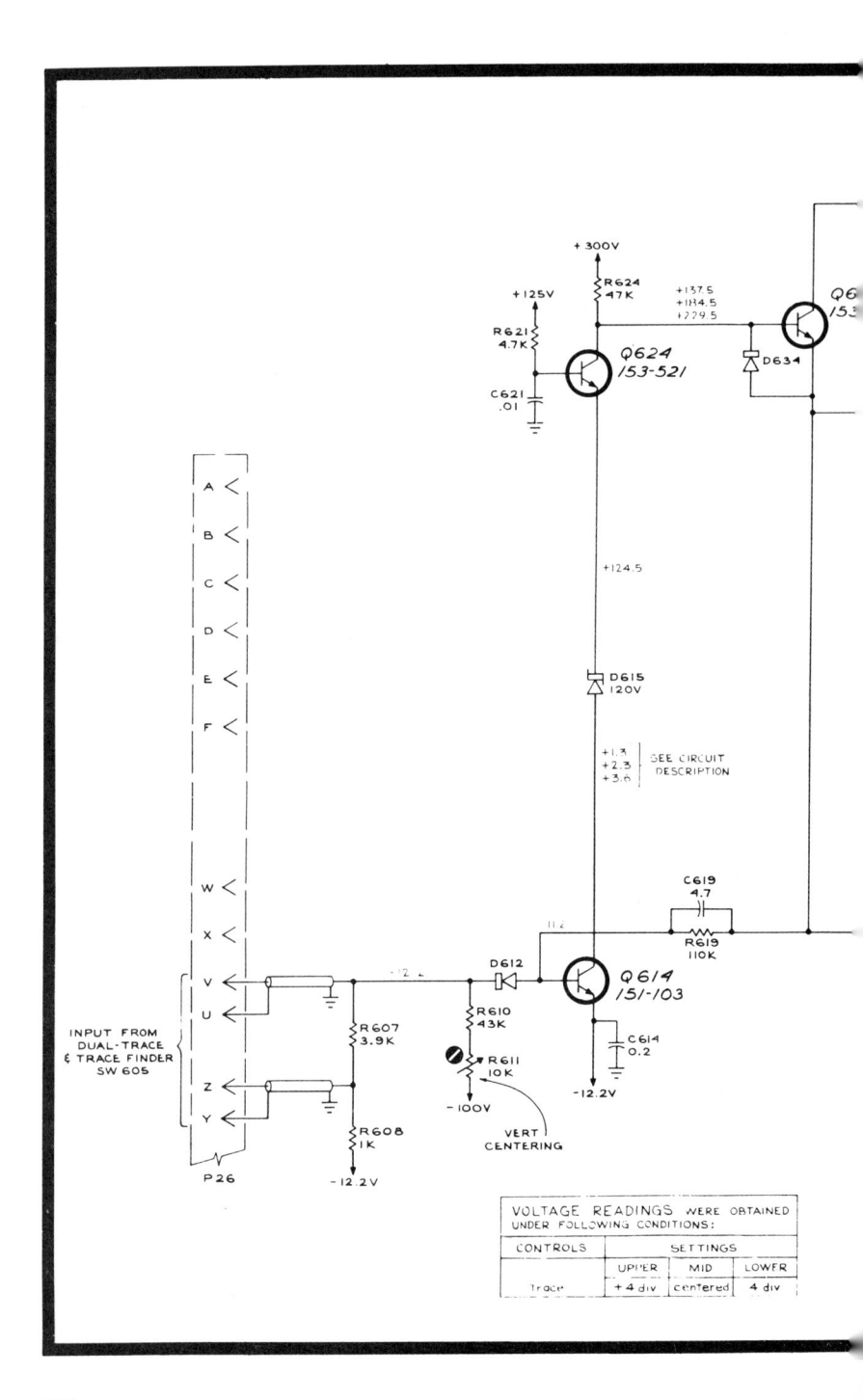

VOLTAGE READINGS WERE OBTAINED UNDER FOLLOWING CONDITIONS:

CONTROLS	SETTINGS		
	UPPER	MID	LOWER
Trace	+ 4 div	centered	4 div

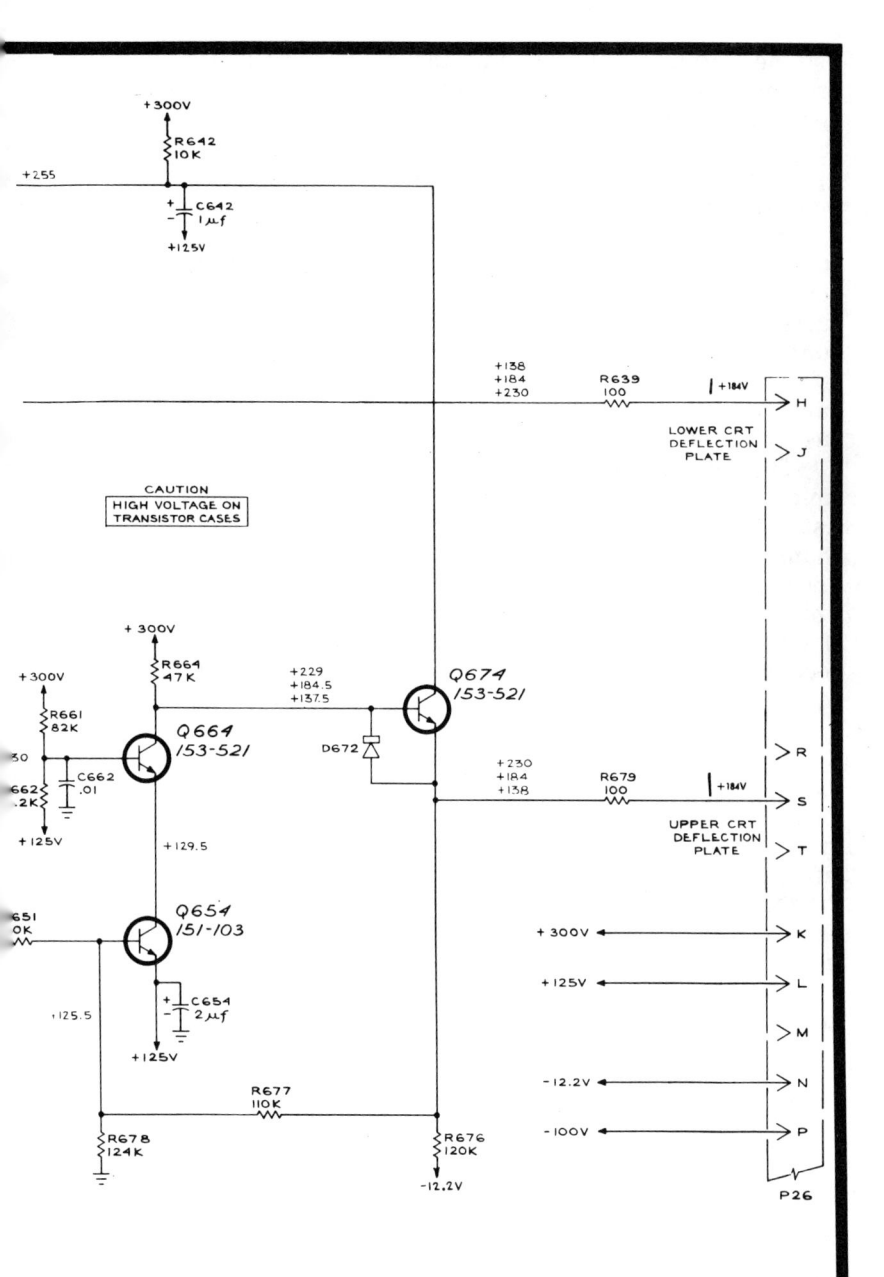

Fig. 8-12. Schematic for quiz (worksheet).

for the cathode follower and the transistor. R_2 will approximate the Z_k of V2 at its mid-point of operation. R_3 will be adequate to give the output stage a gain near 20. (G_m x R_3 equals 22.5.)

The voltage gain action will go something like this: Put in a positive-going signal one unit in amplitude. The collector of the transistor follows in an amplified fashion, but the change of current through Q1 is so small that V_{be} does not change, nor does V1 bias, so e_{in} is seen across R_2, limiting the signal current through V2. Consequently, V2 cannot see a driving voltage any greater than Z_{k2}/R_2. This will be the gain of the effective transistor op amp stage that V2 feels. This voltage gain times the G_m and R_3 product for V2 will be the total circuit voltage gain.

The voltage gain of this stage can be expressed in a much simpler way, though, if we say that the emitter of the transistor follows the cathode of V1, and, since V_{be} does not change, the base of the transistor follows the emitter. We now find that we have unity gain across R_2. R_2 is in series with R_3, which will receive (by rule of thumb) 72 percent of the change of current generated. The total voltage gain of this stage becomes R_3/R_2 multiplied by the plate efficiency (30 x .72 equals 21.6) which really ends up being the more accurate system.

Note that the amplitude of signal seen on the collector of the transistor with respect to ground follows the op amp gain formula $(R_1 + R_2)/R_2$, but since the cathode of V2 also moves in the same direction, this magnitude of signal is something different than what V2 actually feels. V2 is only conscious of the change in voltage between its grid and cathode, thus it only feels the change of voltage developed across the effective RC of the op amp.

This is one of those circuits where the active devices have to go just about completely sour before the circuit stops working correctly. In this respect, it's beautiful.

QUIZ

Now I'm going to let you find out if you've really understood this chapter. Use Fig. 8-12 as your worksheet and answer 10 questions that follow. Then check your work against the values in Figs. 8-13 and 8-14, and the answers on the last page of this chapter.

(1) Determine the DC operating point of each transistor in Fig. 8-12.

	V_{ce}	I_c
Q614		
Q624		
Q634		
Q654		
Q664		
Q674		

(2) Is there any basic difference between the circuit made up of Q614 & Q624 and the circuit consisting of Q654 & Q664? (please be specific)

(3) What specific kind of amplifier does the configuration of Q654, Q664 & Q674 represent?

(4) If I drive this circuit (input at connection point "z") with a 1 volt signal, how large a signal can I expect to see at connection point "s"?

(5) What is the peak to peak voltage gain of this amplifier from point "z" to points "H" & "s"?

(6) What is the output impedance seen looking back into points "H" & "s" if we assume that all transistors have a nominal β of 50?

(7) I am told that if I switch the input from point "v" to point "z", I will cut the voltage gain of this circuit in half. What must be the output impedance of the circuit that I use to drive this one with?

(8) What is the open loop voltage gain of the circuit made up of Q614, Q624 & Q634 or the circuit made up of Q654, Q664 & Q674?

(9) What is the input impedance seen looking into the base circuitry of Q614 or Q654?

(10) If I see a 100 volt signal at points "H" & "s," what will I see at the bases of Q614 and Q654 respectively?

276

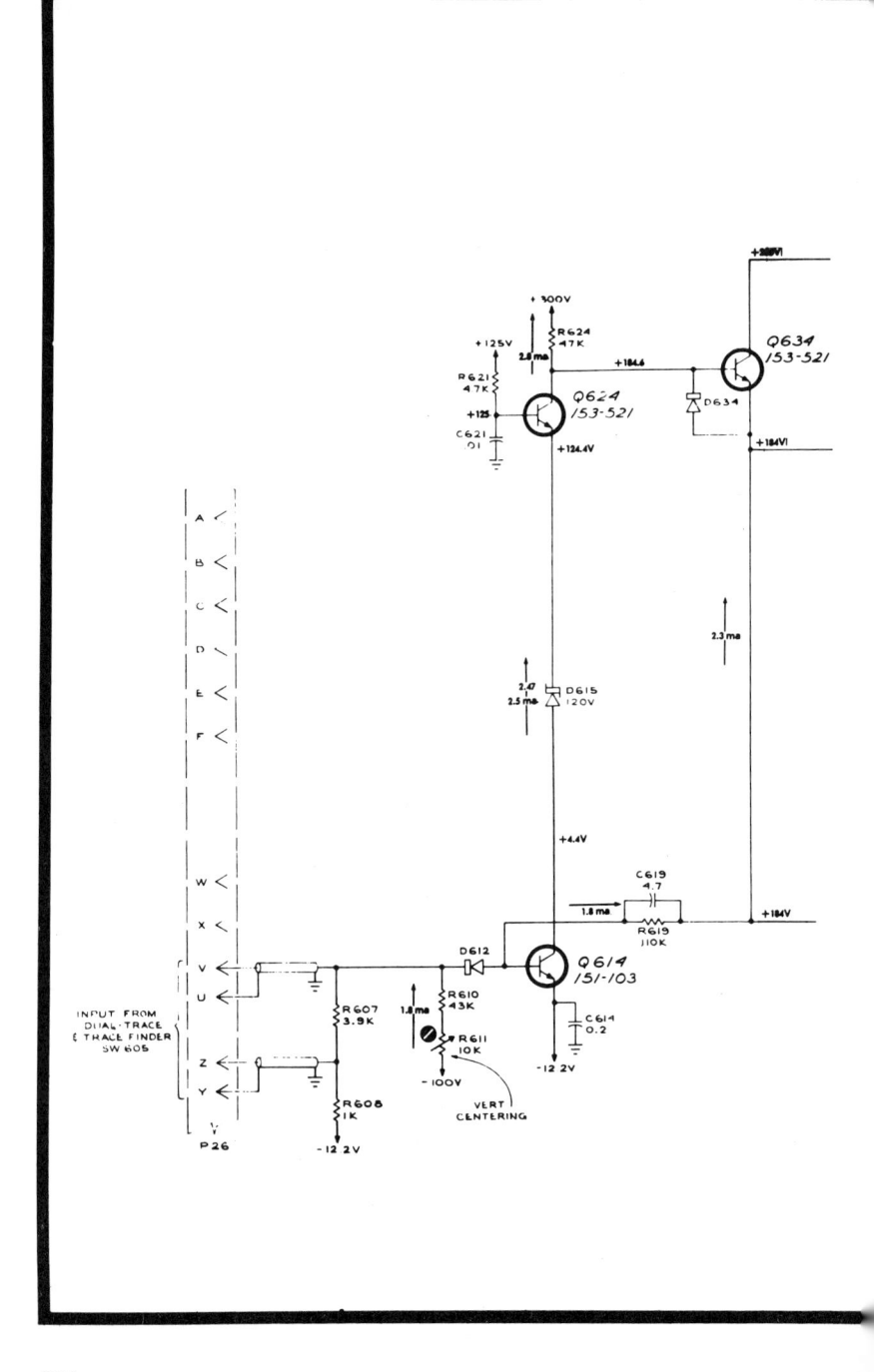

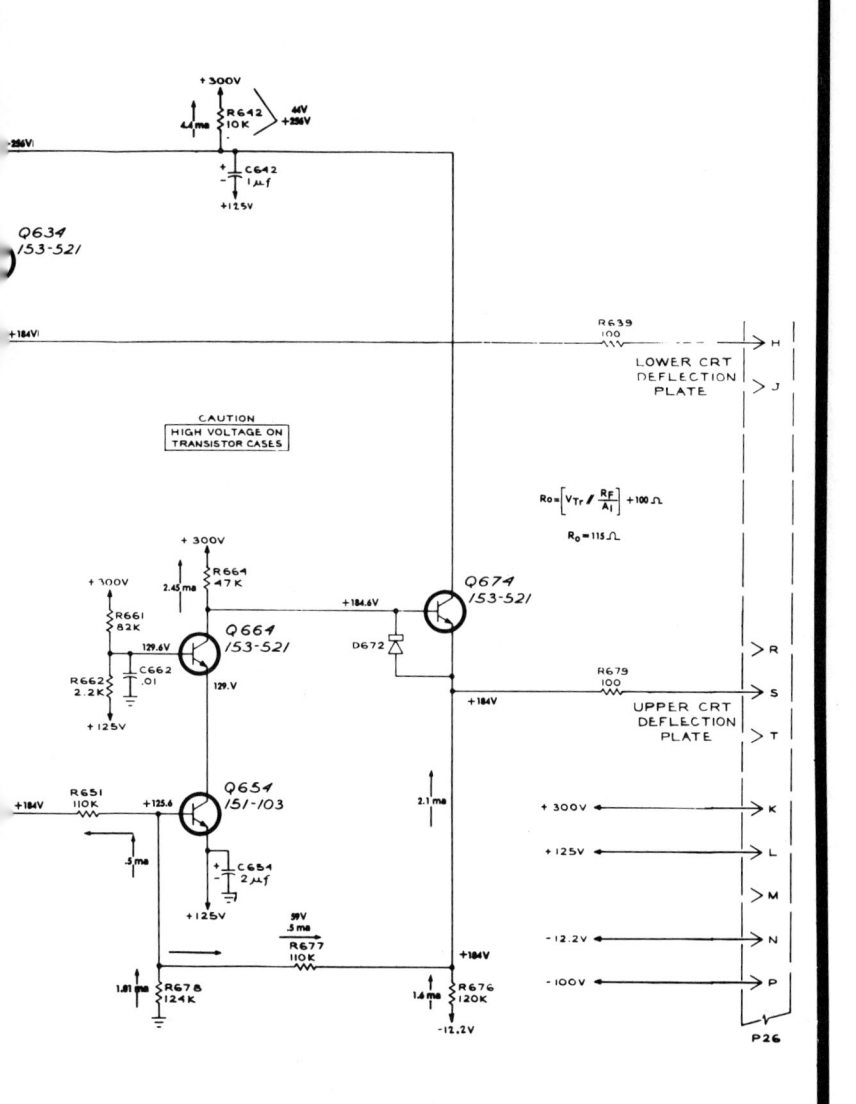

Fig. 8-13. Schematic for quiz with values. (Courtesy Tektronix, Inc.)

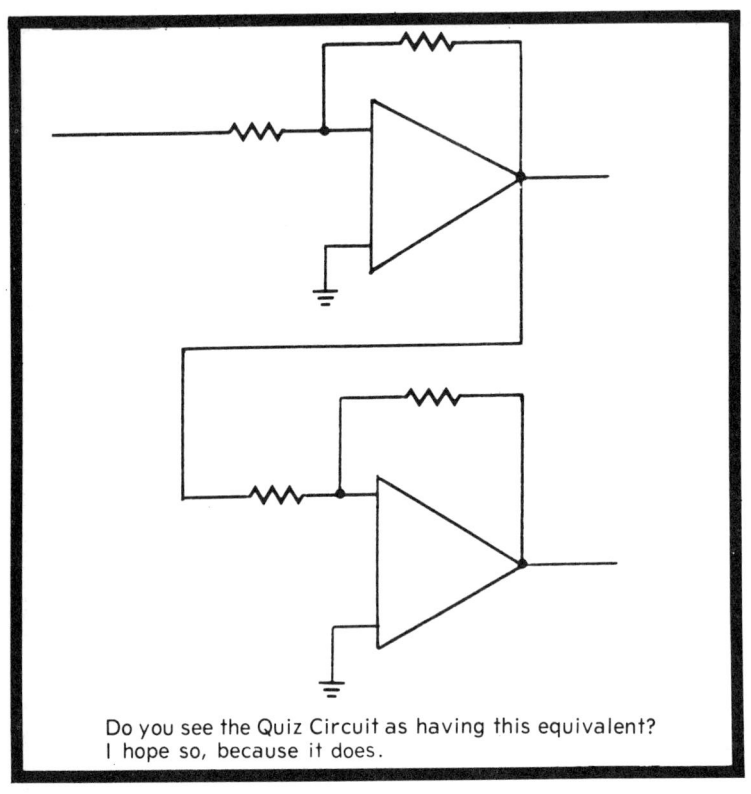

Do you see the Quiz Circuit as having this equivalent?
I hope so, because it does.

Fig. 8-14. Op Amp equivalent for Fig. 8-12.

ANSWERS TO QUESTIONS

(1) Determine the DC operating point of each transistor in Fig. 8-12.

	V_{ce}	I_c
Q614	16.6V	2.47 ma
Q624	60V	2.47 ma
Q634	72V	2.3 ma
Q654	4.6V	2.45 ma
Q664	55V	2.45 ma
Q674	72V	2.1 ma

(2) Is there any basic difference between the circuit made up of Q614 & Q624 and the circuit consisting of Q654 & Q664? (please be specific)

DC levels different.
(AC action & reaction the same.) $Q654\ V_{ce} = 4.6V$
$Q614\ V_{ce} = 16.6V$

(3) What specific kind of amplifier does the configuration of Q654, Q664 & Q674 represent?

op amp

(4) If I drive this circuit (input at connection point "z") with a 1 volt signal, how large a signal can I expect to see at connection point "s"?

28V

(5) What is the peak to peak voltage gain of this amplifier from point "z" to points "H" & "s"? $28 + 28 = 56 = A_{VP/P}$

(6) What is the output impedance seen looking back into points "H" & "s" if we assume that all transistors have a nominal β of 50? $50 \times 50 = 2,500$

$$e_F\ R_o = (r_{tr})^2;\ Op\ Amp\ R_o = \frac{R_F}{A_I} = \frac{110K}{2.5K} = 44\ ohms,$$

from 150 + 100 (R639 + R679)

(7) I am told that if I switch the input from point "v" to point "z," I will cut the voltage gain of this circuit in half. What must be the output impedance of the circuit that I use to drive this one with?

3.9K

(8) What is the open loop voltage gain of the circuit made up of Q614, Q624 & Q634 or the circuit made up of Q654, Q664 & Q674?

$$A_{VO1} = \frac{R_L}{r_{tr}} = \frac{47K}{15} = 3,100$$

(9) What is the input impedance seen looking into the base circuitry of Q614 or Q654?

$$R_{in} = \frac{R_F}{A_{VO1}} = \frac{110K}{3100} = \frac{110K}{3.1K} = 35\ ohms$$

(10) If I see a 100 volt signal at points "H" & "s," what will I see at the bases of Q614 and Q654 respectively?

$$\frac{100}{3100} \simeq 30\ mv$$

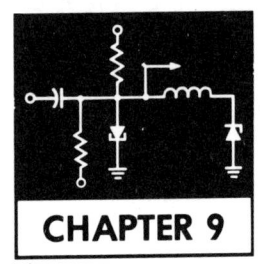

More on Feedback (Oscillators)

CHAPTER 9

So far all we have considered is the forms of feedback used to turn a large-voltage-gain amplifier into (what is called) an Operational Amplifier.

Feedback, of course, can be used to change an amplifier or amplifiers into a whole lot of things. And, each different use of feedback seems to come out under a different name for some reason or other. This was particularly true during the early days of radio. A man would discover a particular point in the output of a specific type of amplifier circuit and use a particular way of returning it to the input to achieve his own special purpose. Then the next thing everybody knew was that that particular circuit became known by this guy's name. It resulted in numerous different circuits that did similar things, with each one known by the name of its developer, but all of them making use of the same basic principles of what we now call feedback.

COAXIAL TRANSMISSION LINE

To gain some real understanding of what will be going on in these circuits, we'll have to take a look-see at one more new item. It's not new to the industry. It's new here because I haven't discussed it before in this text. That item is the Transmission Line, and there's a lot more to it than what I mention about it here. And, I'm going to stick primarily to the coaxial transmission line.

If we have a conductor (piece of wire) going down the exact middle of a piece of metal tubing, we have a piece of coaxial transmission line. A single conductor with plastic covering and a braided flexible shielding on top of that is also a coaxial transmission line.

Electrically, when the outside shielding is grounded the piece of wire going down the middle looks like something different than it did before. If we put a step waveform into one end of this transmission line it will only accept so much current. In other words, it now represents a certain amount of impedance to this sudden change of voltage. This is called its Surge Impedance characteristic. The surge impedance is a constant for this specific piece of transmission line or any other that is made exactly like it regardless of its length.

Since any piece of wire has the basic characteristic of some inductance, this is one of its inherent characteristics.

But, our piece of wire is now a constant distance away from ground with a dielectric (insulator) in between. This says it will act to some degree like a capacitor, and this capacitor is to be thought of as being broken down into a whole bunch of small capacitors distributed down the whole length of the line.

There is, of course, a certain amount of real resistance to this long piece of conductor, but it's usually so small that we can get by with ignoring it. (However, when this coaxial line is extremely long, we can't ignore it.) Anyhow, for our purposes, we'll forget it.

This then says that our transmission line looks like the circuit shown in Fig. 9-1.

Z_0 = Surge Impedance $\cong \sqrt{\dfrac{L}{C}}$

When $F_{in} \ll F_c$ then Z_0 really = $\sqrt{\dfrac{L}{C}\left[1-\left(\dfrac{F}{F_c}\right)^2\right]}$

However, we can use the first formula in most cases.

T_S is the time per section for a signal going down this line and the formula is: $T_S \cong \sqrt{L \times C}$

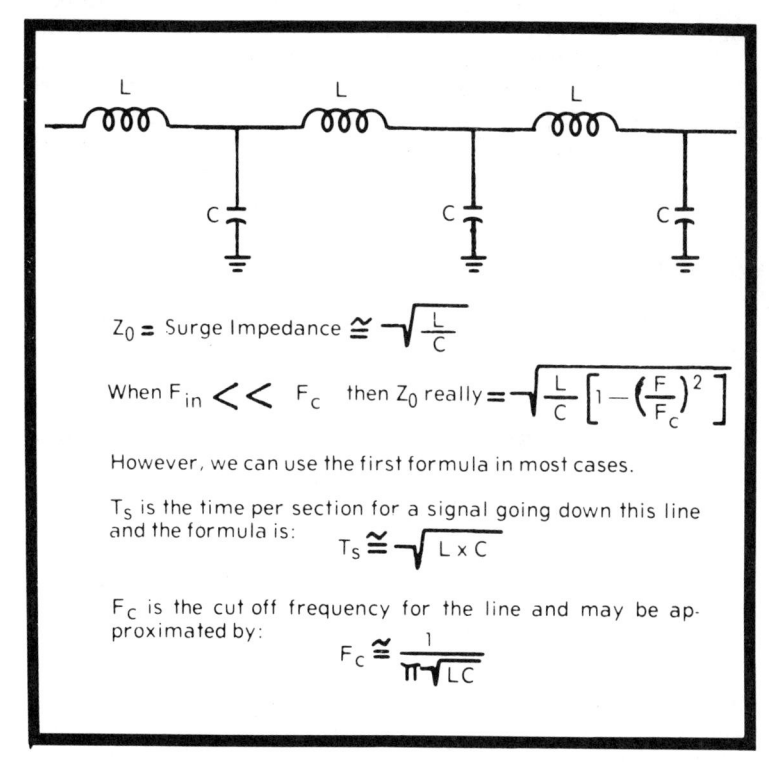

F_c is the cut off frequency for the line and may be approximated by: $F_c \cong \dfrac{1}{\pi\sqrt{LC}}$

Fig. 9-1. Coaxial transmission line representation.

Now, we normally think of a signal going down a piece of wire at the speed of light. In a piece of transmission line, it isn't necessarily so. You see, the inductive characteristic first opposes the flow of current. Then, as the field stabilizes, it lets the current flow. This current charges the capacitance characteristic to the input voltage amplitude as the next section of inductance opposes the flow of current, etc.— on down the length of line.

In other words, there's going to be a finite period of time, a delay, before that change of voltage reaches the other end, and it won't be determined entirely by the speed of light.

Of course, since our coaxial transmission line has a surge inductance characteristic and a surge capacitance characteristic to ground, we have a series tuned circuit for some specific frequency, called the F_c or Cutoff Frequency. Quite obviously, this F_c characteristic has to be higher than the frequency we wish to send down this line, but for our purposes F_c will be so much higher than anything we'll consider, we can forget it too.

The thing I want you to understand is that different lengths of Coaxial Transmission Line can be used to delay the arrival of a sudden change of voltage to some other point by different periods of time. And, lines used for this purpose are called "Delay Lines." Along with this, we must understand that manufacturers make what are called "Lumped Constant Delay Lines" (which look like metal-cased electrolytic capacitors) which are rated by their delay capability.

OP AMP BECOMES OSCILLATOR

Let's say that we had a lumped constant delay line that had a 0.2 microsecond delay characteristic and an F_c of 100 MHz.

And, let's put it between the collector and base of a common-emitter transistor circuit. (We'll also add a resistor in series with it to keep V_{CC} from showing up directly on the base and blowing the transistor sky-high.) See Fig. 9-2. This lets the 27K resistor take place of R_f if you think of it as a variation of an operational amplifier. (We won't use any R_i with regard to this circuit because we don't intend to put any signal in.)

Now, let's see what's going to happen.

We turn on V_{CC}. The collector suddenly has a voltage on it and this change of voltage is a signal applied to one end of our delay line. Point two microseconds later, this positive voltage shows up on the base resistor and turns the transistor on hard.

The collector voltage falls as the transistor saturates. This is a negative signal into the transmission line and point two microseconds later this shows up on the base resistor and turns the transistor off. The collector goes back up to V_{cc} in voltage and the whole process goes on and on.

What's that you asked? "So what?"

Well, we have used feedback (through a different device) to change an amplifier circuit into an Oscillator circuit! This circuit will change voltage at its collector at a rate determined entirely by the delay line. To go through one complete cycle of oscillation, the transmission line will have to deliver two changes of voltage back to the base. This will take 0.4 microseconds time. Consequently, we will have one second divided by 0.4×10^{-6} cycles in one second, or a frequency of 2.5 MHz from this oscillator.

The output waveform won't be a sine wave, and I seriously doubt if it will even approach being a fair square wave. It will, however, have a fundamental frequency plus numerous harmonics (probably odd harmonics). And, if you are going to

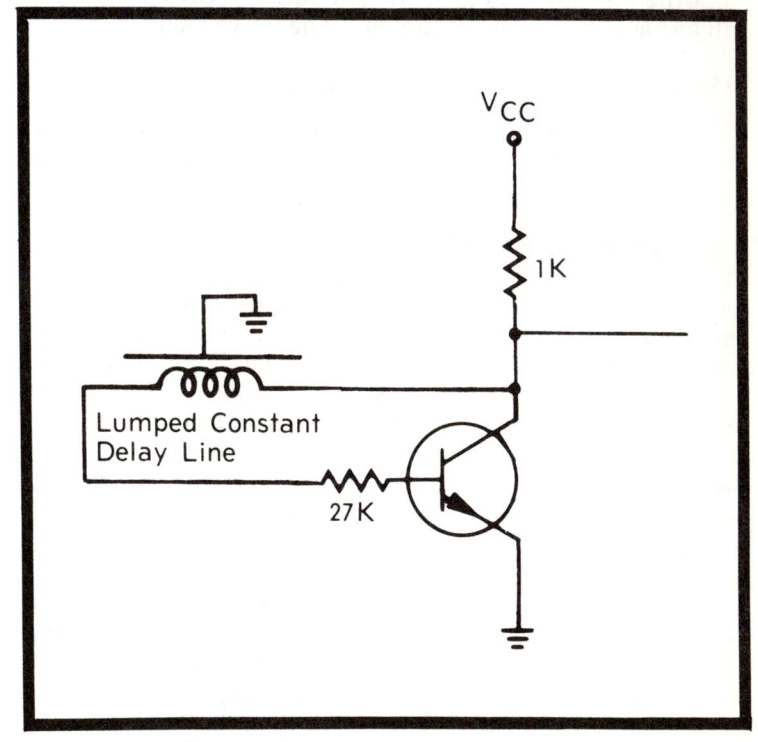

Fig. 9-2. Op Amp becomes an oscillator.

see this waveform on an oscilloscope, it would be best if the scope had a vertical response of 5 MHz at least and a sweep speed in excess of 1 microsecond per division.

Perhaps it would be better if I suggested that you build an oscillator with a real low frequency that you could follow with your voltmeter. That's possible too, but we'll have to change to Zener diodes and RC time constants to do it.

It takes a rather fancy feedback network to make this oscillator put out something closely approaching a sine wave, but since the output is slow, I think you can fully understand it.

Let's start out with what is called a standard operational amplifier, i.e., one with a fairly high open-loop gain characteristic. An integrated circuit (IC) op amp would be a real good deal for this. It would probably have one lead connected to the Substrate as the negative voltage lead or ground lead, a lead for V_{cc} to be connected to (probably for +15 volts), a plus signal input, a negative signal input and a plus signal output (Fig. 9-3). To start with you'll connect a 100K resistor from the plus output back to the plus input and then connect a 10K pot from the plus input to ground. Our fancy feedback circuitry will go between the plus output to the negative input.

SINE WAVE OSCILLATOR EXAMPLE

To make the circuit work, you'll connect the negative voltage power lead to ground and apply +15 volts to the lead for V_{cc}.

If your value for "R" is 1 Meg and the value for "C" is 1 uf this oscillator should put out a signal that can be tuned to look like a sine wave at some setting of the 10K pot. To work best, the back-to-back Zeners should be kind of lousy zeners with very rounded (soft) knees where they turn on. (You don't want a sharp-cornered Zener for this circuit.) If you can get a reject, it'll probably work best. These Zeners should have a turn-on characteristic in the neighborhood of 10 volts or so to give us a peak-to-peak output waveform of something a bit less than 10 volts. The frequency of this oscillator is determined by the formula $\frac{1}{2\pi RC}$. And, if you'll figure it out, you'll find that it should generate one cycle in about 6 seconds. This you ought to be able to watch with your voltmeter.

"How does it work?" What say we turn it on and see. That Twin "T" RC network is right tricky. Let's say that the voltage at the output starts to move positive in response to the application of V_{cc}. The first immediate source of current is going to be from the capacitors, with C on the right (output) supplying most of it at first. This capacitor gets current from

C on the left and as a voltage develops across R/2 it begins to get some from here as well. This interrupts the feedback and allows the output to change a little faster; but this change starts to take electrons away from the top of 2C, which, in turn, changes voltage and demands some current from the negative input through R on the left, which slows down the output change somewhat. Finally it reaches a point where there is enough difference in voltage between output and negative input to start to turn on the Zener diode and the full change per unit time on the output is just about all fed back; and this not

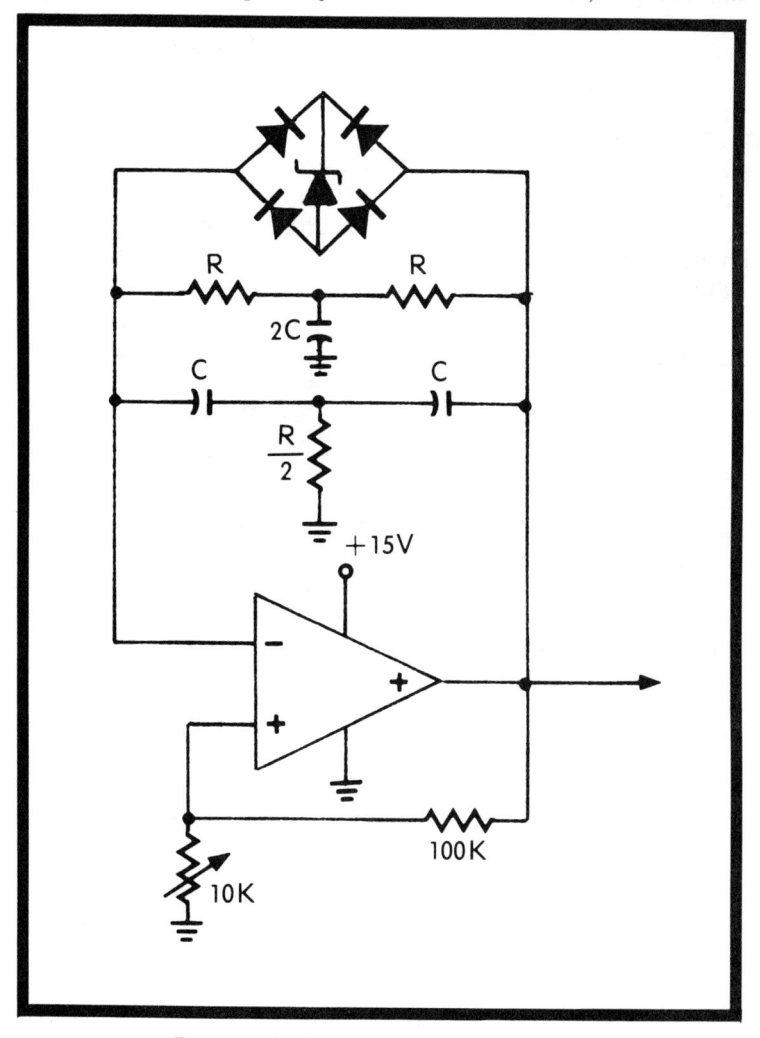

Fig. 9-3. Op Amp sine-wave oscillator.

only brings things to a screaming halt, it actually allows the negative input to move so far positive as to demand that the output go in the negative direction.

For that matter, 2C and C on the left will continue to collect electrons from the negative input right up to the point where we get zero volts across R on the right. And only then will this change of voltage start to diminish. This negative trend will continue to the point where current has reversed in the filter network and we have a voltage difference (output to input) big enough to turn on the Zener diode again, which will not only stop the negative trend of the output voltage but turn it around to go positive again. This just keeps on going back and forth and back and forth (or, if you will, up and down—up and down—up and down). And we have built ourselves a real slow sine-wave oscillator this time, again using time-delaying characteristics of electrical parts in the feedback loop of an amplifier.

"But," you say, "All this is great except you're using modern concepts, and neither of these oscillators seems to be named after anybody."

That's true. But modern parts and modern technology in many cases have made understanding these things a lot easier. For instance, today through the use of modern oscilloscopes with current-sensitive probes as well as voltage-sensitive probes, it is possible to see the current and voltage relationships across a capacitor. Thus, we can prove visually that current leads voltage when we apply a sine wave to a capacitor (we can also prove that current lags voltage when we apply a sine wave signal to an inductance). This makes these things a lot easier to believe. Theory is fine, but "seeing is believing" has been an appropriate saying for a lot longer than I have lived.

Likewise, I can tell you that if you hook up a capacitor and an inductor in parallel, ground one common end of this circuit and for the very briefest moment touch a positive voltage to the other common end, you will start a current bouncing back and forth in this circuit. The capacitor will first charge in one direction and then in the other and continue back and forth until the actual resistance in the circuit has dissipated all the power you applied to it in the initial change of voltage you created in it.

Would you believe me?

I seriously doubt it. I think it's necessary to prove something like this. So, let's take one of those named circuits that I mentioned in the beginning of this chapter and see if we can't figure out what it's going to do.

FRANKLIN OSCILLATOR

The Franklin Oscillator circuit will suit our purpose very nicely. See Fig. 9-4. It consists of two vacuum tubes, each operated as a plate-loaded amplifier, each giving us a signal inversion; by this I mean that if the signal on the grid is positive, the signal on the plate will be negative and vice-versa. Thus, if the grid of V1 is going negative, the plate of V1 will go positive. This positive signal is AC-coupled to the grid of V2 and V2 plate will go negative. This negative signal is fed back to the already negative-going grid and tank circuit of V1. This succeeds in putting back into the tank circuit whatever power was used up by the circuit on the grid of V1 and whatever real resistance actually in the tank circuit itself.

The feedback capacitor and the coupling capacitor to V1 can be very small (somewhere in the neighborhood of a couple of picofarads) and thus won't louse up the frequency of the tank circuit, and the thing will still work beautifully.

What frequency will this thing work at? Oh, I just touched on that very lightly back in Chapter 1. So maybe we'd better talk about that for a bit here.

This circuit is resonant at the frequency where X_L and X_C are equal. Remember? As frequency goes up, X_L goes up in impedance and X_C goes down. Consequently, there is bound to be one frequency where the two impedance graph lines will cross and this is where X_L and X_C are equal to each other. This allows us to say (as a starter):

$$2\pi fL = \frac{1}{2\pi fC}$$

Now, if we transpose (move) the f in the X_C formula to the left and all elements of the X_L formula except f to the right we get:

$$ff = \frac{1}{2 \cdot 2\pi\pi \, L \, C}$$

This gives us a statement that's easy to handle. Take the square root of both sides to obtain the formula for resonant frequency:

$$f = \frac{1}{2\pi \sqrt{LC}}$$

Since this is the frequency at which the tank circuit will oscillate, this is the frequency of our Franklin Oscillator Circuit.

OP AMP DRAWING OF THE FRANKLIN OSCILLATOR

A

FRANKLIN OSCILLATOR CIRCUIT

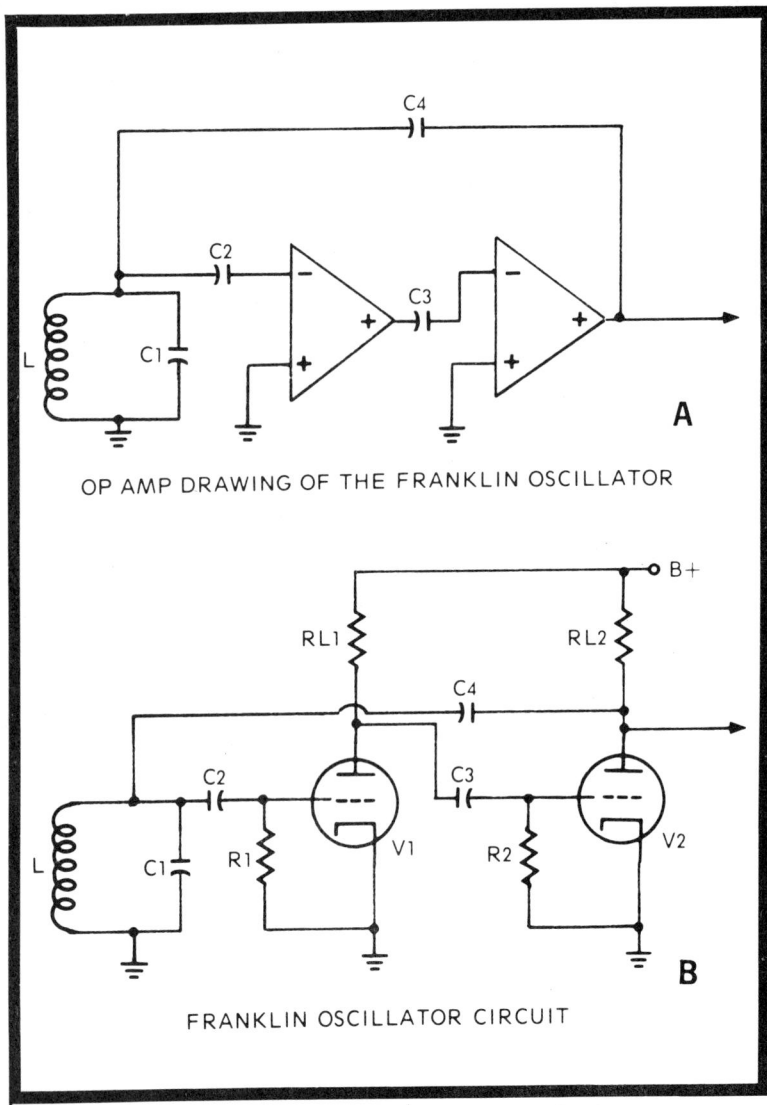

B

Fig. 9-4. Franklin oscillator and equivalent Op Amp.

A word of CAUTION is due here before I go on with any further information. If I were you, I **wouldn't** build this circuit and turn it on except under the test laboratory conditions and under the strict guidance of a good instructor. It is conceivably possible for this circuit to radiate a signal field strength that could get you into loads of trouble with the Federal Communications Commission.

The little transistor, lumped constant delay line oscillator doesn't have the power capability (as long as you are not using a power transistor) to cause any drastic interference, and the low frequency of recommended operation of the Op Amp Sine-Wave Oscillator I discussed with you eliminates it from the class of circuit that can get you into trouble. But, I must admit that this Franklin Oscillator has the capability of interfering with licensed stations and thus the ability to get you into trouble if you build one and turn it on. So my advice is DON'T unless you are under the immediate guidance of a qualified and licensed instructor and he knows what you are doing and when.

This sort of eliminates my proof of how a Tank Circuit works for you, but for your own safety, it's best that way.

Anyhow, I think you can see how the circuit works and in particular if you think of this circuit in modern logic terms as shown in the bottom drawing of Fig. 9-4.

Regardless of the type of active device we use, whether it be transistor, FET or vacuum tube, there is always a natural capacitor in the device from the "effective plate" back to the "effective grid." This natural capacitance can give us feedback, and in many cases is counted upon to do so. The Tuned Plate-Tuned Grid Oscillator is a good example. See Fig. 9-5.

TUNED PLATE—TUNED GRID OSCILLATOR

In this TP-TG oscillator, we have a tank circuit connected to both the plate and the grid of the tube. In this circuit, it is the plate tank that actually sets the frequency of operation since it apparently works best with the grid tank tuned to a slightly lower frequency than that to which the plate is tuned. The much greater power in the plate tank being slightly fed back to the grid tank through the plate-to-grid capacitance tells the grid tank when to switch its direction of voltage change.

Where we wish to leave the control of frequency to just one tank circuit and still get by with one vacuum tube in the oscillator circuit, this natural feedback tendency can be helped along a little bit, such as in the Armstrong Oscillator Circuit, Fig. 9-6.

ARMSTRONG OSCILLATOR

In this type of oscillator circuit, we are counting on the AC-coupled small non-resonant coil in the plate circuit to electromagnetically feed back some of the plate power into the grid tank coil at just the right times to keep oscillations from

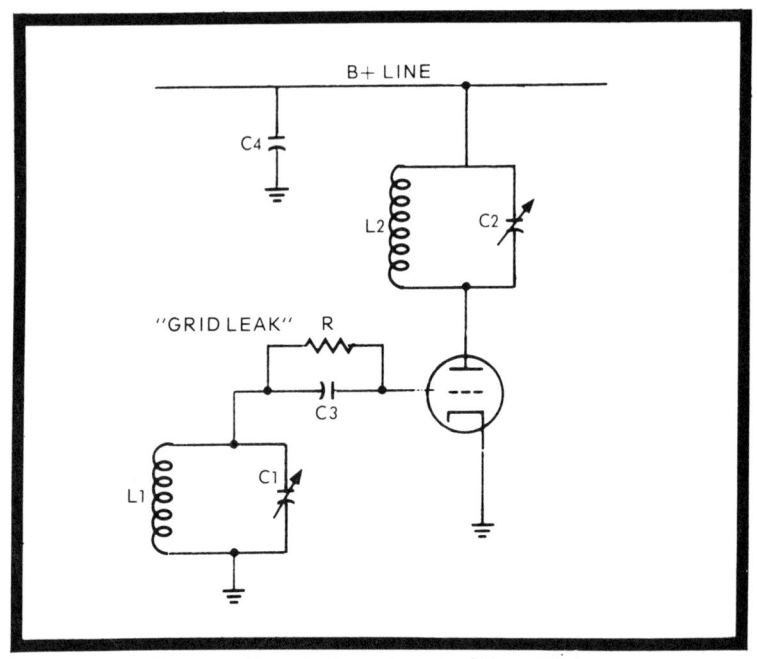

Fig. 9-5. Tuned plate—tuned grid oscillator.

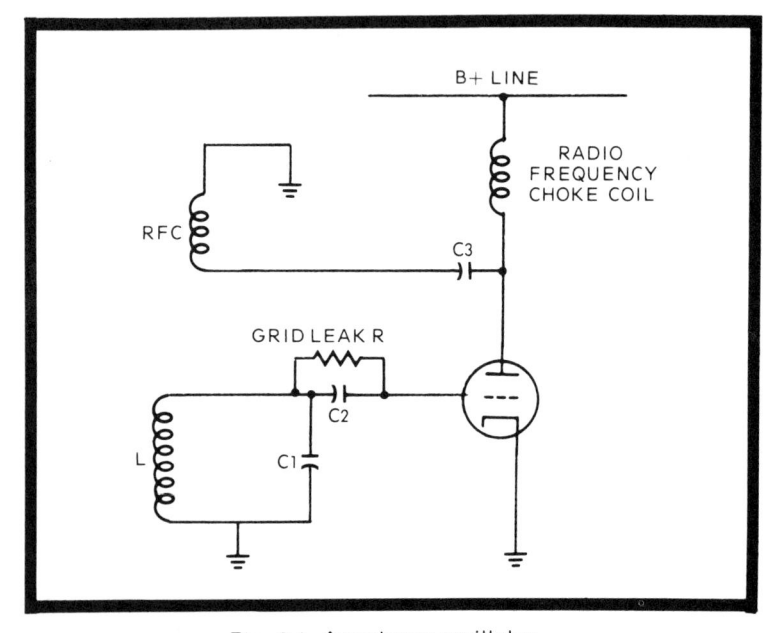

Fig. 9-6. Armstrong oscillator.

dying out. The amount of feedback is controlled by the size of the capacitor and how close the plate coil is to the grid tank coil L.

And, of course, there's the guy who designed these two coils all on the same winding (since they tie together at ground). His name was Mr. Hartley, and the Hartley Oscillator is another time-honored circuit. See Fig. 9-7.

I think the most important thing to remember in figuring out how these circuits work is in how we think about the coil (or inductor). We have to remember that a coil is like a generator of electrical current. Electron current flow in a coil

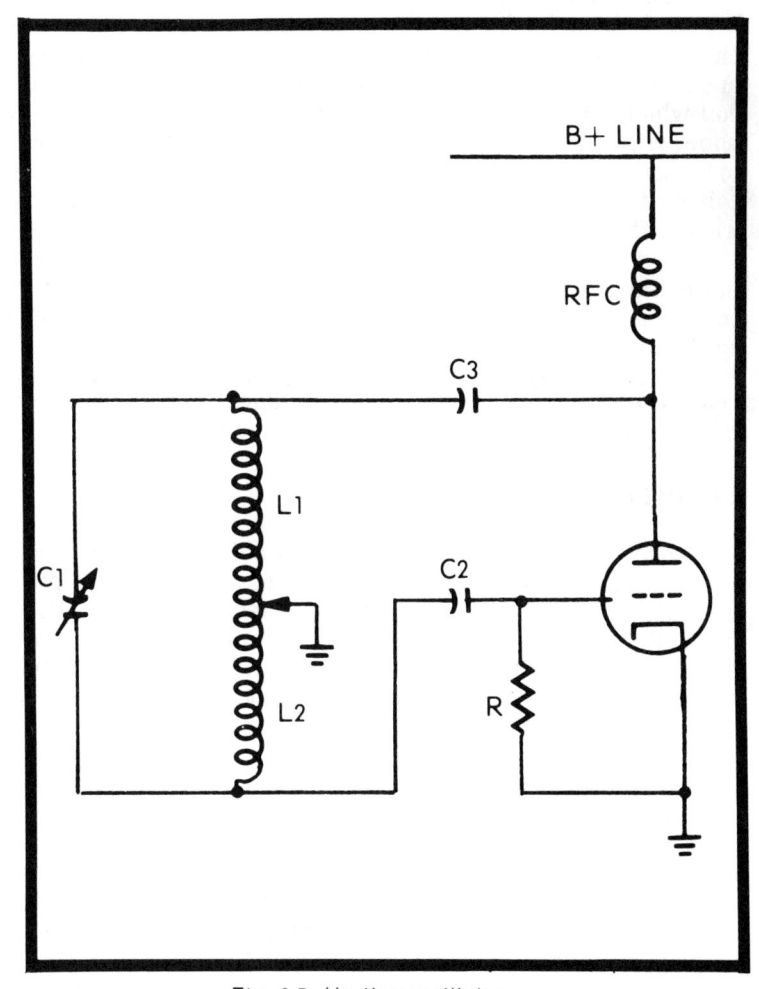

Fig. 9-7. Hartley oscillator.

is not from negative to positive as it is in a resistor. The coil is like a momentary battery. It spits electrons out one end (which becomes negative) and goes positive on the other. So, within the coil, the electron current will actually flow from positive to negative, and if you forget this, you're going to get all loused up. Also, nothing has to tell a tank circuit when to change current direction. It'll do it all by itself. That happens to be one of its characteristics.

HARTLEY OSCILLATOR

Now let's see how the Hartley oscillator works. We turn on B+ and the tube begins to draw plate current. Electrons charge the right-hand side of the coupling capacitor negative and it spits electrons off its other plate which try to go into the coil which refuses to take them and becomes negative on its upper end. This, in turn, says the other end (which is attached to the grid) goes positive, reinforcing the action until the tank coil switches in its own sweep time and the so-called "Fly wheel effect" takes over.

I don't exactly like the term "Fly wheel effect." It's much more like a "zip-zap" or "flip-flip" effect as far as I am concerned, but time and usage of the term overrule my opinion. So, we're stuck with it.

The frequency of the Hartley oscillator was determined by one half of the total inductance (as long as it was center-tapped) and the capacitor in the tank circuit.

COLPITTS OSCILLATOR

Then there was Mr. Colpitts, who split the capacitance rather than the inductance, and of course, we have the Colpitts oscillator circuit. See Fig. 9-8. The frequency of this circuit is determined by the value of one of the capacitors in the tank (the other one matching it) and all of the inductance. Each capacitor acts between ground and the inductance; and since each end of the inductor sees the same capacitance, it works as though there were only one capacitor in the circuit.

Now, please understand—all of these oscillators that depend upon an LC tank circuit to control frequency are a little bit unstable. Some stick to their frequency better than others, but all of them drift a bit with changes in temperature, changes in humidity, changes in barometric pressure, etc. And the need to have more stable frequency control brought out the use of piezoelectric crystals. These pieces of Quartz or Tourmaline crystals will oscillate (cause electrons to move back and forth) at their own mechanical resonant frequency. If you change the shape or put a mechanical strain on these crystals, they will cause an electron current to move. And if you apply an AC signal to a crystal that is equal to its own resonant frequency, it will actually vibrate at that frequency.

The frequency at which a piece of one of these crystals will oscillate, being its mechanical resonant frequency, is largely determined by its thickness. Of course, there are other things that affect the frequency as well, like just what part of the crystal the piece was taken from. If a crystal is cut exactly parallel with the Z axis of the mother crystal, it will still shift frequency slightly with temperature change and is said to have a temperature coefficient. Pieces cut at different angles

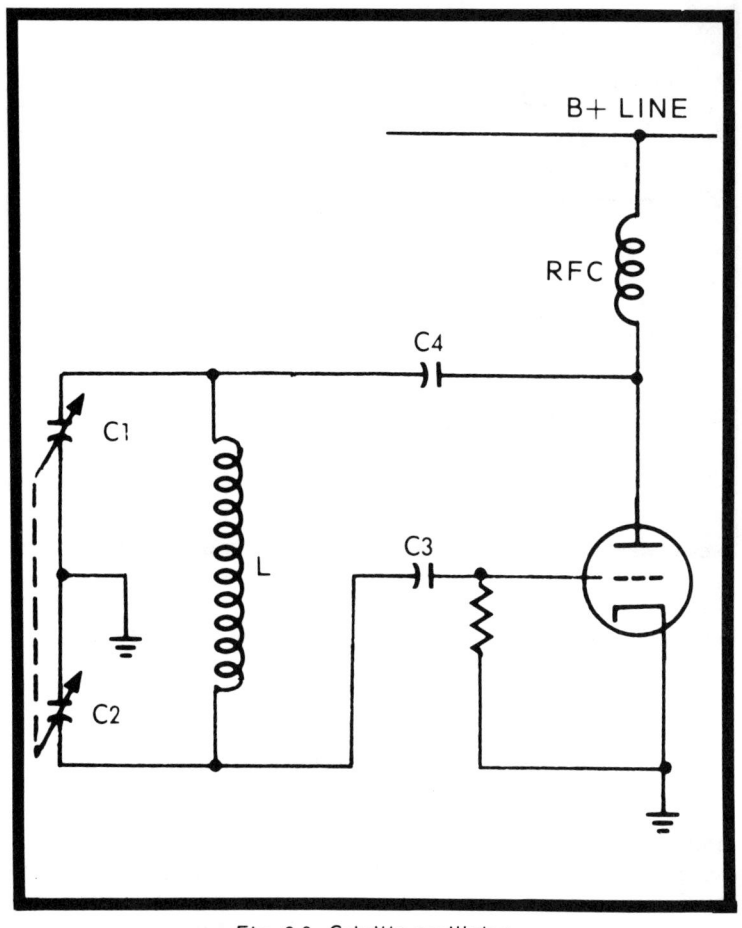

Fig. 9-8. Colpitts oscillator.

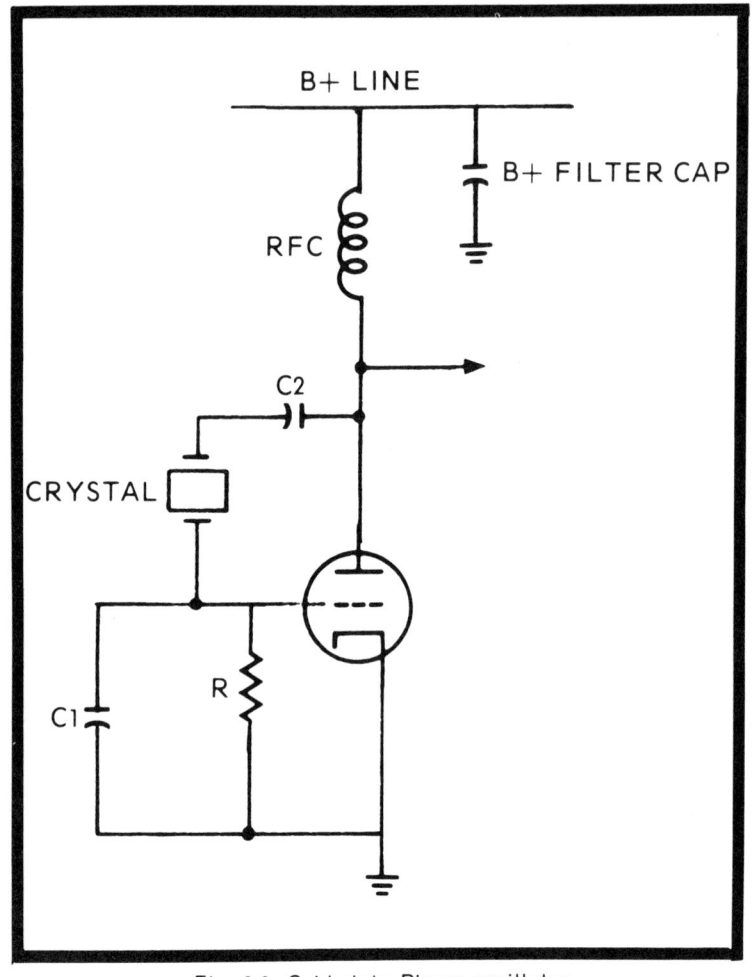

Fig. 9-9. Grid-plate Pierce oscillator.

with respect to the Z axis have been found to have temperature coefficients approaching zero and are much preferred and are quite common.

GRID-PLATE PIERCE OSCILLATOR

Crystals are used much the same as "tank circuits" and in many of the same places. Take the Grid-Plate Pierce oscillator" in Fig. 9-9 for instance. In this circuit, we have the crystal taking the place of the delay line in our first little oscillator, and it works very much the same.

TRIODE-TYPE CRYSTAL-CONTROLLED OSCILLATOR

Crystals may also be used in the grid circuit such as the one shown in Fig. 9-10 (the Triode-type crystal-controlled oscillator). Note particularly that the plate has a tank circuit that is tuned to the frequency of the crystal. This will either be "link coupled" or "capacitance coupled" (as shown) to a tank circuit on the grid of the next stage. And this next stage will have a tank circuit connected to its plate, as well.

Gee, doesn't this sound familiar. Sounds just like a tuned-grid tuned-plate oscillator, doesn't it? Yes, and it'll act just like one if we don't do something to stop it. So, here again we use feedback, but for the opposite purpose. This time to **prevent** the circuit from going into oscillation. See Figs. 9-11 and 9-12.

NEUTRALIZATION

In these two cases, we tune the coupling capacitor in the feedback loop so that just enough of the opposite phase signal

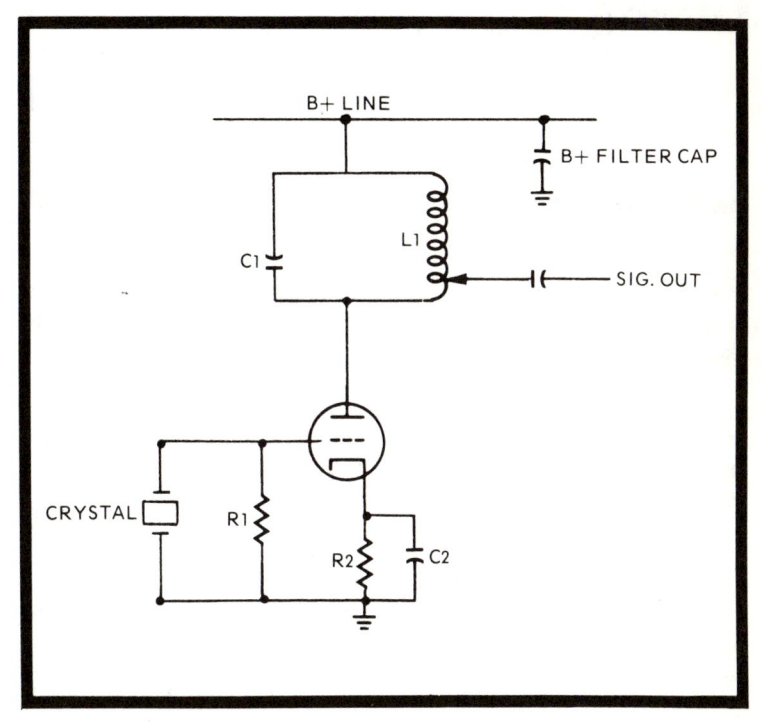

Fig. 9-10. Triode-type crystal-controlled oscillator.

295

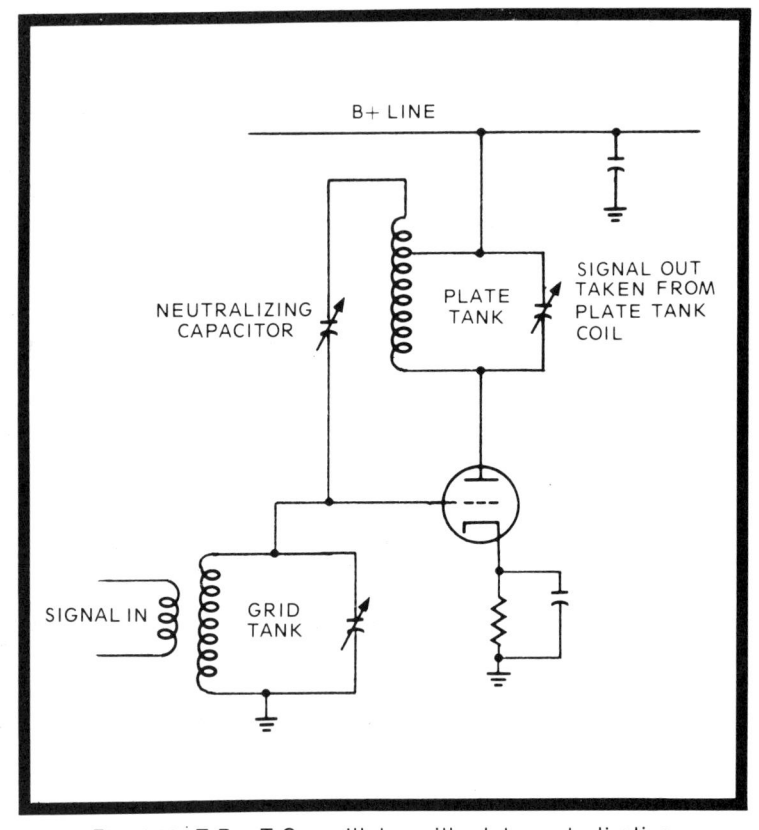

Fig. 9-11. T.P.—T.G. oscillator with plate neutralization.

is fed back to eliminate the natural feedback and allow the amplifier to handle just the signal that comes from the oscillator. This use of feedback is called neutralization.

Note that we can take the signal off the top of the plate tank and feed it back to the grid and call it "plate neutralization," or we can take the signal right off the plate and feed it into the bottom of the grid tank and call it "grid neutralization."

Now then, there's a lot more to be said about oscillators, but I don't intend to try to say it here. A lot more learned men than I have written whole books on the subject. If I have succeeded in getting just some of the basics across to you and tickled your curiosity about them a little bit, I will have reached my goal.

But before we drop this bit of subject matter let's go back to the piezoelectric effect. It can be a lot of fun.

GROWING YOUR OWN CRYSTALS

Quartz and tourmaline crystals are not the only ones that exhibit the piezoelectric effect. Rochelle-salt crystals can do this too, and you can grow these yourself.

If you add 130 grams of Rochelle Salts to 100 cubic centimeters of water and mix it until it's all dissolved and then add about 9 grams more of Rochelle Salts to this mixture (you'll probably have to heat it up now to dissolve this—and you probably had to heat the original up to get it to dissolve too), you'll have what is called a saturated solution. Put a cap on the jar and let it sit for a couple of days. You'll probably notice some crystals growing on the bottom.

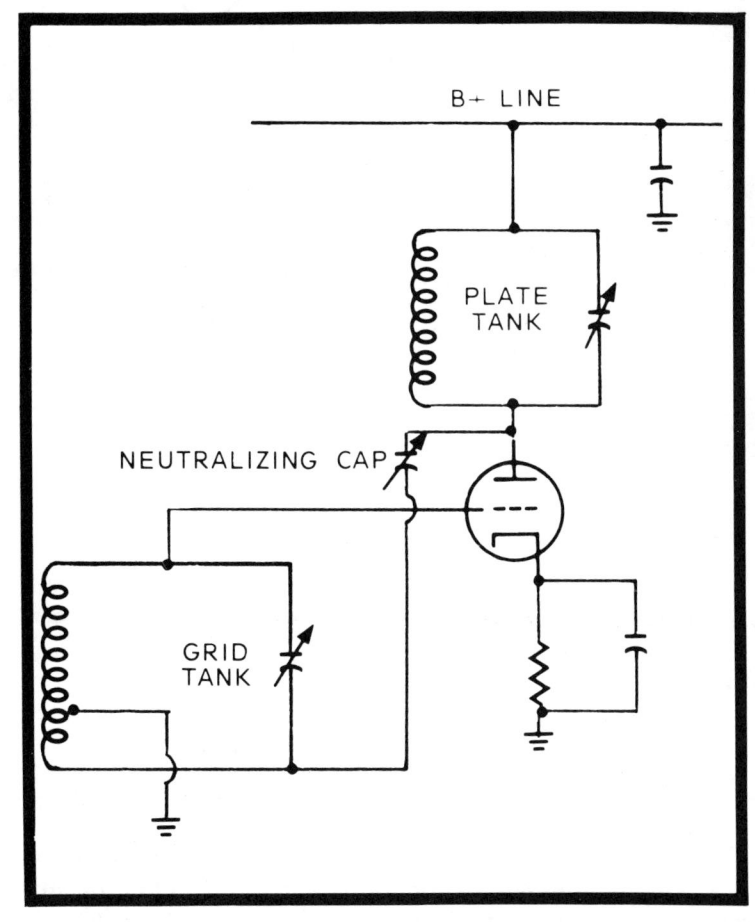

Fig. 9-12. T.P.—T.G. oscillator with grid neutralization.

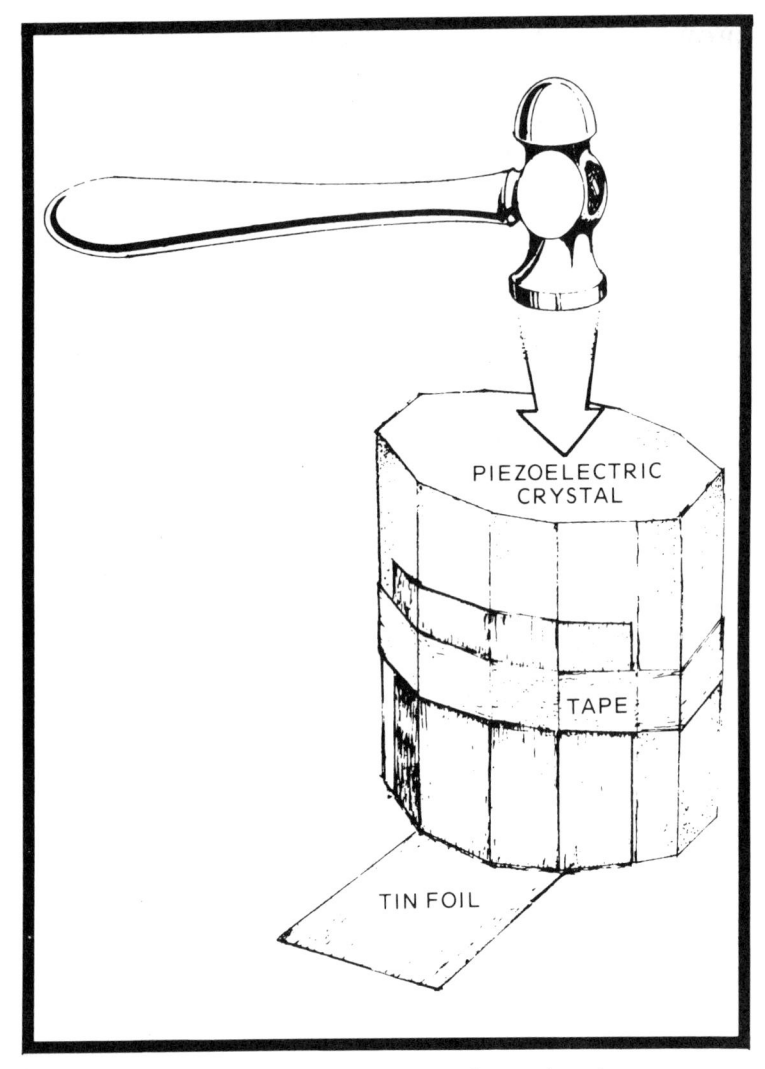

Fig. 9-13. Piezoelectric crystal experiment.

Fish one or two of these crystals out and heat the solution up again a bit and make sure all the rest of the crystals have gone back into solution. Now let the mixture cool down slowly.

While it's doing this, take one of the crystals you fished out and tie it onto the end of a piece of thread.

After the mixture has cooled, hang that little crystal down in the middle of your solution (a slit in a piece of cardboard will hold the upper end of the thread). Now, leave this alone for a couple of days. The crystal at the end of the thread should grow to a usable size. When it has, take it out and you have a piezoelectric crystal.

Stick a couple of pieces of tinfoil on opposite sides of this crystal with a thin layer of vaseline and tape them tightly in place with some plastic tape. Be sure to leave ends of the tinfoil long enough to make electrical connection (preferably with alligator clips).

Now, I have heard that if you strike this crystal on top lightly with a small hammer you can light a small neon bulb. But, it didn't work for me. I hooked it up to an oscilloscope, though, and got a 2-volt signal out of it. See Fig. 9-13.

These crystals aren't very strong, and they don't last very long. They also deteriorate in air and are very difficult to save. But this is fun and does illustrate a point very well.

Power, Power Supplies, and Safety

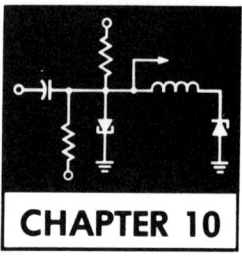

CHAPTER 10

In this chapter we will explore various configurations of power supply circuitry from special viewpoints. To most readers, this may be a review, to others, brand-new. But it is my hope that all readers will learn a little something new, as for example, RC time constants in power supplies.

SAFETY

Too often, people get careless as they gain experience in the trade. Let's review some basic safety items.

TEN ELECTRO COMMANDMENTS

1. BEWARE the lightning that lurketh in an undischarged capacitor lest it cause thee to bounce upon thy buttocks in a most undignified manner.
2. CAUSE thou the switch that supplieth large quantities of juice to be opened before groping, that thy days may be long in this earthly vale of tears.
3. PROVE TO THYSELF that all high voltage circuits on which thee worketh, do not worketh, and have been discharged, lest thee be elevated well above ground, and thy nose light to show others the way.
4. TARRY THOU NOT amongst those fools that engage in intentional shocks for they are surely nonbelievers and are not long for this world.
5. TAKE CARE THOU tampereth not with safety devices as this incureth the wrath of the manager and bringeth the fury of thy supervisor upon thy head and shoulders.
6. WORK THEE NOT on energized equipment for if thou doest so, thy friends may be buying drinks for thy widow and consoling her in ways not generally acceptable to thee.
7. VERILY, VERILY, I say unto thee, never worketh on equipment alone, for electrical cooking is a slothful process, and thou might sizzle for hours in thine own fat

before thy Maker sees fit to end thy misery and drag thee into the fold.
8. TAKE CARE THOU useth the proper method when thou taketh the measure of high voltage circuits so that thou dost not incinerate both thee and thy test meter, for verily, though thou hast no account number and can be easily replaced, thy meter dost have one and as a consequence brings much woe unto accounting.
9. TRIFLE THOU NOT with radioactive tubes and substances, lest thou commence to glow in the dark like a lightning bug, and thy wife be frustrated and have no further use for thee except for thy paycheck.
10. COMMIT THOU to memory all the words of the prophets which are written down in the training manual and the calibration procedure, which giveth out straight dope and consoleth thee when thou hast suffered a ream job from thy supervisor.

As much as we like to chuckle at such things as the TEN ELECTRO COMMANDMENTS, don't let your chuckle prevent you from taking them seriously. In all their lightheartedness, there are times when each can save your life. It doesn't take much in the way of current to kill you, and it is CURRENT that kills—not voltage. BUT, don't let this make you think voltage is safe. It isn't. It moves current.

What's that? How much current does it take?

Well, let's put it this way: In a hospital operating room where there is a great deal of electrical equipment and doctors are working inside of people, ten microamps is thought of as being deadly. A single pulse of 10 ua of current through the heart at the wrong time can cause the heart to go into fibrillation (this is where the heart quivers rather than pumps). Once the heart goes into fibrillation, it can't snap itself back out again. It takes a strong DC jolt to clamp the heart muscle rigidly in one position for a moment and then let it go hopefully back into a normal beat, but it takes a Doctor to do this.

OOPS! Hold it! I know, the literature you read said it took a whole lot more current than 10 ua to even let you feel it, a thousand times more in fact. But you have to remember that was current entering one hand—going through your whole body—and then out your other hand. All of you is conductive, and this current distributes itself through all of you. Under these conditions, the current density in your heart (10 ua) hasn't been reached yet until the current level entering your whole body is in the neighborhood of 100 ma. The article you probably read said the range of current from 100 ma to 200 ma was deadly, and it is. But, believe me—you know it at current levels way below this. You start feeling it at about 8 ma and you can't let go at 12 ma. So, the difference between just feeling it and getting a pretty good shock isn't very much.

There are two things the TEN ELECTRO COMMANDMENTS left out as far as I'm concerned. These are:

11. MAKE HASTE SLOWLY. You'll get it fixed sooner and safer.

12. NEVER GRAB for a falling tool. It might be your soldering iron and you just might catch it.

Oh? You've heard the story that sixty cycles is the most dangerous frequency we could possibly have chosen for power distribution throughout our land? Is it true?

Well, I don't really know. I don't know how it compares with—say, fifty cycles. But I do know that the acceptance threshold of the human body to electricity at ten cycles is higher than it is at sixty, and this acceptance threshold is higher at one hundred cycles than it is at sixty. I can't tie it down any closer than that.

Knowing something about the frequencies involved in this situation can lead to a lot of speculation though. For example, we know that:

(1) The human heart beats at about one cycle per second.

(2) In fibrillation, the heart vibrates at a frequency slightly higher than ten cycles per second.

(3) The human nervous system cannot respond to anything faster than twenty cycles per second (that's its upper cut-off frequency).

Now then, if the heart in fibrillation is going as fast as the nervous system can respond (20 Hz), then 60 Hz is its 3rd harmonic and reinforces the oscillation every one and a half cycles if it's in phase. Also, if the heart in fibrillation is going at something just a bit faster then ten cycles per second, say 12 Hz, then 60 Hz is the 5th harmonic and reinforces the oscillation every two and a half cycles when in phase. I'm not going to try to prove this and I hope you don't either, but it does show possibilities, I must admit.

But getting down to the subject matter at hand, namely, Power Supplies, I'm going to take it for granted that you know how a transformer works. That a ratio of the number of turns of wire in the secondary (right hand side of the transformer in all figures) to the number of turns of wire in the primary is what is called the Turns Ratio and determines the second voltage when the primary voltage is known. See Fig. 10-1.

$$\frac{\text{No. of Secondary Turns}}{\text{No. of Primary Turns}} = \frac{\text{(Turns Ratio)}}{\text{Primary Voltage}}$$

This ratio may be greater than one (in the case of a step-up transformer), or it may be less than one (in the case of a step-down transformer).

I am also going to take it for granted that you know that the current ratio will be the inverse of the turns ratio. That is:

$$\frac{\text{No. of Secondary Turns}}{\text{No. of Primary Turns}} = \frac{\text{Primary Current}}{\text{Secondary Current}}$$

But please understand that it is Ohm's Law along with the secondary voltage and the resistance across the secondary that determine how much current the primary will accept from the line. The power it will accept from the line (that plug in the wall) will always be a little bit more than the power used by the secondary load resistance. The transformer is not a perfect power-transferring device. There is always a little power lost in the transfer from primary to secondary. However, unless you are in the business of designing transformers, this really doesn't bother you very much. The voltage lost can always be made up for with an extra turn or so in the secondary. However, don't ever demand that any set of turns carry more current than the transformer rating. Always remember that it is the load across the secondary that will determine (via the inverse equation) how much current will be accepted by the primary. Don't let this burn up the transformer.

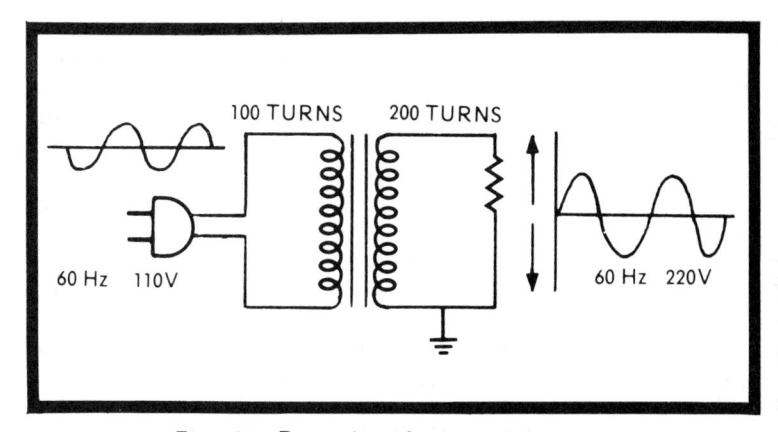

Fig. 10-1. Power transformer schematic.

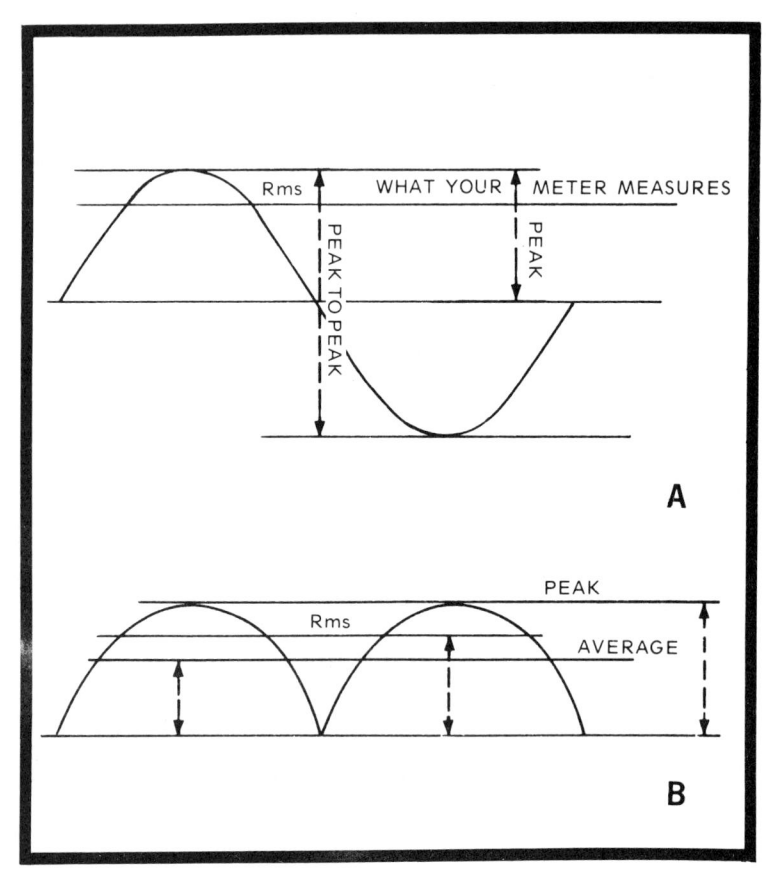

Fig. 10-2. AC measurement of voltage (peak, average, RMS, peak-to-peak).

I am also going to take it for granted that all you need is a reminder that when you take your AC meter and put it across a 110-volt line and it reads 110 volts, it is giving you the rms (root-mean-square) value of voltage on the line. This merely says that this voltage will push the same amount of current through a given resistor as 110 volts DC will. This is in spite of the fact that the average voltage on the AC line is only nine-tenths (0.9) of the rms value, and the peak voltage is 1.414 times the rms reading. The peak-to-peak voltage difference on the AC line is of course just double the peak value, or 2.828 times the measured (rms) value of voltage. Yes, I mean that the peak to peak voltage on the 110 volt AC line is;

110 x 2.828 equals 311 volts

Maybe this chart and Fig. 10-2 will help the situation:

Known	Average	RMS	Peak	Peak-to-Peak
Average	---	1.11	1.57	3.14
Rms	0.9	---	1.414	2.828
Peak	0.637	0.707	---	2.0
Peak-to-Peak	0.32	0.3535	0.5	---

HALF-WAVE RECTIFIER

Now then, if we add a diode to our circuit like you see in Fig. 10-3, we have what is called a half-wave rectifier circuit long as it is handling 60 cycle, 110 volt, AC (when such a circuit is handling RF signals, it's called a detector circuit). Anyway, our half-wave rectifier changes AC into pulsed DC. The electron current in the secondary is always going the same direction, but it does so in pulses (the diode prevents electrons from going in the other direction). If we add a large enough capacitor across the load, we might get something like that seen in Fig. 10-4. This, you will note, sort of fills in the valleys between the half wave pulses and gives us a steadier DC electron current flow, but there still is a ripple to the voltage level even though current flows 100 percent of the time (this, of course, depends on the idea that the RC time constant is long with respect to the time between pulses). And all this sounds great until we realize that the poor diode now is not only handling the load current each time it turns on, but it also is handling the current that flowed into the capacitor during the time it was turned off. If the capacitor is large enough to keep the ripple small enough to be within reason, the peak current that the diode has to handle can be several times the magnitude of the load current alone. If the diode can't handle these peak currents, it's going to get burned up. (In the case of a vacuum tube diode, it gets the cathode blown off.) So, for power supplies that are going to handle much current at all, this isn't very satisfactory. We're going to have to change it.

FULL-WAVE RECTIFIER

In the days of vacuum tubes (they're not gone yet), designers put two of these half-wave rectifiers back to back (doubling the number of turns in the secondary and requiring a center tap) to get the circuit you see in Fig. 10-5. This reduced the time between pulses greatly, allowing the use of a

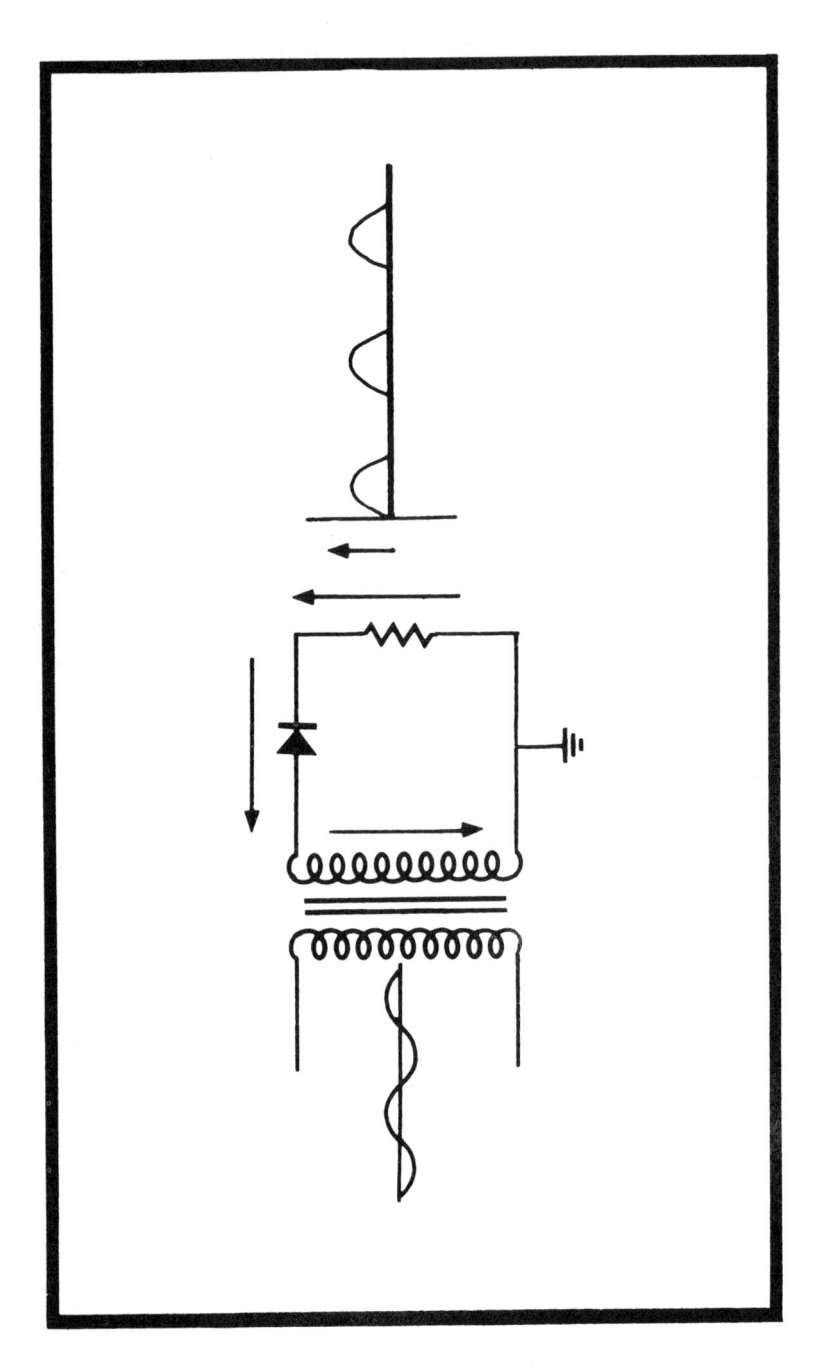

Fig. 10-3. Half-wave rectifier.

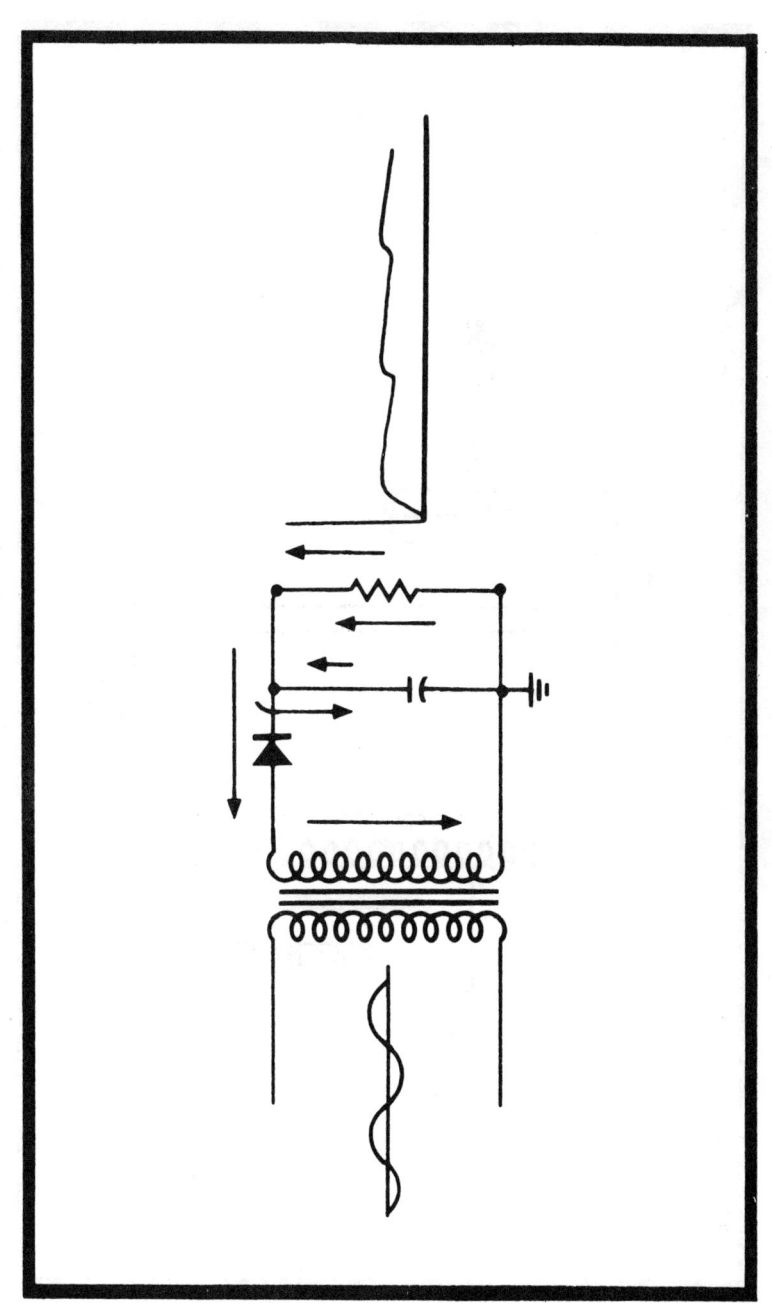

Fig. 10-4. Fig. 10-3 with shunt capacitor added to "fill-in" between the half-wave peaks.

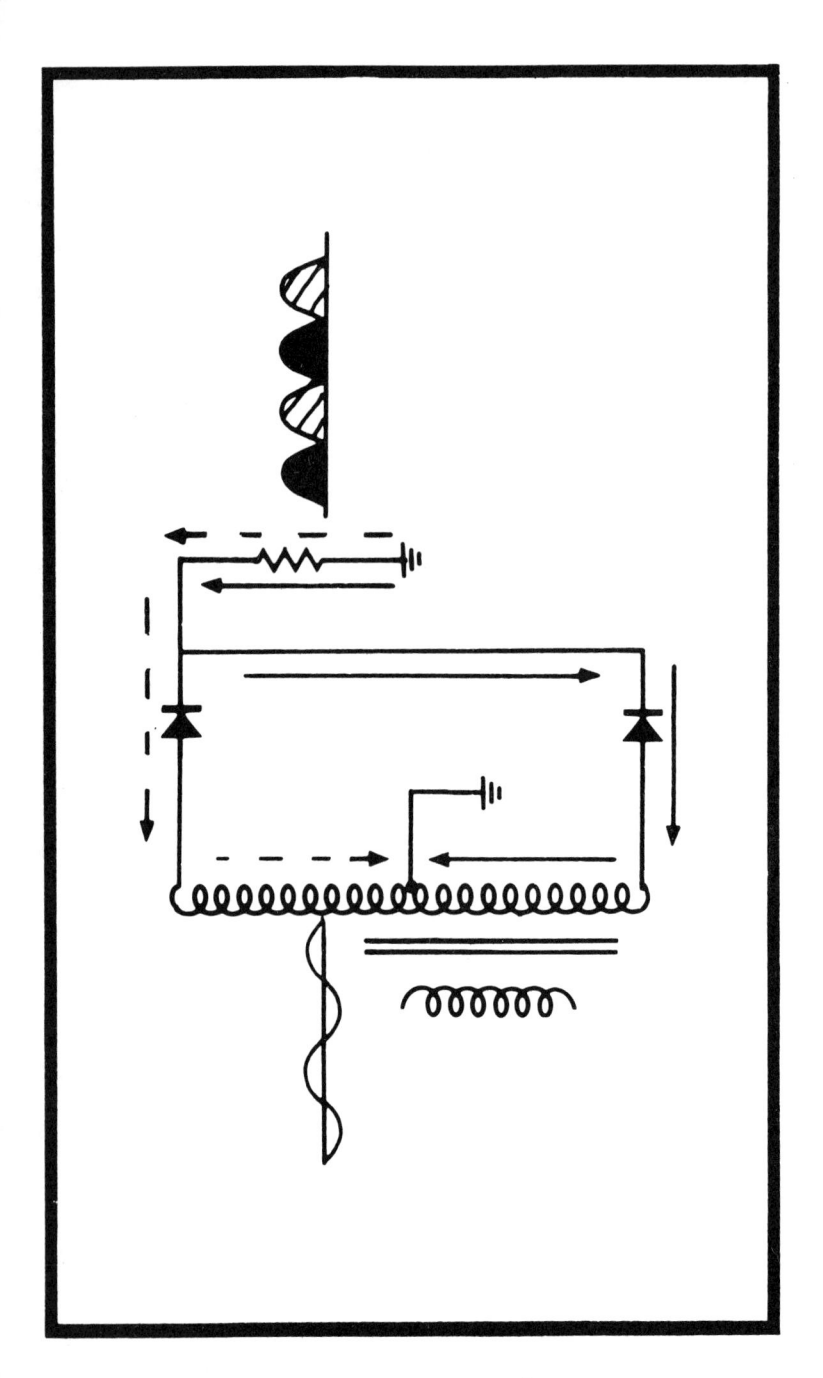

Fig. 10-5. Full-wave rectifier.

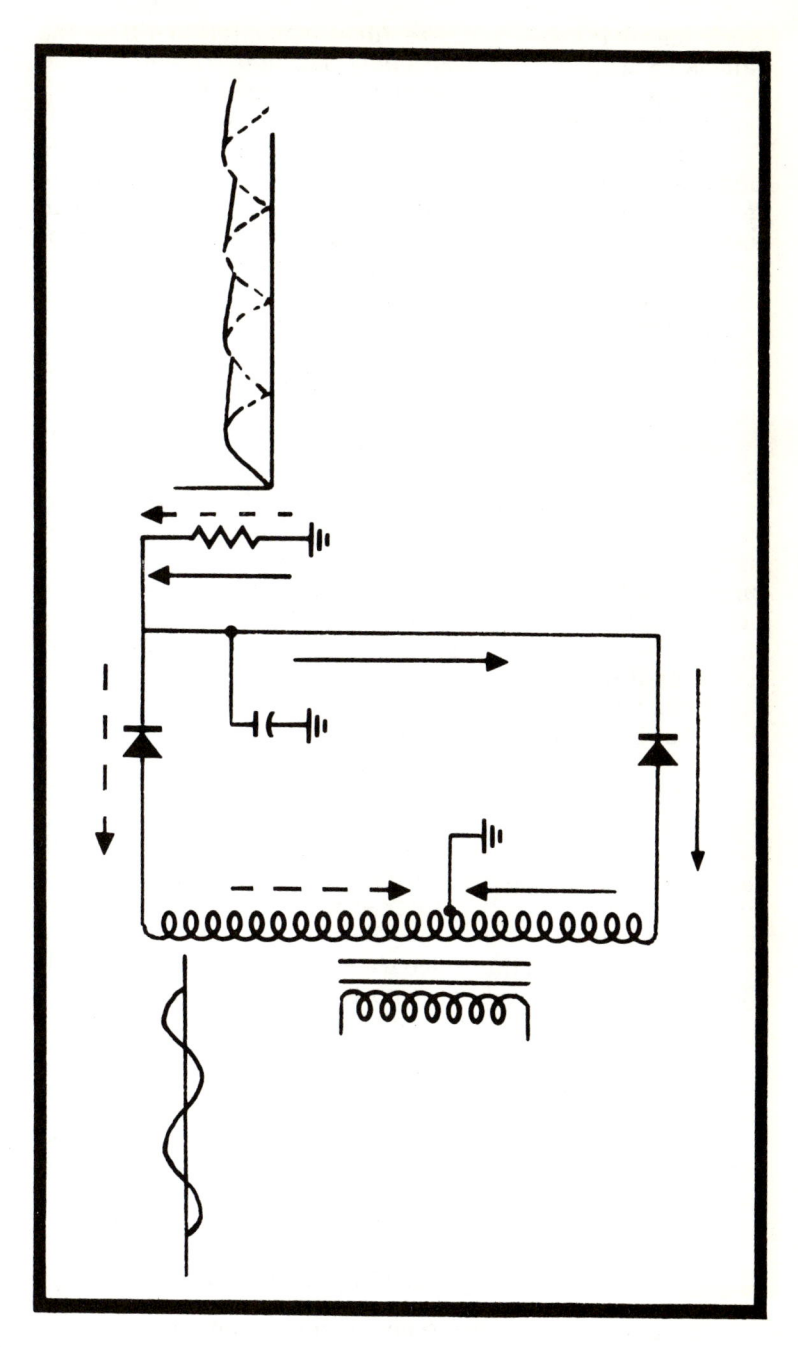

Fig. 10-6. Fig. 10-5 with filter capacitor added.

much smaller value and size filter capacitor to fill in the valleys, (Fig. 10-6). The diodes only had to be rated for peak currents that were twice the amount of current that was being drawn up through the load.

I am personally convinced that it took the use of solid-state diodes and the desire to conserve on both weight and space to develop the circuit I show you in Fig. 10-7. Once more you will note that the full secondary is used during both half cycles of the primary signal and we are back to the turns ratio bit that we started with, along with smoothed full-wave rectification.

It's interesting to note here that if my transformer has a 1:1 turns ratio, I ought to be able to build a DC 140-volt supply with just a few volts of ripple with this circuit. Let's look at it closer.

If we have a frequency of 60 cycles per second (60 Hz), then each full cycle lasts about 16.7 msec. Each half cycle then lasts for about 8.35 msec. Now, if we allow for a 6 volt ripple, the change of voltage time back to peak voltage level (the time during which the diode has to conduct the extra electron current from the capacitor) will be about 1 msec long. (This is a rough estimate, but plenty close enough to work with.) And—all of a sudden—we find that we have enough information to determine the size of the capacitor:

We have the "change of voltage": dV,
"change of time": dT
"load current": I
And we want to find: C.

Do you recognize the formula that it fits? I sure hope you do. It's: dV/dT equals I/C. And if you are willing to allow 18 volts of ripple, you can roughly double the dT estimate and get by with a much smaller filter capacitor, about one third smaller in fact.

This tells us then that since we do not have diodes that will conduct infinite currents any more than we have capacitors that have infinite value capacitance, we will always have some ripple content in the DC output voltage of such a supply. This is what we refer to when we speak of as "unregulated voltage."

Now, the need was and is for regulated voltages, that is, DC voltages that only have a few millivolts of ripple. This gave rise to what was called the series regulator type of supply (there was a shunt regulator supply also).

SERIES REGULATOR AS OPERATIONAL AMPLIFIER

The Series Regulator type of supply might be better thought of as a DC Operational Amplifier type of supply; see

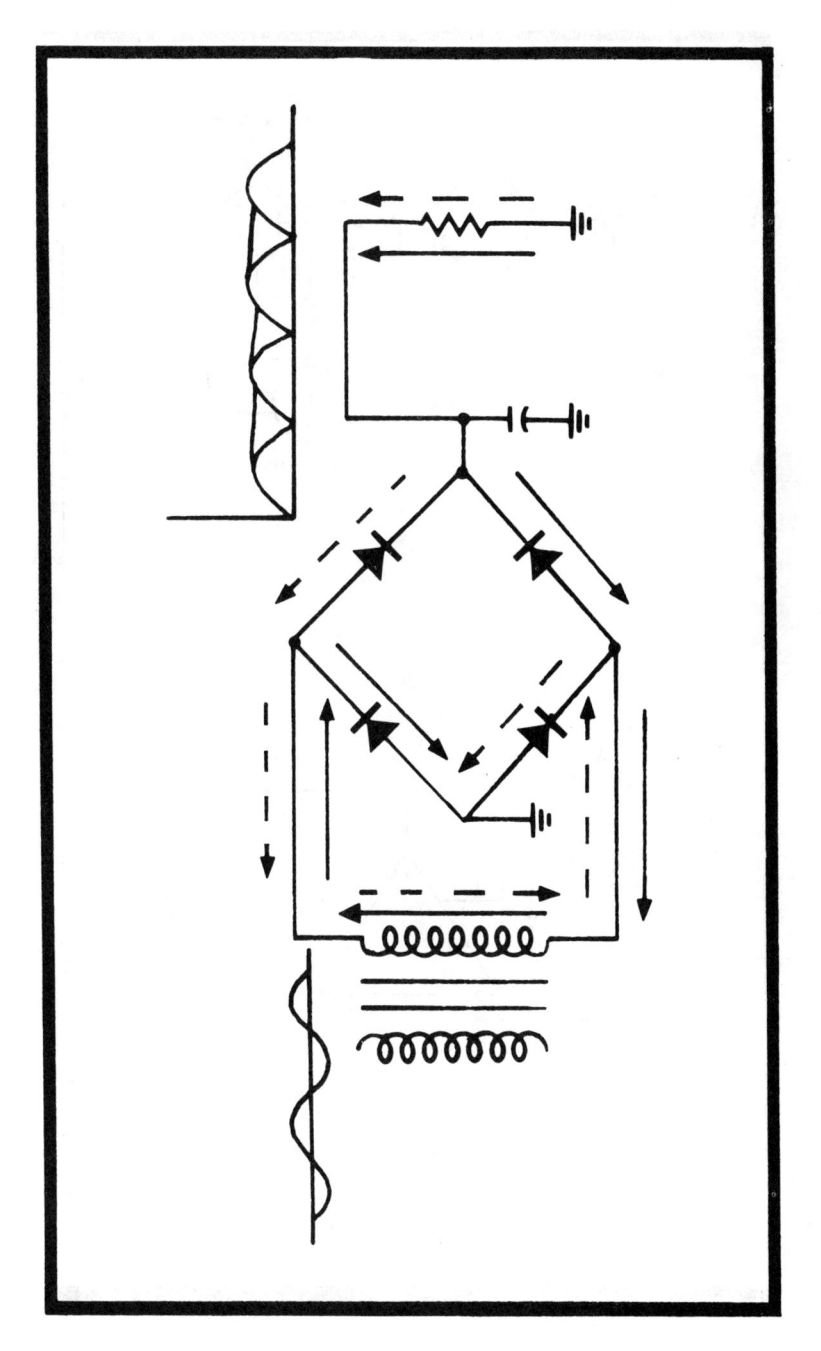

Fig. 10-7. Full-wave bridge rectifier.

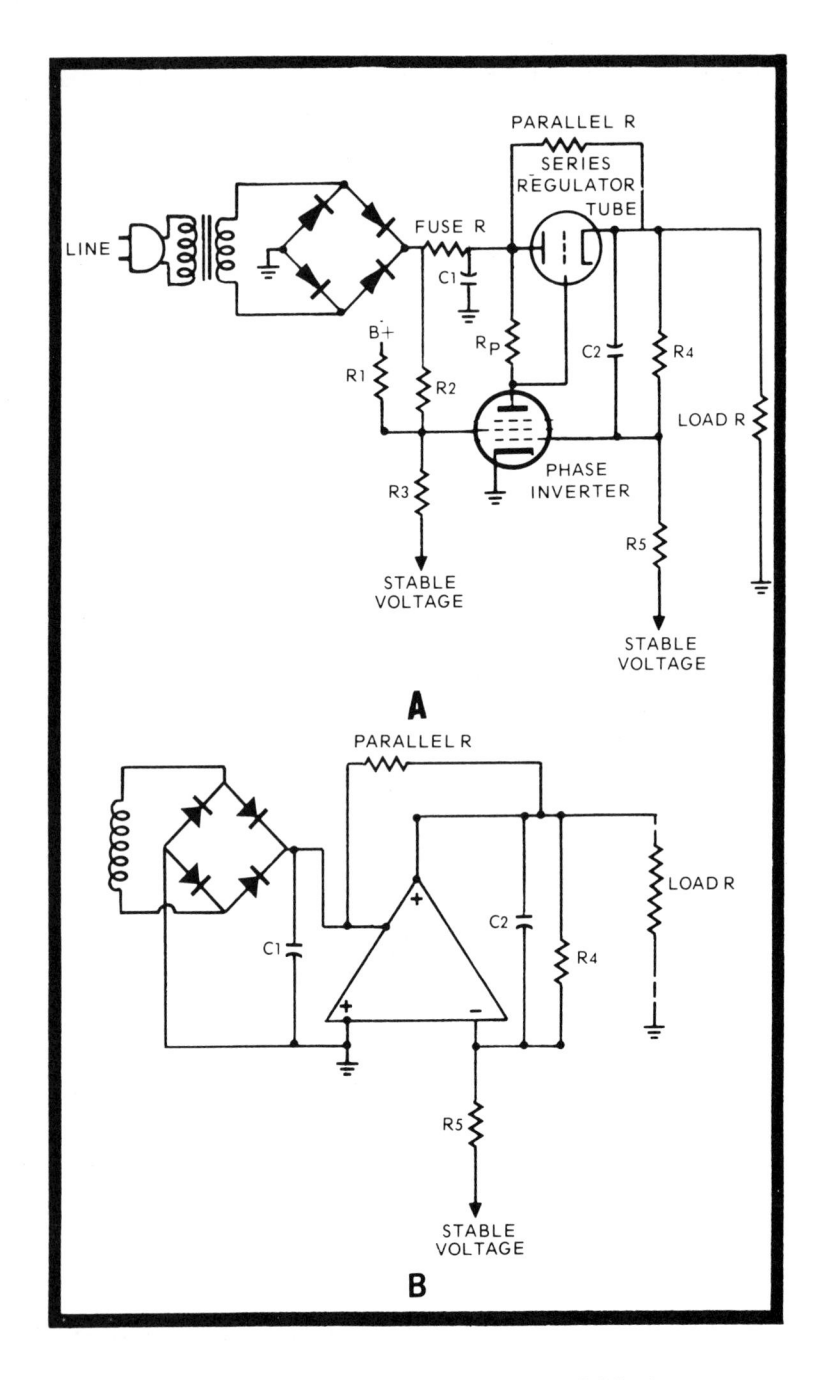

Fig. 10-8. Relationship of series regulator to DC Op Amp.

Fig. 10-8 where A has been redrawn at B to point out that the two circuits shown in this diagram are the same. The circuit using the triangle in place of the Phase Inverter and Series Regulator Tube shows the parts critical to the understanding of how this circuit works. I should point out here that the unregulated DC voltage across C1 has to be enough greater than the regulated supply voltage at the output of the Op Amp to give the series regulator tube adequate B+ to operate on. This difference in voltage also appears across the Parallel Resistor.

The Op Amp in Fig. 10-8B operates just as any Op Amp does. If the input DC at the bottom of R_5 is negative (—), then the output will be positive (+) to the magnitude determined (remember?) by R_f over R_i or in this case R_4 over R_5 times the input voltage (as long as the reference voltage of the Op Amp is ground). If the supply of this nature happens to be stacked on top of another supply (i.e., the + input to the Op Amp at some regulated DC level other than ground), then R_4

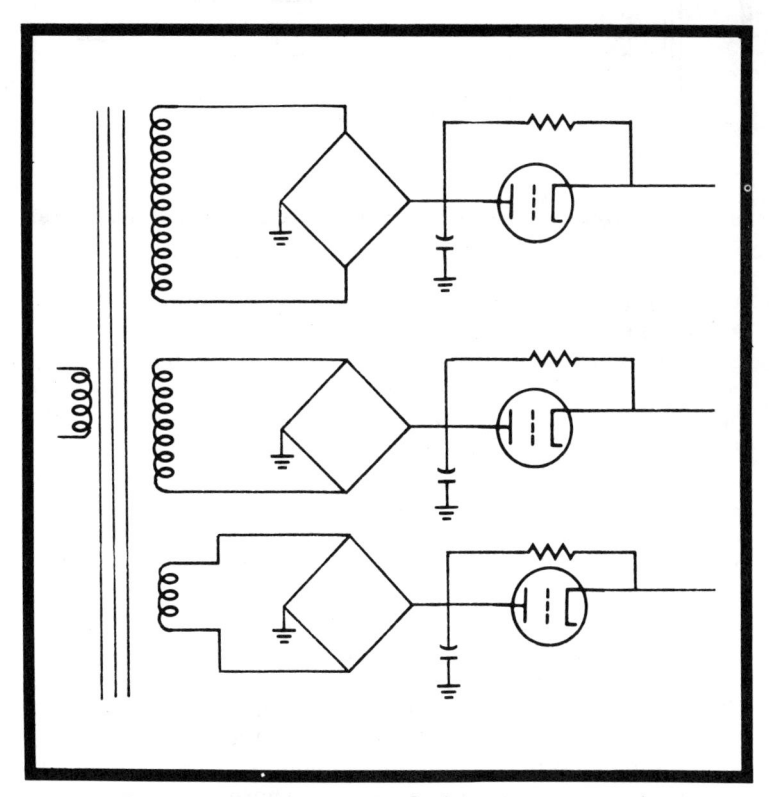

Fig. 10-9. Independent series Op Amps in power supply.

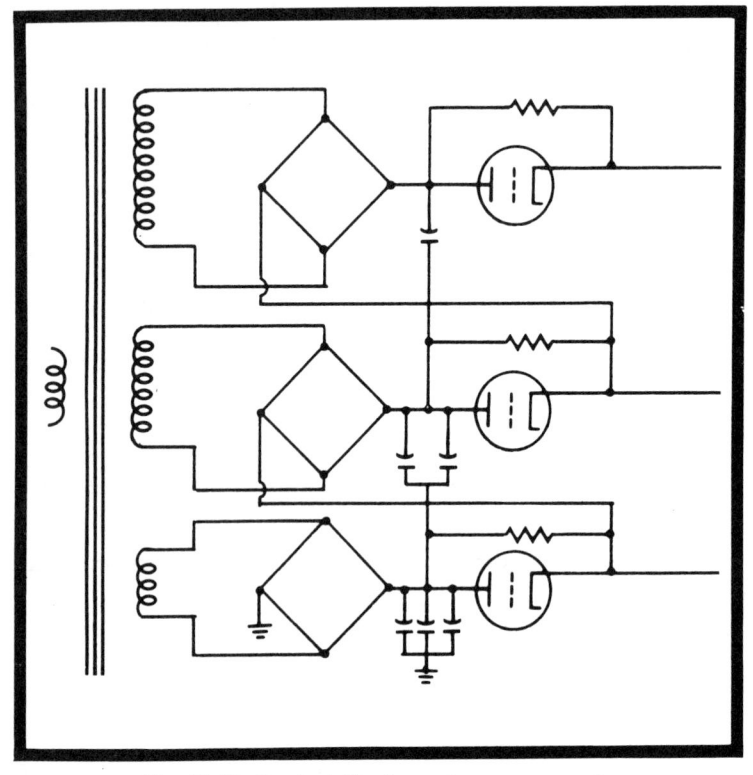

Fig. 10-10. Stacked Op Amps in power supply.

over R_5 multiplies the voltage across R_5 and then is added to the reference voltage.

If you will refer to Figs. 10-9 and 10-10, you'll get the idea of what I mean. Very seldom do we find a set of secondaries like those shown in Fig. 10-9, each operating independently of the others and each with its own reference to ground. This would lead to an excessively large transformer and they're expensive as well as heavy. Usually, we find one supply stacked on top of the other, like those shown in Fig. 10-10. This obviously allows us to get by with a smaller transformer, but it has its share of problems when it comes to the filter capacitors. The bottom supply, for instance, has to filter not only for its current but the other two supplies as well. The middle one has to filter for its current plus that of the third (the top one), and the top one is the only one that has to filter for its own current only. This means that the I in our formula dV/dt=I/C can get pretty big when it comes to the poor guy on the bottom, so you have to use discretion here also.

This combination of filter capacitor and series regulator might readily be compared with the suspension system of your car, the filter caps being comparable to the springs on each wheel mount taking out the major bumps in the road, and the series regulator being comparable to the shock absorbers which complete the job of giving you a smooth ride. Of course, our idea is to give the electrons a smooth ride in the B+ bus and maintain a steady pressure here so the electrons respond only to the input signal stimulus and to nothing else.

To see more of what I meant about this R_f/R_i business with the DC Op Amps in these power supplies containing the series regulators (so called), see Fig. 10-11. This is a power supply that contains many of the circuits of which I've been talking. Note the +100V supply. It has a ground reference (V664 cathode is at ground). R_f is R650 (333K), and R_i is R651 (490K). R_f over R_i then is 333/490. This times the minus 150 (which is the DC input on the end of R_i) along with a change of sign, gives us the +100 volts of this supply.

Again, check the +225V supply. V694 cathode is at ground, and R_f/R_i will be R680 and R681 or (333/220) x —150 equals +225V.

The +350V supply works out in the same fashion with R710 and R711 doing the job for us.

The +500V supply is a little different though. Note that its reference voltage is the +350V supply (V754 cathode is at +350V). R_f is R740 and R_i is R741 and their ratio (220/720) has to multiply the voltage across R_i which is 500V (i.e., 350 + 150 equals 500). This ratio times 500 gives us +150V which has to be added to our +350 reference voltage to get the supply voltage out with respect to ground. And, as you can see, 350 plus 150 gives us our +500 for the op amp regulated supply.

As good as they are, these supplies have their limitations too. They can't carry any more current than the cathode of the series regulator unless they are modified a bit (as these are). Note that each series regulator tube has a resistor in parallel with it. This resistor carries the extra current. There are times when there is more than one series regulator too. The +350-volt supply has such a combination; two cathode as well as two parallel resistors are required to carry the current load on this supply.

What? Oh, yes—thanks for reminding me. "What's the system for determining the size of the parallel resistors?!!!"

That's quite a story and I think you'll like it. Turn to Fig. 10-12 and I'll explain it to you.

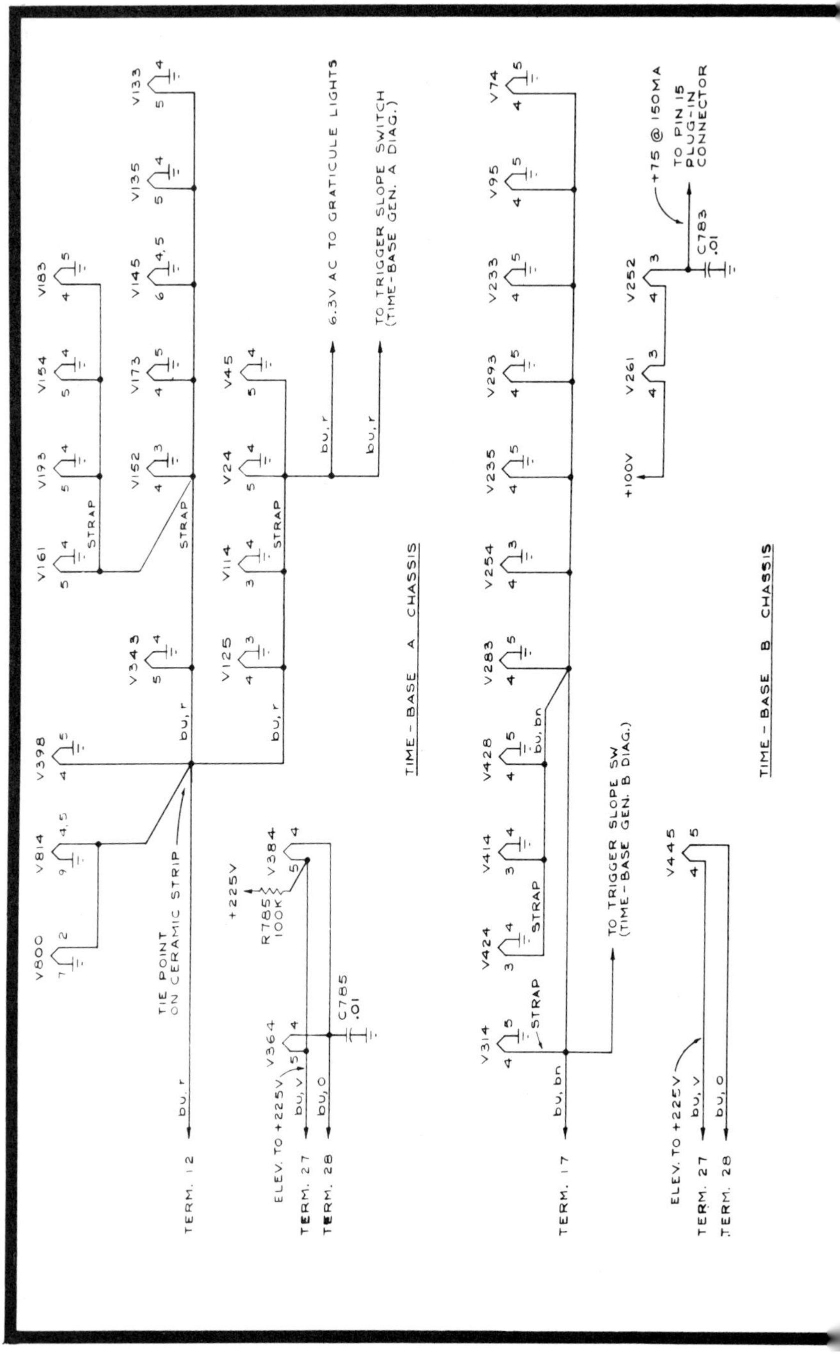

TIME - BASE A CHASSIS
TIME - BASE B CHASSIS
V800 V814 V398 V343 V161 V193 V154 V183 V152 V173 V145 V135 V133
V125 V114 V24 V45
V364 V384
V314 V424 V414 V428 V283 V254 V235 V293 V233 V95 V74
V261 V252 V445
TIE POINT ON CERAMIC STRIP
STRAP
bu, r
bu, bn
bu, v
bu, o
6.3V AC TO GRATICULE LIGHTS
TO TRIGGER SLOPE SWITCH (TIME-BASE GEN. A DIAG.)
TO TRIGGER SLOPE SW (TIME-BASE GEN. B DIAG.)
+225V
R785 100K
C785 .01
C783 .01
+100V
+75 @ 150MA
TO PIN 15 PLUG-IN CONNECTOR
ELEV. TO +225V
TERM. 12
TERM. 27
TERM. 28
TERM. 17

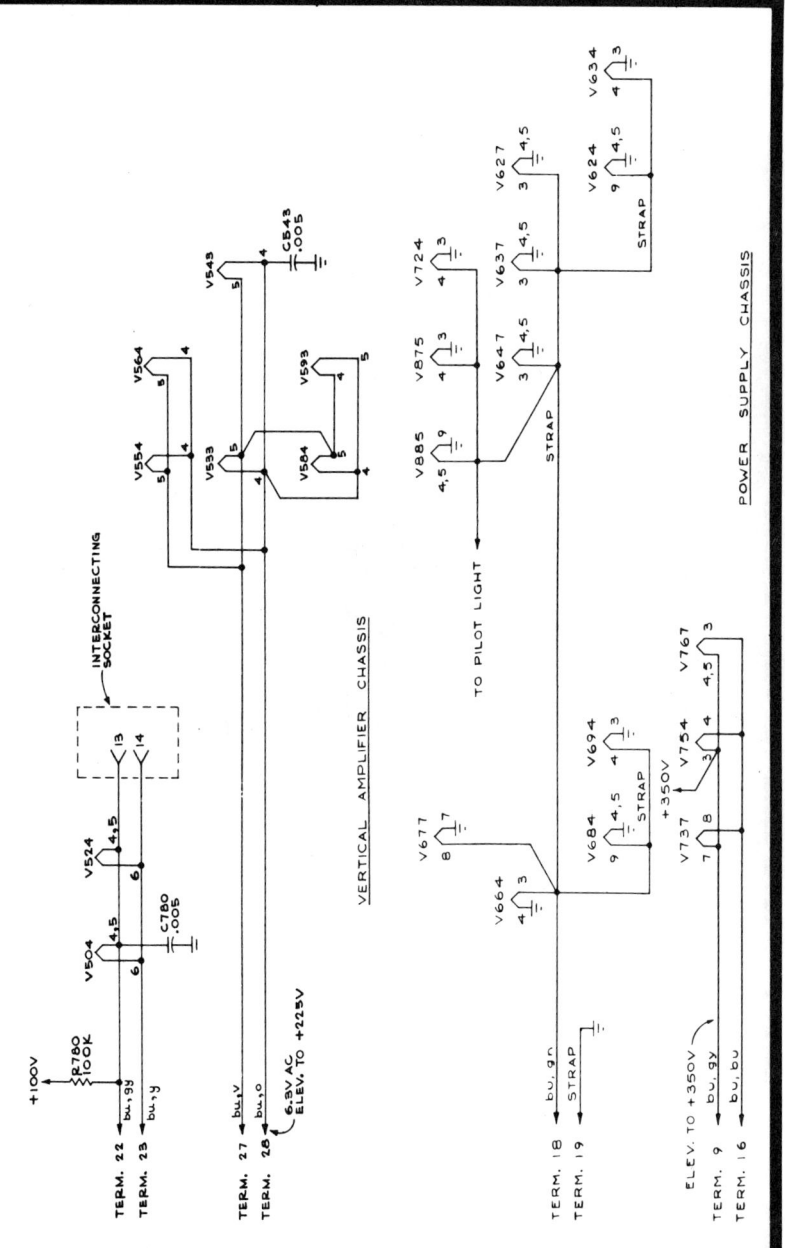

Fig. 10-11A. Commercial supply illustrating Op Amp usage (part 1 of 2 parts) (Courtesy Tektronix, Inc.).

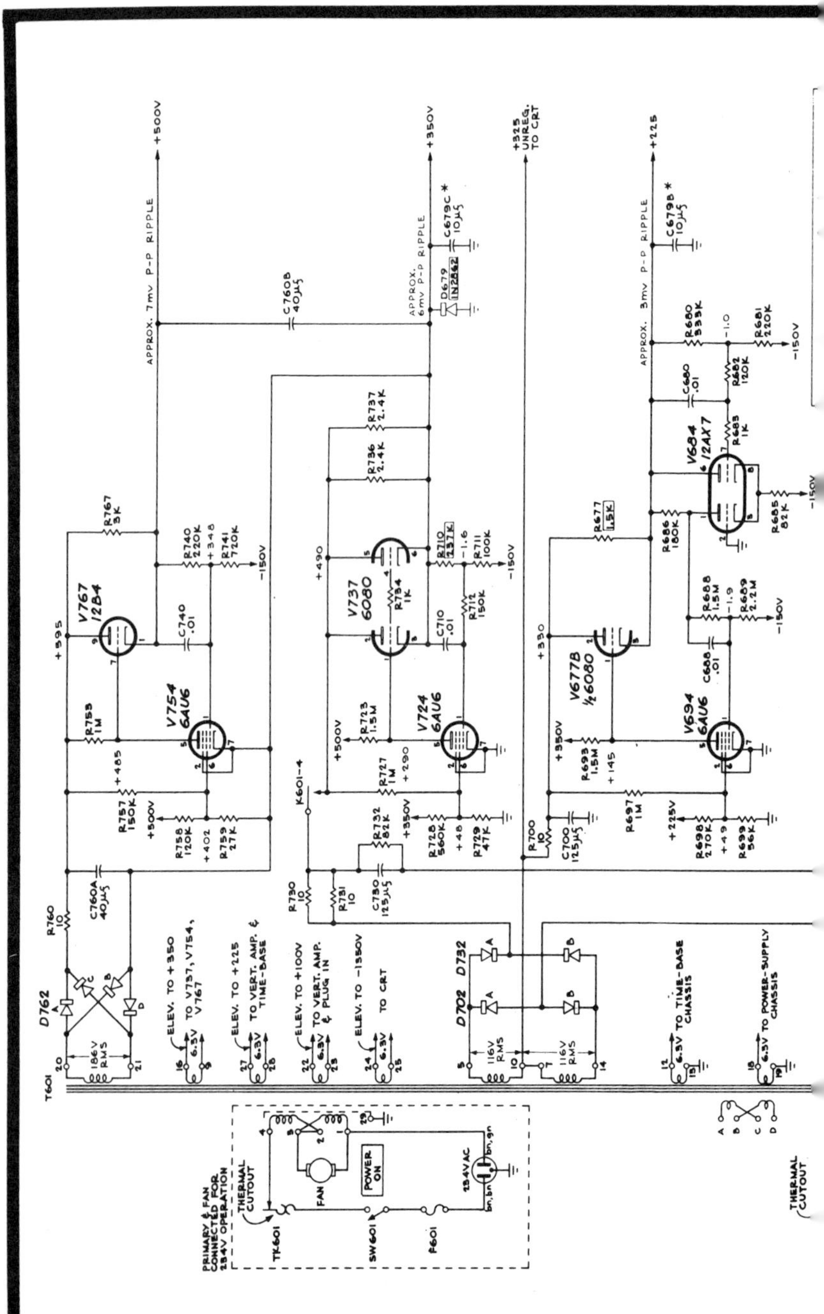

+500V
APPROX. 7mv P-P RIPPLE
+350V
APPROX. 6mv P-P RIPPLE
+325 UNREG. TO CRT
+225
APPROX. 3mv P-P RIPPLE
C760B 40µf
C679C * 10µf
C679B * 10µf
D679 1N2862
R680 333K
R681 220K
R682 120K
C680 .01
R683 1K
V684 12AX7
R685 82K
R686 180K
R677 1.5K
R688 1.5M
R689 2.2M
C688 .01
V677B ½6080
V694 6AU6
R693 1.5M
R697 1M
R698 270K
R699 56K
C700 125µf
R700 10
R737 2.4K
R736 2.4K
R710 237K
R711 100K
R712 150K
C710 .01
R734 1K
V737 6080
V724 6AU6
R723 1.5M
R727 1M
R728 560K
R729 47K
R732 82K
C730 125µf
R730 10
R731 10
K601-4
R767 3K
R740 220K
R741 720K
V767 12B4
C740 .01
V754 6AU6
R755 1M
R757 150K
R758 120K
R759 27K
C760A 40µf
R760 10
D762
D702
D732
186V RMS
116V RMS
T601
ELEV. TO +350 TO V737, V754, V767
ELEV. TO +225 TO VERT. AMP. & TIME-BASE
ELEV. TO +100V TO VERT. AMP. & PLUG IN
ELEV. TO -1350V TO CRT
6.3V TO TIME-BASE CHASSIS
6.3V TO POWER-SUPPLY CHASSIS
PRIMARY & FAN CONNECTED FOR 234V OPERATION
THERMAL CUTOUT
TK601
SW601
F601
FAN
POWER ON
234VAC

Fig. 10-11B. Commercial supply illustrating Op Amp usage (part 2 of 2 parts) (Courtesy Tektronix, Inc.).

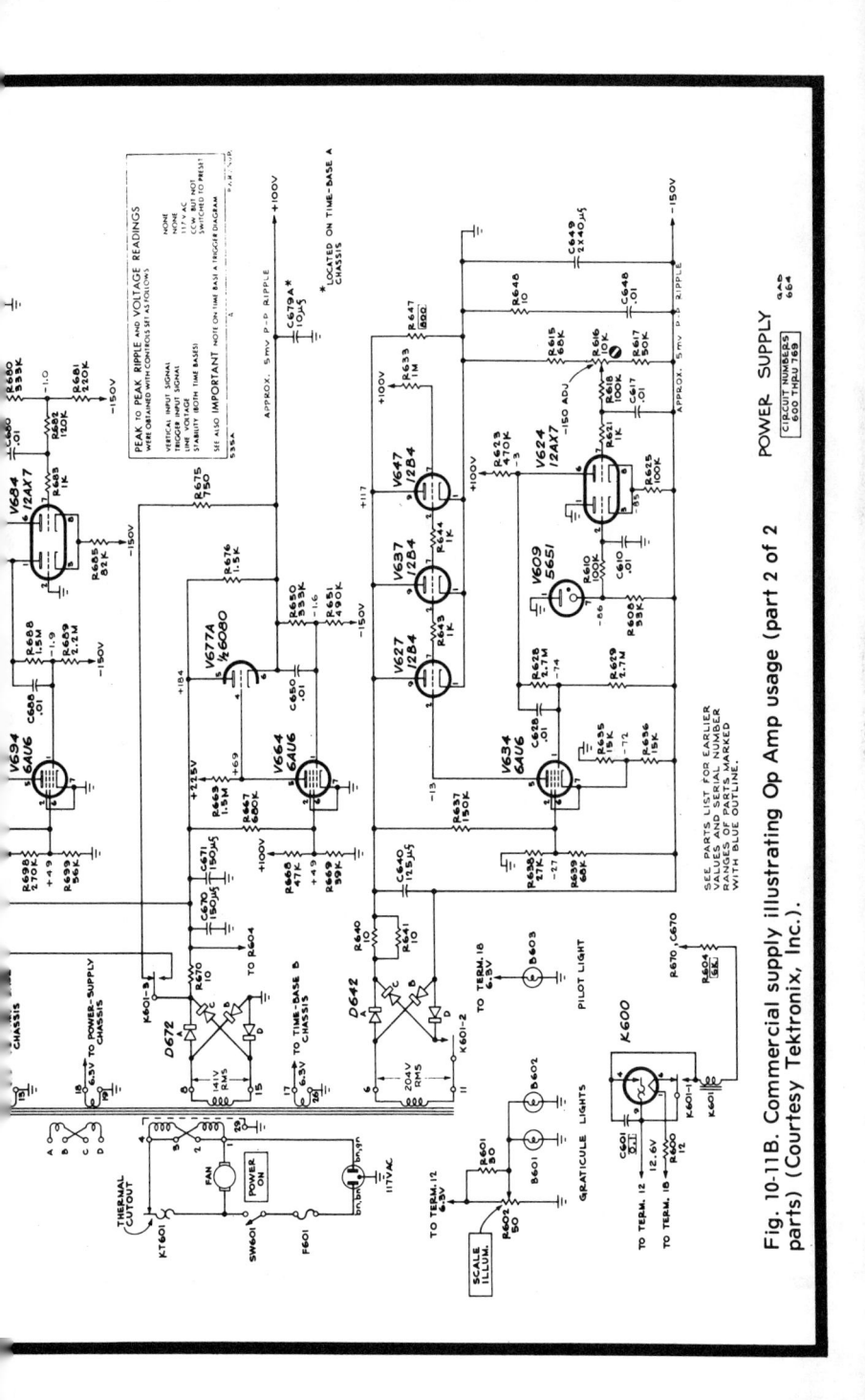

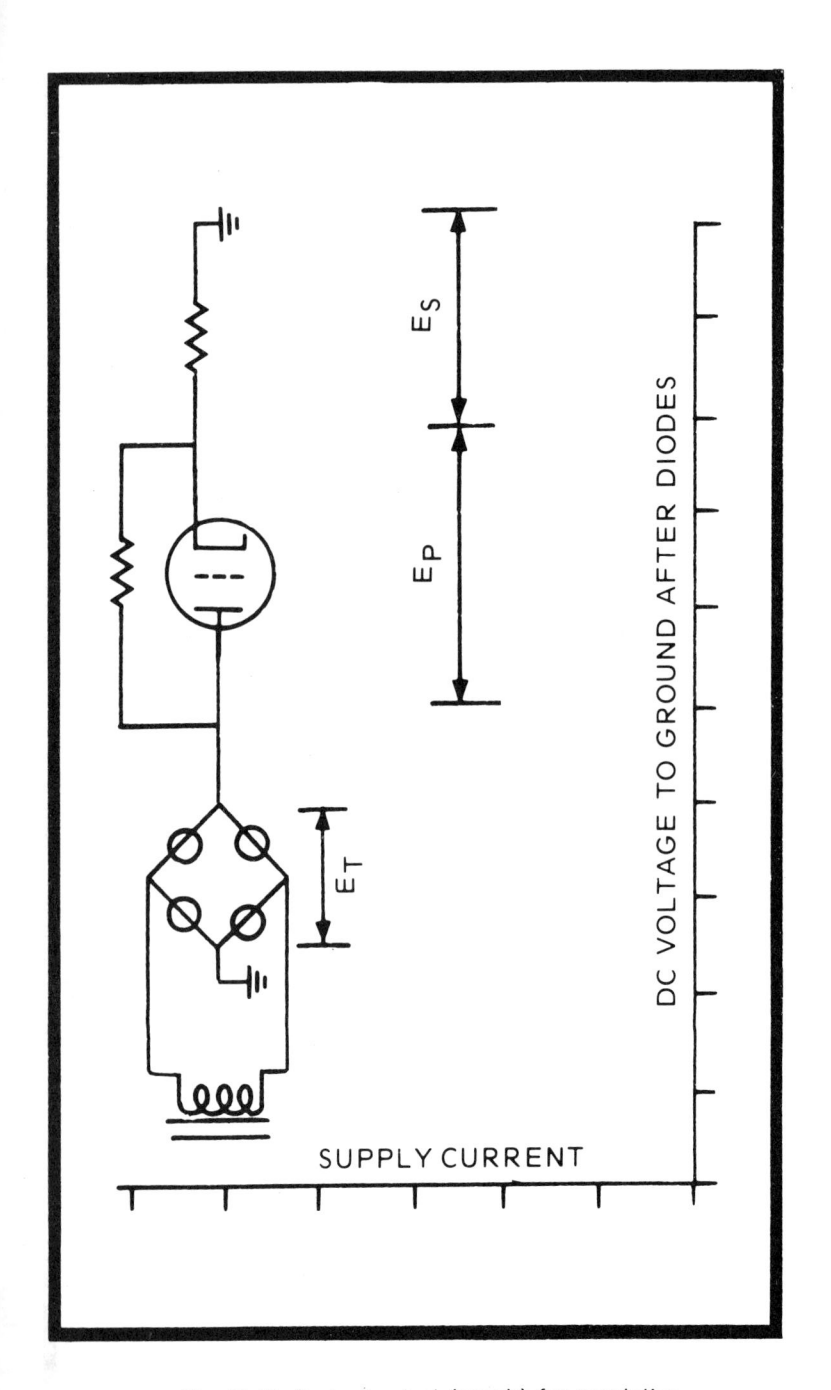

Fig. 10-12. Designing tool (graph) for regulation.

"DESIGNING" A POWER SUPPLY

Remember I told you that the power supply was the last thing that was designed for any piece of equipment? Well, this will illustrate why this is so. You see, we need to know just how much current this supply will be called upon to handle. You'll see why in a minute.

Fig. 10-12 has the axis of a graph on it. The horizontal is for the DC voltage when full-wave rectified and capacitor-filtered. This secondary level of voltage will shift if the input 110V AC changes its value. You have probably noticed that during peak usage times, the line voltage is a bit lower than 110 volts and at times of least usage, the line voltage may be as high as 120 volts (possibly higher). And, since much electronic circuitry can't operate properly under this type of voltage shift, we eliminate it in this fashion. Assume that you want power supply regulation to be accurate over a range of input (line) voltage swing. Say for instance you want the supply voltages to be accurate over a swing of input voltage from 95V AC input up to 125V AC input from the line. Since our filtered secondary voltage will be close to the peak voltage of the secondary, we will take the measured secondary voltage and multiply it by 1.414 to get the DC level on the plate of the series regulator. For a 1:1 transformer then, it would be 95 x 1.414 (for the lower end of regulation) and 125 x 1.414 (for the upper end of regulation). Our supply then must regulate from a DC level of +134 volts up to about +178 volts.

Mark these two points on the horizontal axis of our graph and draw vertical lines up from these two points.

Now assume some amount of current required of the supply (for this example only). This you will draw across the graph from about the fourth or fifth mark up on the vertical axis. This represents the amount of current this supply must handle regardless of the input voltage from the line, and the area between high- and low-line secondary voltage becomes the region of regulation.

Now, assuming that our supply is to be a +100 volt supply, we will note this point on the horizontal axis of our graph also.

You see, the idea is to have the series regulator tube and the resistor in parallel with it carry the full amount of current the supply has to handle throughout the full range of regulation. At high-line conditions, the tube should carry only 1/10 the total current and the resistor 9/10 of the total. So, from a point 9/10 up the vertical at high-line conditions, you will draw a straight line down to the 100-volt mark on the horizontal axis. Since, R=V/I we must have a resistor equal to

321

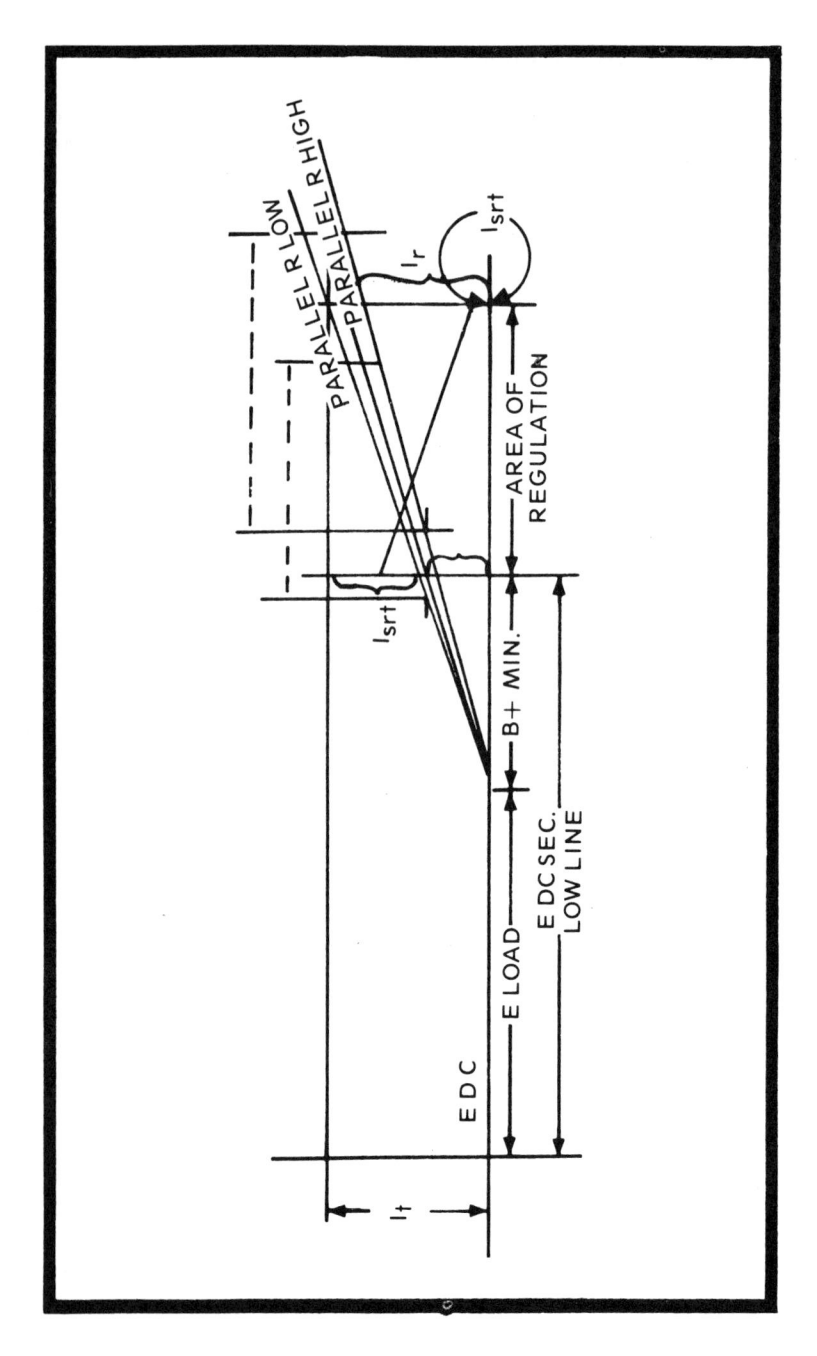

Fig. 10-13. Resultant graph when Fig. 10-12 is filled in.

78 volts divided by 9/10 of the lead current. The series regulator tube will have maximum bias on it and carry minimum current, and its curves will tell you just how much bias will be needed. Bias must be able to change to the level necessary for operation with only 34 volts B+ and that amount of current it will have to carry at low-line conditions. Since—as I pointed out long ago—plate voltage and plate current will determine bias as long as it is an operating point the tube is capable of (within the active region of the tube curves), then everything will work OK.

Fig. 10-13 shows you what your graph should now look like, as well as the results of having a parallel resistor of the wrong value. Note that a bigger resistor than your design calls for pushes the area of regulation to a higher voltage, while a parallel resistor that is too small lowers the region of regulation.

I haven't said anything about two of the parts in the sample series regulator circuit (Fig. 10-8A). These are C_2 and the "Fuse R."

C_2 is sort of a cut-and-try type of part used to eliminate any glitches generated in the B+ bus by the circuitry that the supply powers. If there's a fast multivibrator in the load, then a fast spike might be generated in B+ and this capacitor would tend to eliminate it. A 0.01 uf capacitor with an adequate breakdown voltage rating should be more than OK for the job.

The Fuse R is a small value of resistance (maybe 10 ohms) with a wattage rating just a slight bit bigger than required of it to carry maybe twice the load current. This guarantees that if there is a partial short in the load, this resistor will burn up rather than either the transformer or the load circuitry. Twice the value of the load current quantity squared times 10 ohms would give you the power rating needed.

For example, say we had a supply that handled 250 ma. Two times .25 is .5 and .5 squared is .25. Now ten times .25 is 2.5 and a 2.5 watt, 10 ohm resistor would do the trick. A 3 watt resistor would also be satisfactory and probably more available. I don't think a 5 watt resistor would offer you adequate protection (it might not blow), but it would be better than nothing.

If you happen to find yourself working on a piece of equipment with such a resistor burned up, it is NOT the major trouble. It's only an indicator telling you which load the actual trouble is in.

Now all of this is great except it is entirely dependent upon having a stable, well-regulated negative voltage (−150V in Fig. 10-11). In the power supply we are about to discuss

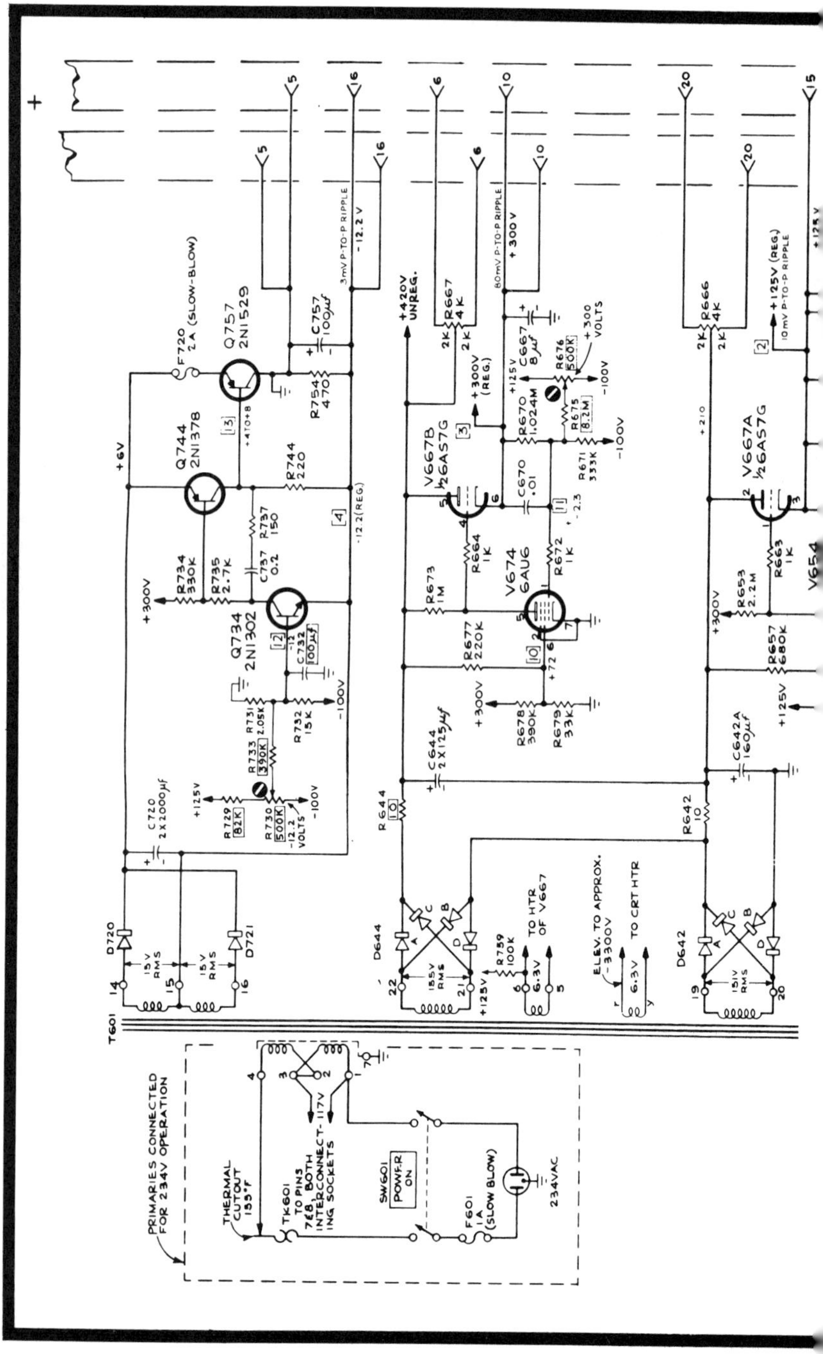

F720 2A (SLOW-BLOW)
Q757 2N1529
Q744 2N1378
Q734 2N1302
+6V
R754 470
C757 100µf
3mV P-TO-P RIPPLE
-12.2V
-12.2 (REG.)
R744 220
R737 150
C737 0.2
R734 330K
R735 2.7K
+300V
C732 100µf
R731 2.05K
R733 390K
R732 15K
-100V
R729 82K
R730 500K
-12.2 VOLTS
+125V
C720 2 X 2000µf
D720
D721
15V RMS
T601
+420V UNREG.
R667 4K
2K
+300V (REG.)
80mV P-TO-P RIPPLE
+300V
C667 8µf
R676 500K
+300 VOLTS
R670 1.024M
R675 8.2M
R671 333K
V667B ½26A57G
C670 .01
R664 1K
R672 1K
V674 6AU6
R673 1M
R677 220K
R678 390K
R679 33K
C644 2 X 125µf
R644 10
D644
155V RMS
TO HTR OF V667
6.3V
R759 100K
ELEV. TO APPROX. -3300V
TO CRT HTR
R666 4K
+125V (REG.)
10mV P-TO-P RIPPLE
V667A ½26A57G
R663 1K
R653 2.2M
R657 680K
C642A 160µf
R642 10
D642
151V RMS
PRIMARIES CONNECTED FOR 234V OPERATION
THERMAL CUTOUT 155°F
TK601
TO PINS 7&8, BOTH INTERCONNECTING SOCKETS
117V
SW601 POWER ON
F601 1A (SLOW BLOW)
234VAC

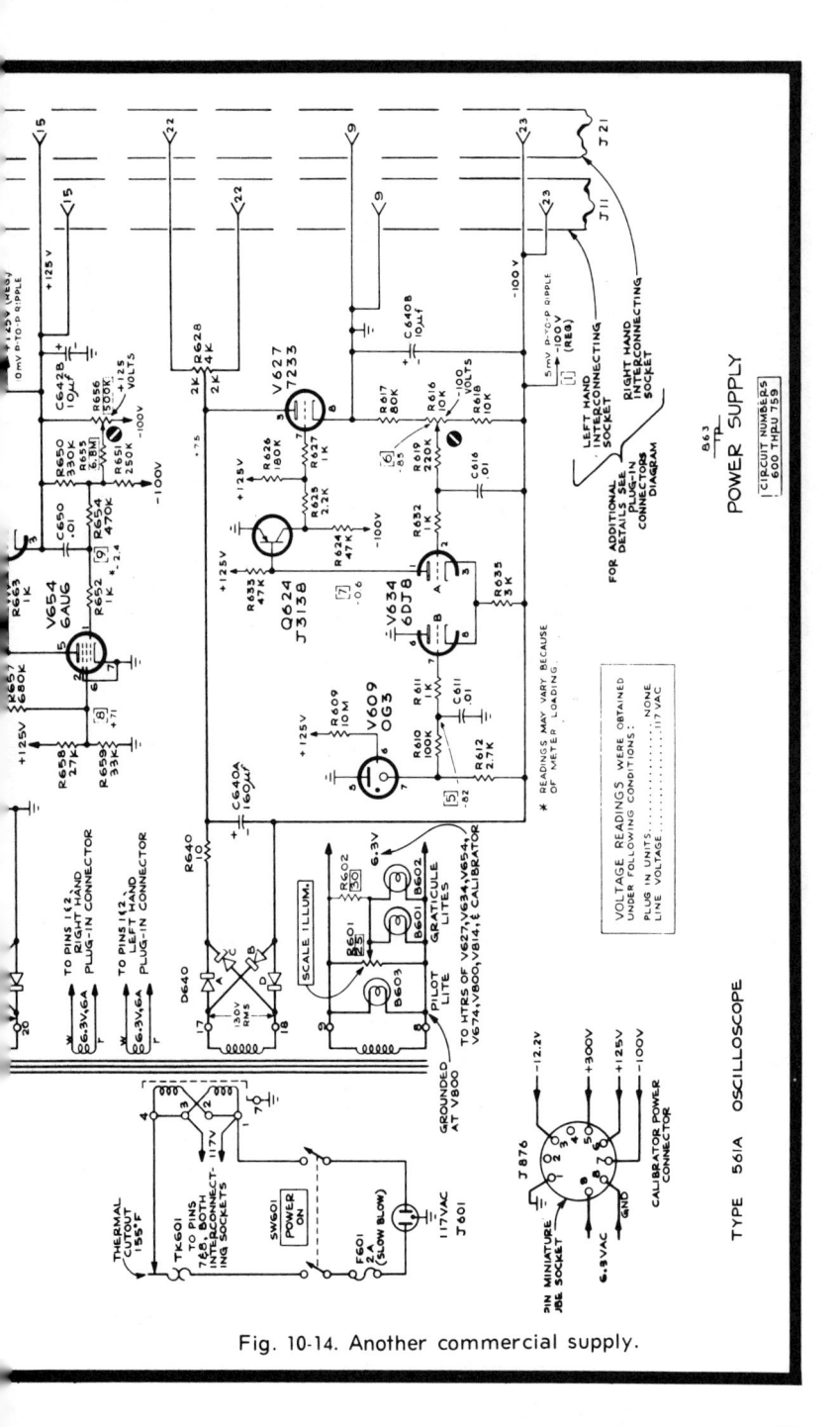

Fig. 10-14. Another commercial supply.

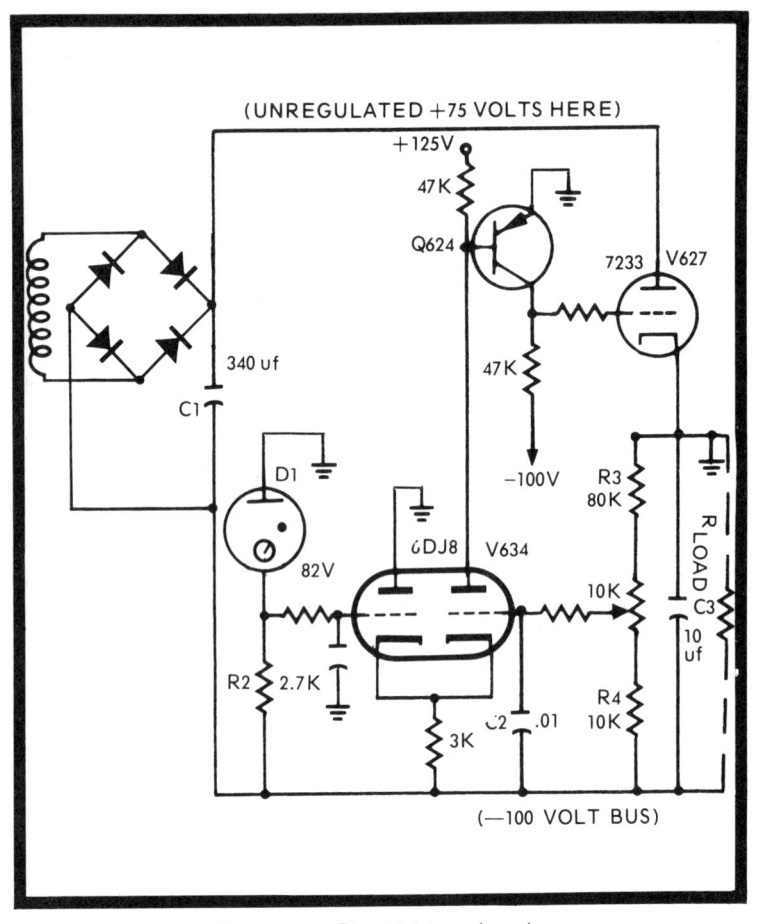

Fig. 10-15. Fig. 10-14 analyzed.

(because its negative supply is simpler) you will note that the negative supply is a —100V supply. See Fig. 10-14. The same basic units are contained in both supplies, though.

Now let's turn to Fig. 10-15 and note that I have isolated the —100V supply and its critical parts. The diode bridge fullwave rectifier and C_1 limiting the ripple to around 6 volts on the unregulated +75 volt side of the supply are conventional and already commented on. The rest is what we have to talk about.

First, we'll look at that part of the supply that is below ground. In other words, we'll look at V634 and the circuitry that surrounds it. This circuit (in essence) is a balanced bridge circuit with the grids on both sides of V634 being at

326

approximately the same voltage. D_1 is an OG3 type Gas Diode that has about the same characteristics as a Zener diode. It will maintain a stable voltage across itself over some given range of current change. The OG3 in this circuit has about 82 volts across it, setting the left grid of the 6DJ8 at —82v with respect to ground. The current that will make D1 do this comes through R_2. A similar 82 volts will be across R_3 (the 80K resistor) and that part of the 10K pot below it that remains above the grid connection. That part of the 10K pot that remains below the grid connection plus R_4 (the 10K resistor) will carry the same current determined by the 82 volts across the upper portion and thus determine the negative voltage level of the minus supply. If we move the adjustment on the 10K pot up, the minus supply will go more negative (more current through a smaller resistor with the same old 82 volts across it). And, if we move the adjustment on the 10K pot down, the minus supply will become less negative (for the opposite reasoning). We must bear in mind though that we are not getting something for nothing. If the negative supply is adjusted more negative, the positive side of the supply becomes less positive, and if the negative supply is adjusted less negative, the plus side of the supply moves more positive. We cannot change the magnitude of the secondary voltage across C_1. We can only move it around with respect to ground. Also, this system works only so long as V634 carries current in both cathodes. If one side (for any reason) should be driven to cut-off, the supply will no longer regulate.

The series regulator in this case is V627. If the supply tries to go too negative, this becomes a negative signal on the right grid of V634 and thus positive on its plate. This positive signal is on the base of Q624 and thus negative-going on the grid of V627—turning it off slightly, saying don't draw quite so much current up through the negative load and thus bring the minus voltage back to normal. If the supply tries to go less negative, V627 feels reduced bias and demands more current up through its load, and the voltage drop across the load gets bigger. Since the grounded end can't move, the negative end **must**, and the —100 volts is returned to normal.

This supply then doesn't have an Op Amp to set its output level of voltage; it has a balanced bridge. It does have the proper number of phase inversions to make the series regulator tube do its job (regulating the voltage set by the balanced bridge). C_3 is just a sort of extra bit of filtering for this —100V supply and it may or may not be needed, depending on what type of circuitry is represented in the load. C_2, of

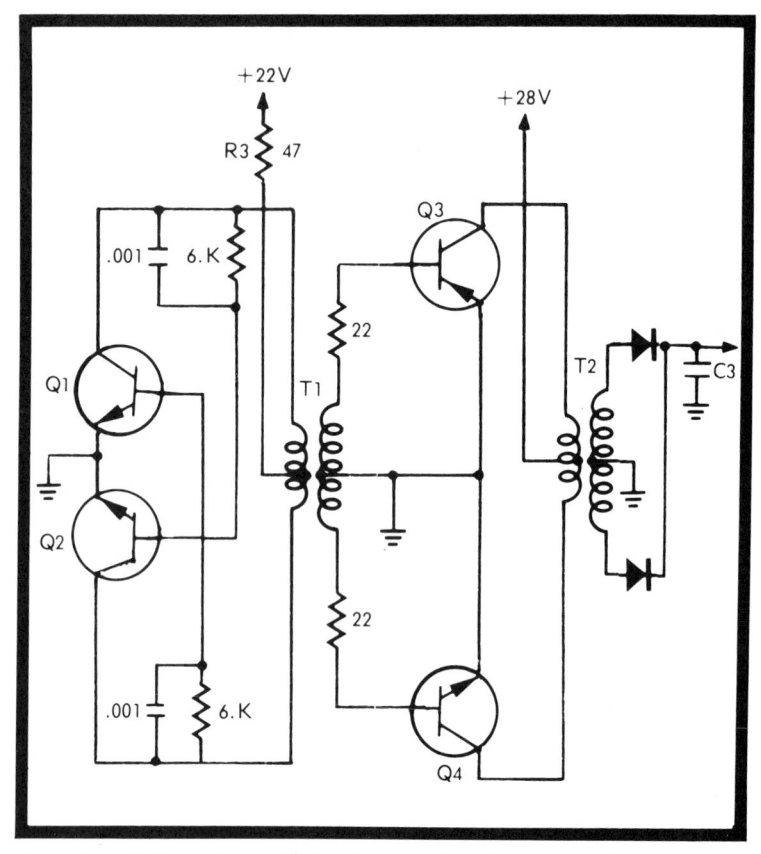

Fig. 10-16. Solid-state multivibrator supply. (Courtesy Tektronix, Inc.)

course, is the high frequency path for spikes that C_3 cannot filter.

There is no parallel resistor to V627 (this series regulator) as such. There is one, however, in the load on the +75 (unregulated) buss and the center-tapped 4K resistor, adding an equivalent 1K to this load (2K in parallel with 2K). This would be figured in the same fashion as we did before.

What with PNP Transistors, P-Channel FETs, and Zener diodes, I suppose positive supplies could be regulated in this fashion too, but I've never seen any. But—believe me—that doesn't mean there aren't any like this. I'm a long way from having seen every kind of circuit that can be built.

Of course, changing AC to DC or one AC level to a higher or lower AC level aren't the only jobs of power supplies, and it's easy enough to change a high DC level to a lower DC level with a resistor and Zener diode. But, very often we find it necessary to change a low DC source of voltage (like lead-acid batteries) to a higher DC level of voltage. Time was when the main system of doing this was with a DC motor coupled to an AC generator or the cheaper method using a mechanical vibrator to do the job, but these have been beaten. Oscillators and multivibrators can now change DC to AC and a transformer can change a low AC to a high DC with solid-state diodes to rectify the transformer secondary voltage to the higher DC level desired. All these are now quite common.

SOLID-STATE POWER SUPPLY

Fig. 10-16 shows an astable multivibrator (multi), Q1 and Q2, powered from a +22V DC source, as the AC generator. T1 steps up the AC voltage and drives two power transistors (Q3 and Q4) which give this supply its current-supplying capability. T2 steps up the AC voltage still more (to whatever is needed) and CR1 and CR2 do the full-wave rectification. C3 is the standard filter.

This system looks beautiful, but it does have its share of difficulties. Yes, we can make the multi operate at a much higher frequency than 60 Hz and thus get by with a much smaller transformer, but when we do this, we run into difficulties turning Q3 and Q4 off at the proper times. Since these two are power transistors, they do store carriers (electrons in my language) in their base regions. This means that the collector signal will be delayed in following the base signal, during the "ON to OFF" part of the cycle in particular. Consequently, there will be a portion of each cycle where both transistors are turned "ON" and no secondary voltage will be generated through T2 (or—at least it will be at a minimum with opposing currents flowing). This in its own turn tells us that this circuit should be designed to operate at less than 1 kHz, or that we'd better expect something much more complicated in the power amplifier part of the circuit if its going to operate at around 13 kHz as this multi is designed to do.

See Fig. 10-17 and note the added circuitry. Q5 and Q6 are now the power transistors and Q3 & Q4 help to control them so that only one of the power transistors is "ON" at a time. The attempt isn't so much to speed up the turn "OFF" time of these power transistors as it is to slow down the "Turn-ON" time of the "OFF" transistor so that it matches the inherent delay naturally built into the "ON" device.

This kind of thinking may appear kind of backwards to you, but I think you'll have to admit that it is effective.

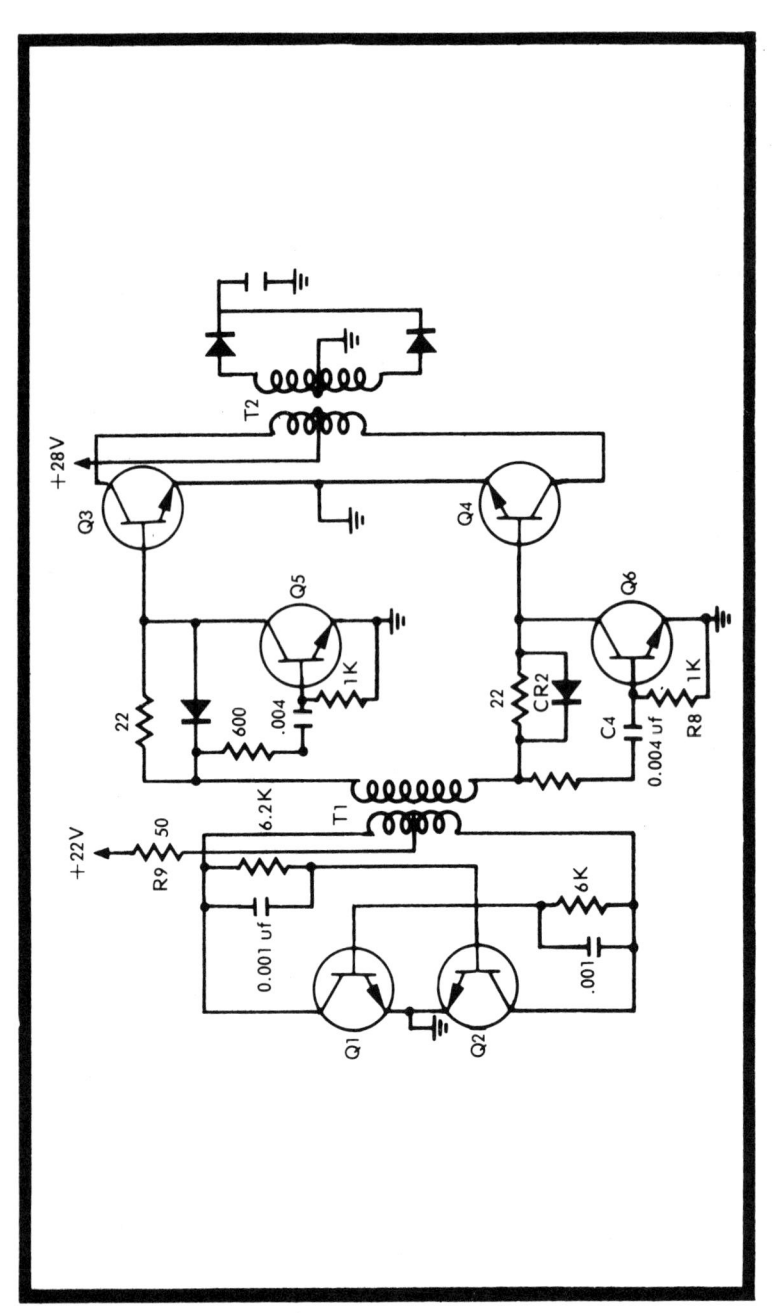

Fig. 10-17. Fig. 10-16 with parts added for high frequency operation. (Courtesy Tektronix, Inc.)

In Fig. 10-18 we have a high voltage power supply used in a Tektronix 545A Oscilloscope. This is used to power the cathode ray tube in the instrument. It's an excellent example of a supply that changes a DC voltage of around +325 volts to +8,650 volts and two sources of —1,450 volts. (A low DC voltage changed to a high DC voltage.) This supply doesn't handle much in the way of current, so it can get by with half-wave rectification and small wire in the secondary windings in particular. V800 (the oscillator) probably is operating around 30 kHz and the circuit is almost a conventional Hartley oscillator. The tank circuit is made up of C808 and the primary of the transformer. The frequency of the oscillator and the low current demand made on this supply allow it to get by with a transformer much smaller than normal.

It is interesting to compare the two negative supplies; the first one using V822 as its rectifier and the second using V862 as its rectifier. In both cases we have the plate of the tube as the output source of voltage. If you can think of a negative voltage as a source of electrons you will realize that the plates are getting nothing but extra electrons every other half cycle, thus recharging the filter capacitors whose other end is at a DC level of voltage.

In the case of V822 and C820, the reference voltage is ground for this explanation.

For V862 and C831, the reference voltage is +100 volts as you can probably plainly see.

Both of these rectifiers operate from secondaries of the same number of turns, enough to develop a secondary voltage across C820 and C831 of something close to 1,450 volts.

Since the reference voltage of V822 and C821 is ground, it becomes a —1,450 volt supply.

In the case of V862 and C831, the reference voltage is +100 volts and this combination becomes a —1,350 volts supply with respect to ground.

Two highly negative supplies—one hundred volts apart.

For that part of the secondary that is made up of V832, V842, and V852, note that the output is connected to a cathode. This means that it will take electrons away from anything connected to it and thus it becomes a positive supply. It's a rather tricky little supply that's been around for quite a while. It's called a Voltage Multiplier. The number of diodes in this supply with their capacitors tell you the number of times the rectified DC is multiplied. This one is a voltage tripler.

When V832 conducts, C832 charges to the secondary DC voltage of about 2,850 volts.

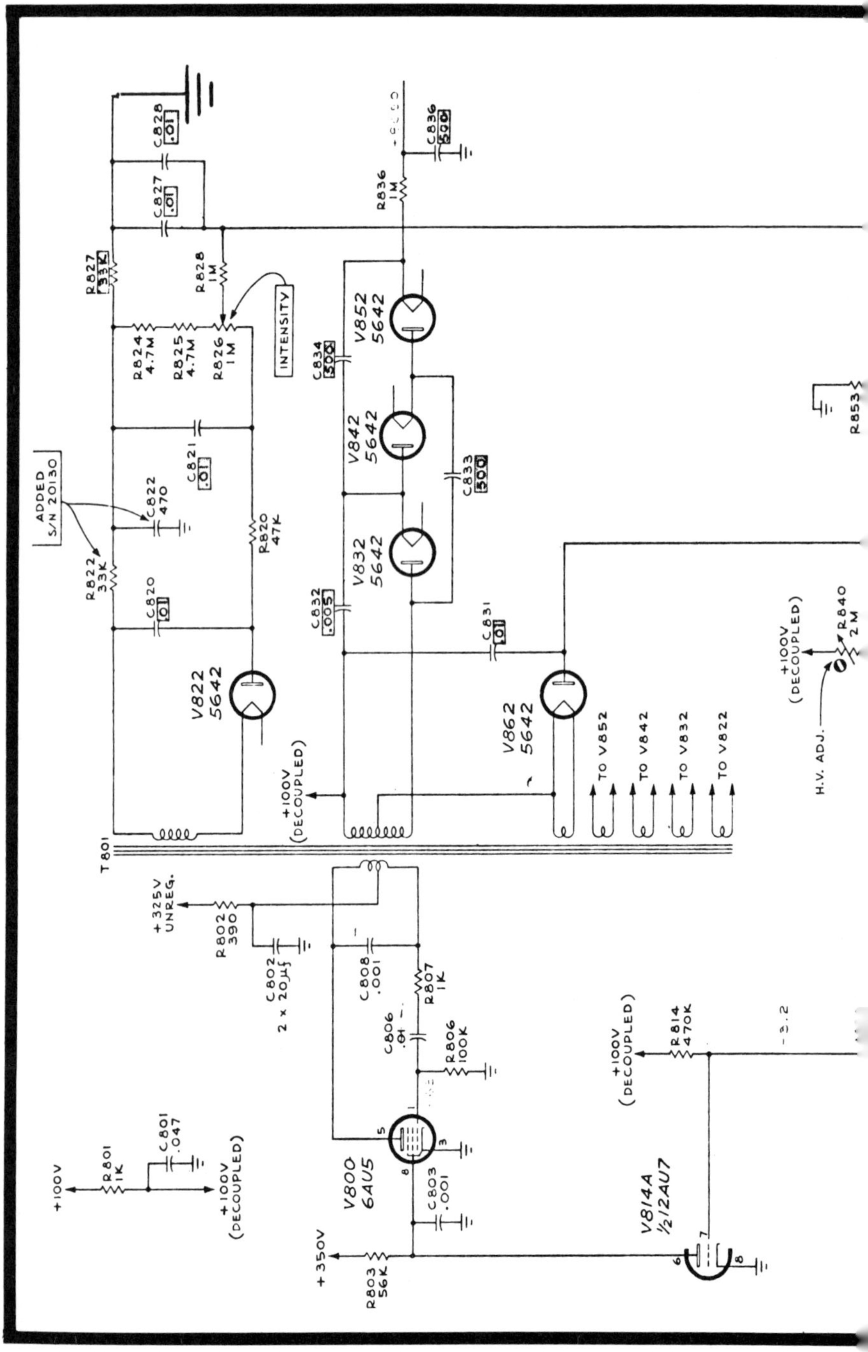

C828 .01
C827 .01
R827 33K
R828 1M
R824 4.7M
R825 4.7M
R826 1M
INTENSITY
C821 .01
ADDED S/N 20130
C822 470
R822 33K
R820 47K
C820 .01
V822 5642
C836 500
R836 1M
V852 5642
C834 500
V842 5642
C833 500
V832 5642
C832 .005
C831 .01
V862 5642
+100V (DECOUPLED)
TO V852
TO V842
TO V832
TO V822
R853
+100V (DECOUPLED)
R840 2M
H.V. ADJ.
T801
+325V UNREG.
R802 390
C802 2 x 20µf
C808 .001
R807 1K
C806 .01
R806 100K
+100V (DECOUPLED)
R814 470K
-3.2
+100V
R801 1K
C801 .047
+100V (DECOUPLED)
V800 6AU5
C803 .001
+350V
R803 56K
V814A 1/2 12AU7

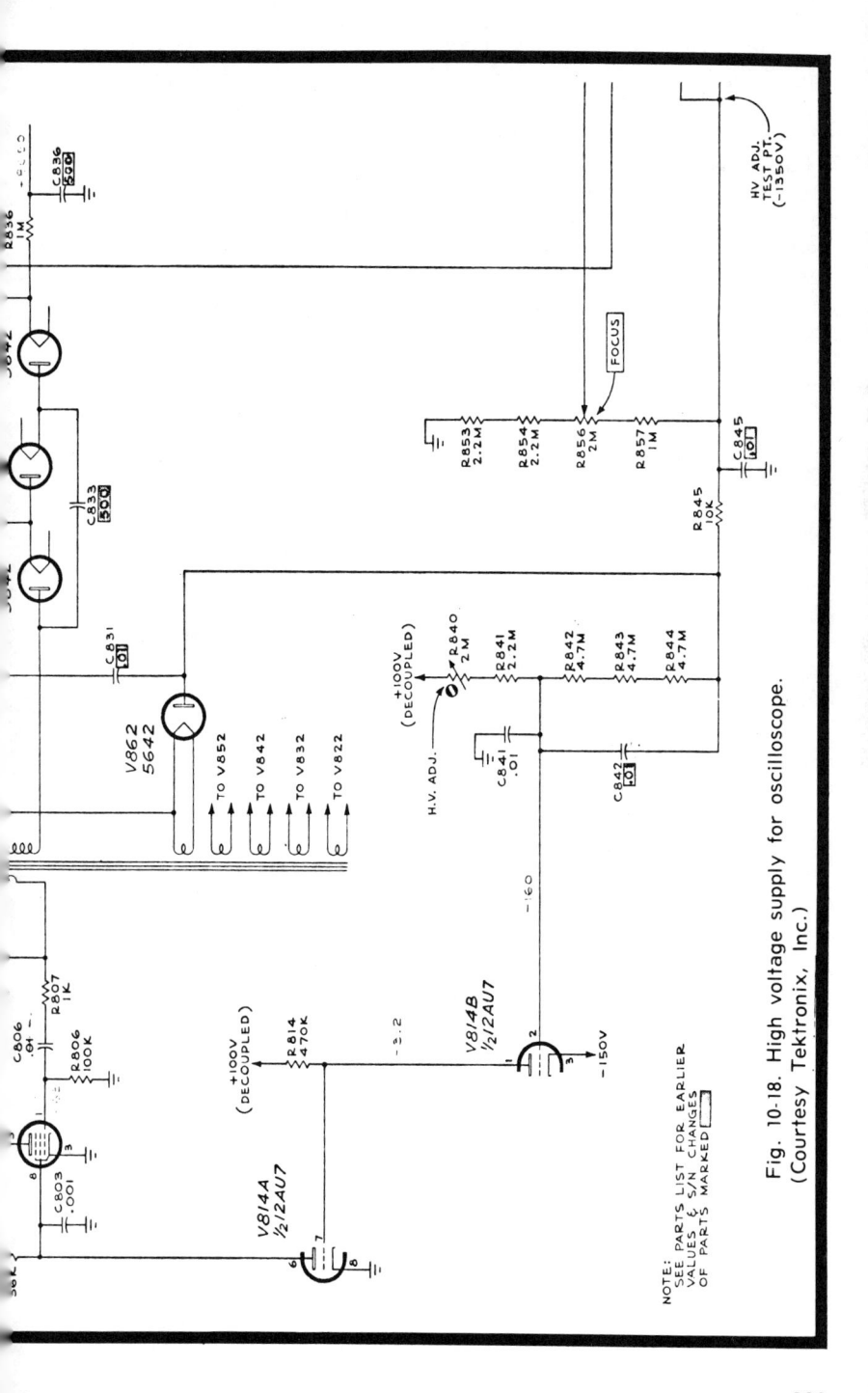

Fig. 10-18. High voltage supply for oscilloscope. (Courtesy Tektronix, Inc.)

333

On the next half cycle, V842 conducts and C833 charges to the secondary voltage plus the charge across C832 which gives C833 a charge of 5,700 volts.

Now when V832 comes back on, V852 conducts as well. The charge across C833 is transferred to C834 and C832 is recharged. This puts the right-hand end of C834 at 8,550V above the supply's +100 volt reference, giving us a +8,650-volt supply as long as we don't allow much current (electrons) to get to it.

This doesn't happen within a couple of half cycles as I have described it, but it can be thought of as happening this way. (It's the easiest way of explaining it that I know of—in other words!)

Since we're on the subject of high voltage supplies, let me urge you NOT to try measuring them without the proper equipment. Normal test leads are not adequately insulated to withstand voltages of this magnitude, and neither are you. A burn from a supply of this nature goes right to the bone and remains sore for weeks. And another thing to remember about such a supply is this: It'll jump right out to get you. You don't have to touch it. All you have to do is come too near it, and its got you. One thing I will guarantee you is this—it won't get you without you knowing it. You'll know all about it when it happens. But then, it's too late. So, be careful.

PAGE MAY BE REMOVED AND MOUNTED ON CARDBOARD

USE THIS CARD AS A SHIELD TO COVER THE FOLLOWING QUESTION WHICH WILL CONTAIN THE PROPER ANSWERS TO THE QUESTION YOU WILL BE READING. THE ANSWERS WILL BE FOUND IN THE LEFT HAND MARGIN OF THE FOLLOWING QUESTION EXCEPT IN CASES OF MULTIPLE CHOICE.

PLEASE FEEL FREE TO USE THE INFORMATION ON THIS CARD AT ALL TIMES AS YOU GO THROUGH THE PROGRAM.

CUT ALONG THIS LINE

UNIT	PREFIX	SYMBOL
10^{12}	tera	T
10^9	giga	G
10^6	mega	M
10^3	kilo	k
10^2	hecto	h
10	deka	da
10^{-1}	deci	d
10^{-2}	centi	c
10^{-3}	milli	m
10^{-6}	micro	u
10^{-9}	nana	n
10^{-12}	pico	p
10^{-15}	fento	f
10^{-18}	atto	a

The Powers of Ten

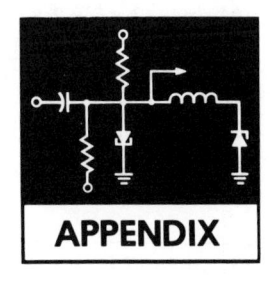

APPENDIX

Page 335 may be removed from the book and mounted on cardboard. It will help when taking the self-test provided in the appendix.

OBJECTIVES

The Student should be capable of:
(1) recalling the power of ten represented by the prefix used in front of the units of measure with the aid of the printed list which he will use as a mask to cover correct answers as he proceeds through the program.
(2) multiplying one number by another when both are stated in terms of powers of ten and get an answer in a proper power of ten terminology.
(3) dividing one number by another when both are stated in terms of powers of ten and get an answer in a proper power of ten terminology.
(4) recognizing those powers of ten which cancel each other.
(5) changing a given number with its associated power of ten to a new statement using a different power of ten without changing the combined numerical value of the original statement.
(6) knowing what procedure to follow when adding or subtracting numbers stated in terms of powers of ten.
(7) reading and substituting the proper power of ten for each related prefix in the poem used as the gating frame (1.1) or a similar poem, as well as finding the one incorrect statement contained and prove that it is wrong.

If you don't know what a POWER is and least of all what a POWER OF TEN is, then begin here. Otherwise go to frame (1.1) and start there.

(1) 2,500 is a rather large number. Let's break it down into its prime factors. We will have 5 x 5 x — x — x — x —.

337

(2) ans. (2 x 5 x 2 x 5 or any sequence of these same numbers)
Another way of writing this would be to write the actual different factors (i.e., 2 & 5) and then use exponents (a small number slightly above vertical center and to the right of the number it is related to) to show how many of each of these factors is used:

$$2{,}500 = 2^? \times 5^?$$

---

(3) ans. ($2{,}500 = 2^2 \times 5^4$)
If we do **not** have to stay with PRIME factors, we could say:

$$2{,}500 = 25 \times 10 \times 10 \text{ or } 25 \times 10^?$$

---

(4) ans. (or 25×10^2)
And, as long as we do not have to stay with PRIME factors, we can also use decimal numbers. This would allow us to say:

$$2{,}500 = 2.5 \times 10 \times 10 \times 10 \text{ or } 2.5 \times 10^?$$

---

(5) ans. (or 2.5×10^3)
We can continue in this fashion saying that:

$$2{,}500 = 2.5 \times 10^?$$

---

(6) ans. ($.25 \times 10^4$)
It should be evident by now that we can factor out as many **tens** as we wish. After all, the only thing that changed was the location of the decimal point in the remaining factor and the exponent that kept track of the number of t____ we did pull out as factors.

---

(7) ans. (tens)
When this number (2,500) is used as a factor in some problem then, we should have a choice as to how we will use it. Perhaps it would be easier to use it in the form of 25×10^2. Perhaps it would be easier to use it as .25 x ____ .

---

(8) ans. ($.25 \times 10^4$)
Sometimes you can see a quantity of 25's easier than 2,500's. For instance, what is 400 x 2,500 = _____ ? To take this problem just as stated is a bit awkward. But, IF we break it down into $4 \times 10^2 \times 25 \times 10^2$, it isn't so hard. It's easier to see that we have 4 x 25, which is 100 times four tens. The answer must be _____ .

338

(9) ans. (1,000,000 in both cases)

Or maybe it would be easier for you to see the problem this way: $400 \times 2{,}500 = 4 \times .25 \times 10^2 \times 10^4$

Let's see, one quarter of four is one and now I must multiply it by s____ tens. This will give me _____ (the same as before).

---

(10) ans. (six) & (1,000,000)

In question No. 8, it was noted that $10^2 \times 10^2$ was multiplication by four tens. In question No. 9, $10^2 \times 10^4$ was the same as multiplying by _____ tens.

---

(11) ans. (six)

At this rate, 10^3 times 10^5 must equal $10^?$.

---

(12) ans. (10^8 or multiplication by **eight** tens)

This gives a rule we can now go by. When multiplying factors of ten that are raised to a whole number power, we simply (add-subtract) their exponents (or Powers) to obtain the exponent of the ten factor in the answer.

---

(13) ans. (add)

This should give us a fairly easy way to tackle a problem like this: $20{,}000{,}000 \times 350{,}000 =$ _____. We can break this down into this: $2 \times 3.5 \times 10^7 \times 10^5$ and the answer is easily seen as 7×10^{12}. Now write out the correct answer in the blank. It will be _____ followed by _____ zeros.

---

(14) ans. (7,000,000,000,000)(7)(12)

Maybe it would have been easier for you to see it as 3.5 twenties. Then you would have wound up with ten to the eleventh power in the answer. Or, maybe you saw two times 35 and wound up with ten to the _____ power. Any one is just as correct as the other since all give the same final answer.

PROGRAMMED INSTRUCTION OF "THE POWERS OF TEN"

(1.1)   Read the following and find the boo-boo.
1   Enee meenee minee mo,
2   What power of ten does the name show?
3   Tera means twelve and giga means nine,
4   While mega means six and kilo means three and this doesn't rhyme.
5   Hecto means two and deka means one.
6   Yes, we have ten to the zero power too,
7   But this just stands for the number one.

8	Now, decimal numbers continue the same
9	With prefixes telling the negative power's name.
10	Deci is ten to the minus one while centi continues to the minus two.
11	Then milli becomes ten to the minus three,
12	And micro a minus six you see.
13	Nano is ten to the minus nine,
14	And pico becomes a milli nano combine.
15	Deci times centi becomes a milli
16	and milli times milli makes up a micro.
17	So, milli and micro combine in giga
18	And a micro micro then is a pico.
19	Then, on the positive side of the fence,
20	Deka times hecto becomes a kilo,
21	And kilo times kilo makes up a mega.
22	So, kilo and mega combine in a giga.
23	And a mega mega then is a tera.
24	Cycles or ohms or farads, watts or meters,
25	The units of measure depend on the measured,
26	And how big or small we'll tell with a symbol,
27	That stands for a power of ten quite nimble.

Now, choose one of the three reactions as being closest to your own reaction to this poem.

REACTIONS:

(a) I understood it all the way through.
(b) I'm confused!
(c) Ha!!—I found a boo-boo!!!!

If you chose (b), continue with frame (1.2).
If you chose (a), chances are that you will benefit by the program, so, continue with frame (1.2).
If you chose (c), you may not have to take this program. If the error you found was in line 17, where it says, "So, milli and micro combine in a giga," you have correctly identified the line with the error. Turn to frame (1.14), and continue there.

(1.2)   10 times 10 times 10 times 10 may be written 10 to the fourth power or 10^4. This simply means that we are using four tens as multiplers. Consequently, we should be able to write 10 x 10 x 10 like this_____.

(1.3) ans. (10^3)

Some number divided by "10 times 10 times 10 times 10" can be said to be multiplied by 10 to the MINUS fourth power or 10^{-4}. Thus, if I were to divide the number 12.34 by "10 times 10 times 10," I should be able to say that I am multiplying it by _____.

(1.4) ans. (10^{-3})

The historical development of our number system for extremely large values has resulted in the marking off of our numbers in groups of three digits. That is, one thousand is written 1,000., one million is written 1,000,000., one billion is written 1,000,000,000., and one trillion is written 1,000,000,000,000. The names "thousand, million, billion, and trillion" have developed. At the same time, a different set of names has developed in scientific circles which represent these same numbers as powers of ten. These new names were much shorter and served a different purpose. They were used as prefixes to basic units of measure. And, they developed with regard to the negative powers of ten just as they did for the positive powers of ten. Because we divide by large numbers as well as multiply by them. Thus:

Multiplication by 1,000. is multiplication by 10^3 or so many **kilo** units.

Multiplication by 1,000,000. is multiplication by 10^6 or so many **mega** units.

Multiplication by 1,000,000,000. is multiplication by 10^9 or so many **giga** units.

Multiplication by 1,000,000,000,000. is multiplication by 10^{12} or so many **tera** units.

Division by 1,000. is multiplication by 10^{-3} or so many **milli** units.

Division by 1,000,000. is multiplication by 10^{-6} or so many **micro** units.

Division by 1,000,000,000. is multiplication by 10^{-9} or so many **nano** units.

Division by 1,000,000,000,000. is multiplication by 10^{-12} or so many **pico** units.

Consequently, the complexity of an extremely large or small number is tied up in its name rather than in its value, which we may have to use as a real number in some scientific equation.

(1.5) ans. (no answers asked for)

Our scientific power of ten prefix system did not start out in groups of three, however. It started out one ten at a time. Thus ten units became a **deka**-unit. If I had ten tens of units, I had a **hecto**-unit. Conversely, if I had one tenth of a unit, I had a **deci**-unit and if I had a tenth of a tenth of a unit, I had a **centi**-unit. Thus, we might say that a tenth of a tenth of a dollar is a _____ dollar. We even abbreviate this more and call it one cent.

(1.6) ans. (**centi**-dollar)

If we carried this system on even further with our decimal monetary system, ten dollars could be called a_____ dollar and one hundred dollars could be called a_____ dollar. We don't use this notation in money, however, but this system does find use in geometry where a ten-sided closed figure is a **deka**-hedron and a one-hundred-sided closed figure is a **hecto**-hedron.

(1.7) ans. (deka) (hecto)

As noted in the conclusion of unit (1.4), the complexity of an extremely large or small number is tied up in its name rather than in its numerical value since the numerical value may have to be used in some scientific equation. The number answer of this equation may be quite simple, but the resultant name must be a **combination** of the **powers of ten** related to the individual parts of the equation. Combining the powers of ten into the proper name is usually the more important of the two parts of the problem. If I say 2 times 200 is 500, I am 25 percent off. But if I get my arithmetic correct and gain or lose one power of ten, I have an answer of either 40 or 4,000 which is a much, much greater error. Consequently, combining my powers of ten is most times the (more-least) important of the two parts of my problem.

(1.8) ans. (more)

Let's see how this combining powers of ten works. If I have a simple multiplication of a **mega**-unit by a **milli**-unit, I must be multiplying my final answer by six tens and dividing it by three (see table of (1.4)). It's obvious that three of the tens on top will cancel with the three tens below the dividing line and the final answer must still be multiplied by the remaining three tens left on top. The final answer must then be in 10^3 units. or_____ units.

(1.9) ans. (kilo-units)

This system of figuring, correct as it is, is a bit awkward. We have to convert negative powers of ten into positive powers of ten by moving them below the dividing line before we can cancel. It is much easier to algebraically add the powers of ten to get the final answer name. Ten to the sixth times ten to the minus three becomes ten to the six-take-away-three power, or ten to the third power known as _____ units.

(1.10) ans. (kilo-units)

Thusly, if **kilo**-units (10^3 units) were multiplied by **milli**-units (10^{-3} units) we would get just plain units in the answer. WHAT units, would be determined by the scientific equation the powers of ten were related to. We would have 100 left for which there is no prefix since there is really no need for one. We are not multiplying this answer by any tens; therefore, we do not need a prefix. If, however, we had **mega**-units times **nano**-units, then the answer would be in terms of _____ units, or still multiplied by 10^{-3} units.

(1.11) ans. (milli)

One of the steps in becoming conversant in the powers of ten is to know the combinations which cancel each other as did **kilo**-units times **milli**-units in the previous step. Fill out the following chart with the appropriate prefix for the power of ten indicated;

kilo (10^3) times milli (10^{-3}) equals units (10^0)
_____ (10^6) x _____ (10^{-6}) units (10^0)
_____ (10^9) x _____ (10^{-9}) units (10^0)
_____ (10^{12}) x _____ (10^{-12}) units (10^0)

(1.12) ans. (mega x micro)(giga x nano)(tera x pico)

You say, "This is great, BUT my numbers don't always come out in appropriate powers of ten like that."

My answer is, "True. They don't. However, when you want them to, change them so that they will cancel."

REMEMBER—knowledge of the powers of ten and how they combine is a tool to be used at the discretion of the student. This knowledge allows you to write any number in several different ways. The power of ten you use with a number merely tells us how many digits we must move the decimal and in what direction it must go to return the number to its original value with relation to the basic unit of measure. A positive power of ten indicates that the decimal should move to the right, while a negative power of ten says to move the decimal to the left. Now, let's see how this works;

500 kilo units is the same as 500,000 units.
.5 mega units is the same as 500,000 units also.

So, .5 mega units, 500 kilo units & 500,000 units are all the same.
Try it with 397,000 units.

   _____kilo units.
   _____mega units.

(1.13) ans. (397 kilo)(.397 mega)

Let's try this with negative powers of ten. Remember—negative moves the decimal to the left the number of digits the power indicates.

       1,000 pico units.
       _____ nano units.
       _____ micro units.
       _____ milli units.
       _____units.

DIRECTIONS:

Continue from here to frame (1.16) above which you will find the correct answers to this frame's questions.

(1.14) Line 17 of the poem says, "milli and micro combine in a giga." Agreed, this is wrong. However, what is the product of milli and micro? Choose the proper prefix for the correct power of ten representing their product.

 (a) pico
 (b) nano
 (c) atto

ANSWERS:

If you chose (a), you do not know your powers of ten as well as you should, so turn to frame (1.2) and continue from there.

If you chose (b), you are quite correct. Please continue with frame (1.15).

If you chose (c), your system of handling the powers of ten is wrong and you should turn to frame (1.2) and continue from there.

(1.15) Quite often we find ourselves working with a power of ten in the denominator of a problem, and it helps to know what the inverse of this power of ten is so that we may move it to the numerator and handle it as a multiplier.

What is the reciprocal of one divided by a number stated in micro units?

(a) one over that same number the quantity times a thousand.
(b) one over that same number times a thousand.
(c) a mega unit divided by the same number without its power of ten.

ANSWERS:

If you chose (a), you should continue with frame (1.16) since you have not chosen the correct answer.
If you chose (b), you have not chosen the correct answer, so you should continue with frame (1.16).
If you chose (c), you have chosen the correct answer and you should turn to frame (1.20) and continue from there.

(1.16) ans. (1.)(.001)(.000,001)(.000,000,001)

So much for multiplication with powers of ten. What happens when we divide one power of ten by another?

What were you taught to do with a fraction made up of two rather large numbers, the first large one in the numerator and the second large one in the denominator? You were taught to break these numbers down into their prime factors. Then, if there was a 3 above the line and another 3 below the line, they could be cancelled. You could cross them out and forget them, since 3 divided by 3 was one and one times anything did not change its value. Well, the powers of ten work the same way. Milli units divided by milli units comes out units. Kilo units divided by kilo units comes out units too. ANY number divided by itself equals one and ten to a power divided by another ten to the same power is a factor of one just as 3 divided by 3 is.

There is another way of thinking about this though. (Ref. (1.4)) You were told that division by 1,000 is the same as multiplication by milli units. At this rate, division by 10^3 must be the same as multiplication by 10^{-3}, and it is. We moved the ten above the division line and changed the sign of the power. Thus division by 10 to the plus 6th power is the same as multiplication by 10 to the minus 6th power.

Now, we have a rule that allows us to move all our dividing powers of ten above the division line (just change the sign of the power and move it). Then proceed as we did with multiplication.

In this fashion, units divided by kilo units gives an answer in _____ units. And, units divided by milli units becomes _____ units.

345

(1.17) ans. (milli)(kilo)

Let's try this with some larger powers of ten. Mega units divided by giga units gives an answer in _____ units (10^{-3} units). Conversely then—micro units divided by nano units will give _____ units (10^3 units). Milli units divided by kilo units gives an answer in _____ units (10^{-6} units).

(1.18) ans. (milli)(kilo)(micro)

Powers of ten are like fractions when it comes to **adding** them or **subtracting** them. You can't add or subtract fractions until you have them in terms of a common denominator. Well, you cannot add or subtract numbers in terms of powers of ten unless the powers of ten are the same. Then the answer comes out in terms of the same power of ten. Milli units cannot be added to micro units. Both numbers must be in terms of the same power of ten. (i.e., each in milli units or each in terms of micro units for instance). One or both numbers must be changed.

Thus; 1,000 micro units + 1 milli unit equals 2,000 micro units or 2 milli units. And 1 mega unit + 100 kilo units equals 1.1 _____ units or 1,100 _____ units.

(1.19) ans. (mega)(kilo)

Do not mix up the sign of the power of ten with the sign in front of the number the power of ten relates to. For instance, a problem tells you to find the algebraic sum of +.5 kilo units and a —500 milli units.

ANSWERS:

(a) 499.5 units is the correct answer.
(b) The correct answer is .495 kilo units.
(c) The answer is, of course, 250 units.

If you chose (a), you changed both terms in the problem to basic units before taking their difference. This is absolutely correct. Continue with frame (1.20).

If you chose (b), you goofed and should review frame (1.18). (hint) Change both numbers in the problem to the basic units thus eliminating the powers of ten from this problem. Then take their difference.

If you chose (c), BOY did you goof!! This is a basic problem in subtraction and you tried to make a multiplication problem out of it. Remember, you cannot add or subtract unlike quantities. And, if the power of ten names are different, you do not have like quantities. Review frame (1.18) and try again. It's nothing new. You do the same thing every time you reach in your pocket, pull out some coins and add them up to see how much money is there. Let's say you have two dimes and three pennies. If you think of them as **Coins**, you have five coins, but you don't have five pennies and you don't have five dimes. If you change their common name to cents, then you know you have 23 cents. It's the same thing. You have to change their names in order to add them up. The two quantities must be both stated in terms of the same name. Now turn to frame (1.18) and give it another whirl.

(1.20)   Read the following;
1   It's an interesting thing to note as we go,
2   That powers of ten can be multiplied so,
3   Deka times deci is ten times point one,
4   Which is, of course, unity—son-of-a-gun!
5   Then hecto times centi is exactly the same,
6   And kilo times milli can go into the game.
7   Mega times micro is giga times nano,
8   And tera times pico are micro—some game!
9   Tera times pico may equal one,
10 But tera times nano is kilo by gum!
11 And, giga times pico is milli—some fun!
12 While micro times kilo is milli too.
13 A micro times giga now becomes a kilo,
14 As also is milli times mega you know.
15 Some fun these prefixes—powers of ten—
16 But, Oh what a headache I've got in the end!!!!

CONCLUSIONS:

With the aid of the cover-card tables showing the powers of ten, you should have been able to substitute the proper power of ten for each prefix mentioned in the poem above.

All lines of the poem above should have made sense except line 8. If you cannot find the boo-boo in line 8, return to frame (1.10) and take this part of the program over again.

INDEX

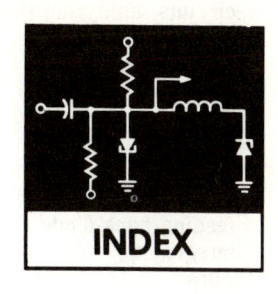

A

Adding a diode	113
Amplifier	
—circuits	197
—complete analysis of an	231
—devices	135
—operational	261
—and frequency response	252
—push-pull	264
Analysis	
—graphical	47
—in detail, author's	237
—triode tube	136
Analysis of transistor and FET amplifiers using "Transresistance"	210
Analyzing time constants of diode circuits	109
Answers	344
Armstrong oscillator	289
Author's analysis in detail	237

B

Back diodes	96
Bandwidth	
—gain times	255
—relating, rise time, and RC	257
Basic laws of electronics	7

C

Calculating input and output resistance, and voltage gain	197
Capacitor, the	42
Capacitance	
—determining stray, and their effects	257
—multiple	45
"Catching" diodes used in both directions	121
Circuits	
—amplifier	197
—equivalent, Norton	24
—equivalent, Thevenin	15
Coaxial transmission line	280
Colpitts oscillator	292
Complete analysis of an amplifier	231
Complex, getting a bit more	117
Components theorem, Millman's	31
Constants	
—time	56
—time, L, R, C and	41
Construction of junction diodes	86
Crystals, growing your own	297

D

DC circuit	
—parallel	12
—series	10
"Designing" a power supply	321
Determining stray capacities and their effects	257
Devices, amplifier	135
Differential amplifier	227
Diode	
—adding a	113

—circuits, analyzing time constants of	109
—circuits, field effect	127
—field effect	97
—Shockley	96
—"Snap"	106
—tunnel	91
—tunnel, differentiator feeding back-diode integrator	124
Diodes	85
—back	96
—junction, construction of	86
—used in both directions, "catching"	121
—zener	90
Directions	344

E

Emitter follower—common emitter op amp	264
Equating tubes and transistors	164
Equivalent circuits	
—Norton	24
—Thevenin	15

F

Feedback in-phase	266
Field effect	
—diode	97
—diode circuits	127
Filter, simple high pass (a review)	109
Formulas, reactance	41
Franklin oscillator	287
Frequency response	254
—and operational amplifiers	252
Frequency-compensated op amp	263
Full-wave rectifier	305

G

Gain times bandwidth	255
Generator, sawtooth	131
Getting a bit more complex	117
Graphical analysis	47
Growing your own crystals	297

H

Half-wave rectifier	305
Hartley oscillator	292
High pass filter, simple (a review)	109

I

Inductor, the	41, 42
Inductors, multiple	43
Initial methodology	231
In-phase feedback	266
Instructions, programmed "the powers of ten"	339

J

Junction diodes, construction of	86

K

Kirchhoff's and Ohm's laws, combining	10
Kirchhoff's laws	9

L

Laws	
—basic, of electronics	7
—Kirchhoff's	9
—Ohm's	7
L, R, C and time constants	41

M

"Metering resistance," the	201
Methodology, initial	231
Millman's	
—composite theorem	31
—theorems	14
More on feedback (oscillators)	280
Multiple	
—capacitors	45
—inductors	43

N

Neutralization	295
Norton equivalent circuits	24
Norton equivalent of vacuum tube as current generator	200
Norton's theorems	14

O

Ohm's and Kirchhoff's laws, combining	10
Ohm's law	7
Op amp	
—becomes oscillator	282
—frequency compensated	263
—transistors	261
Operational amplifiers	261
—series regulator as	310
Oscillator	
—Armstrong	289
—Colpitts	292
—Franklin	287
—Hartley	292
—op amp becomes	282
—sine wave, example	284
—more on feedback	280
—triode-type crystal-controlled	295
—tuned grid	289
—tuned plate	289

P

Parallel DC circuit	12
Paraphase amplifier	227
Pentodes and triodes	154
Polarities, reversing	117
Power, power supplies and safety	300
Power supply	
—"Designing" a	321
—solid-state	329
Powers of ten	337
Preliminaries and self-test	234
Programmed instruction of "the powers of ten"	339

Push-pull amplifier	227
Push-pull op amps	264

Q

Quiz	274

R

Reactance formulas	41
Reactions	340
Rectifier	
—full-wave	305
—half-wave	305
Relating bandwidth, rise time, and RC	257
Response, frequency	254
Reversing polarities	117

S

Safety	301
—power, power supplies and	300
Sawtooth generators	131
Self-test and preliminaries	234
Series DC circuit	10
Series regulator as operational amplifier	310
Shockley diode	96
Simple high pass filter (a review)	109
Sine wave oscillator example	284
"Snap" diode	106
Solid-state power supply	329
Step-function waveforms	51

T

Theorem, Millman's composite	31
Theorems, Von Helmholtz Thevenin, Norton, and Millman	14
Thevenin equivalent circuits	15
Thevenin equivalent of vacuum tube as voltage generator	199
Thevenin's theorems	14
Time constants	56
—L, R, C, and	41

351

—of diode circuits, analyzing	109
Transistor op amp	261
Transistors and tubes, equating	164
Transmission line, coaxial	280
Triode plate-loaded stage voltage gain formulas	197
Triode tube analysis	136
Triode-type crystal-controlled oscillator	295
Triodes and pentodes	154
Tuned plate-tuned grid oscillator	289
Tunnel diode	91
—differentiator feeding back-diode integrator	124

V

Von Helmholtz theorems	14

W

Waveforms, step-function	51

Z

Zener diodes	90